W9-CUK-509

FEMTOSECOND
BEAM SCIENCE

FEMTOSECOND
BEAM SCIENCE

edited by **Mitsuru Uesaka**
University of Tokyo, Japan

ICP

Imperial College Press

phys

0142270122

Published by

Imperial College Press
57 Shelton Street
Covent Garden
London WC2H 9HE

Distributed by

World Scientific Publishing Co. Pte. Ltd.
5 Toh Tuck Link, Singapore 596224
USA office: 27 Warren Street, Suite 401-402, Hackensack, NJ 07601
UK office: 57 Shelton Street, Covent Garden, London WC2H 9HE

British Library Cataloguing-in-Publication Data
A catalogue record for this book is available from the British Library.

FEMTOSECOND BEAM SCIENCE

ISBN   1-86094-343-8

Printed in Singapore by World Scientific Printers (S) Pte Ltd

Preface

The main purposes of this book are to summarize the recent developments and advances in femtosecond beam science and to make them accessible to researchers, engineers, technicians, graduate students and postdoctoral researchers. Femtosecond beam science is a development of femtosecond laser spectroscopy and pump-and-probe analysis, for which Prof. A. H. Zevail won the Nobel Prize in Chemistry in 1999. Femtosecond beam generation comprises visible lasers, electrons, X-rays, ions, THz radiation, positrons, neutrons and so on. This science is expected to lead to the development of technology to realize dynamic microscopy, that is, the visualization of atomic motions, chemical reactions, protein dynamics and other microscopic dynamics. Advances have already realized the visualization of atomic motions and enabled the visualization of phonons thermal expansion and shock wave propagation by advanced time-resolved X-ray diffraction, at a time resolution of 10 ps. These achievements will act as a bridge to the development of femtosecond X-ray sources and fourth generation synchrotron light sources. Dynamic microscopy promises to be one of the most important issues in dynamic nanotechnology in the near future. For this reason, we felt it was necessary to provide an overview of femtosecond beam science to assist current and future researchers in the field, and hopefully to accelerate the progress of the field.

In this work, the theories, state-of-the-art techniques and recent achievements in femtosecond beam science are described by worldwide leaders in the field. In early 2002 I was asked by Imperial College Press/World Scientific to publish a monograph on this topic. I proposed the title of "Femtosecond Beam Science" and sent in a proposal with preliminary contents. After it was reviewed and approved in August 2003, I decided to ask prominent pioneers, who are at the forefront of this fascinating science, to cooperate in producing this volume and make it of much higher quality than I could manage alone. They responded perfectly with excellent contributions within only a few months. I deeply appreciate their understanding, effort and hard work. I expect this book will be useful especially for young researchers and graduate students. To make it easy to understand for these readers, I asked all co-authors to keep to the following style.

- Start by explaining their particular subject as simply as possible, for example by using a schematic drawing that shows how it works.
- Briefly review recent advances worldwide.
- Feature at least one of their more important results.
- Even if femtosecond pulses are not yet used in their particular area, foresee future uses of femtosecond pulses.

I would like to thank Dr. T. Imai of the University of Tokyo for acting as editor for this volume and Dr. K. Dobashi and Ms. H. Zhao for their editorial assistance.

M. Uesaka

Contents

Chapter 1

Introduction

M. UESAKA
Nuclear Engineering Research Laboratory
University of Tokyo,
22-2 Shirane-shirakata, Tokai, Naka,
IBARAKI 319-1188, JAPAN.

Femtosecond beam science consists of the generation, measurement and application of a variety of ultrashort beams. Synchronized femtosecond beams are used to visualize ultrafast microscopic phenomena. The science is also applied to advanced compact radio frequency (RF) accelerators.

Chapter 2 covers femtosecond beam generation. Femtosecond laser pulses are produced by chirped pulse compression. Analogous magnetic bunch compression of electrons generates a femtosecond electron beam in high quality linear accelerators (linacs). In synchrotrons, femtosecond electron/synchrotron radiation (SR) pulses are produced by slicing with a wiggler and femtosecond laser. Momentum compaction control and strong longitudinal focusing can realize femtosecond electron and SR pulses. Femtosecond and picosecond hard X-ray pulses can be produced by compact systems based on terawatt (TW) laser–plasma interaction and inverse Compton scattering between short electron and laser beams. The configuration of a femtosecond beam generation system and the relevant physics is depicted in Fig. 1.1. Large intense X-ray free electron lasers and energy recovery light sources are in development. Tabletop TW lasers have recently enabled the generation of several kinds of ultrashort beams from laser–plasma, as schematically shown in Fig. 1.2. Laser plasma cathodes can produce femtosecond electron beams. Picosecond ion, THz radiation and positron pulses can also be generated via the TW laser–gas/cluster/solid

1

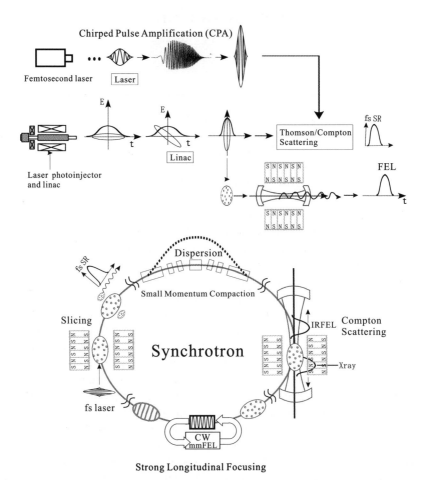

Fig. 1.1   Femtosecond particle beam generated by femtosecond laser in a particle accelerator.

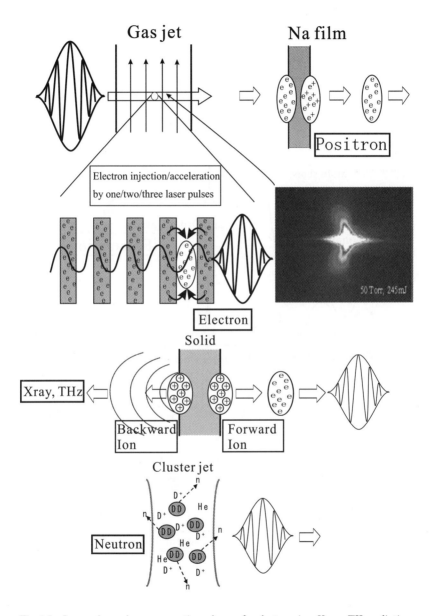

Fig. 1.2   Laser–plasma beam generation schemes for electron, ion, X-ray, THz radiation, neutron and positron pulses.

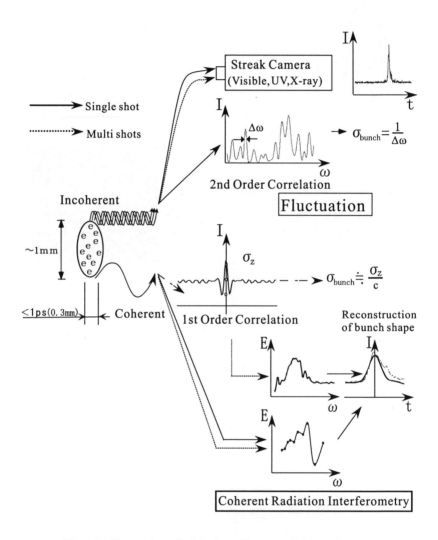

Fig. 1.3   Diagnostic methodologies of femtosecond electron beams.

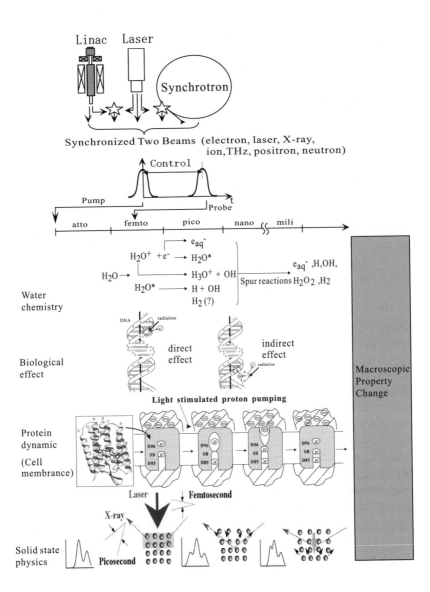

Fig. 1.4 Applications of femtosecond beam pump-and-probe analysis.

interaction. The TW laser–D_2 cluster interaction initiates nuclear fusion and yields picosecond neutron pulses.

The major diagnostic methodologies for femtosecond radiation and electron beams are explained in detail in Chapter 3. The physics of these methodologies is depicted in Fig. 1.3. An overall comparison among the methodologies for electron beam diagnosis is given. A newly developed jitter free X-ray streak camera using a GaAs optical switch is described. Several new promising ideas are also referenced. Details of a synchronization system between femtosecond lasers, linacs and synchrotrons are given. This system consists of an RF control system, timing stabilizer, passive mode-locked laser, RF amplifier and linac. Timing jitter between laser and electron pulses, the sources used in the system and the method of suppression are described. The influence of the environment, such as temperature, humidity, laser transport line, dust, etc., is discussed. New synchronization schemes and advances are also described.

The applications of femtosecond beams are summarized in Chapter 4. In pump-and-probe analysis with two synchronized beams, the pump pulse induces a reaction and the delayed probe pulse extracts a signal of the state at specified time steps. This is summarized in Fig. 1.4. State-of-the-art pulse radiolysis systems for radiation chemistry using synchronized femtosecond lasers and linacs (including the use of laser photocathode RF guns) and recent results are described. Recently, X-ray pulses have been used as probe pulses in time-resolved X-ray diffraction. Laser plasma X-ray sources and SR have also been used. Phonons, thermal expansion and shock wave propagation in semiconductors have been visualized by pump-and-probe analysis using laser plasma X-ray sources and third generation synchrotron light sources. Furthermore, time-resolved Laue diffraction analysis of phase transitions and fast motion of photoactive proteins is under way at third generation synchrotron light source facilities, which will lead on to future experiments with fourth-generation synchrotron light sources. Femtosecond intense lasers are being used to develop a new chemistry in intense laser fields. Finally, computer simulations of ultrafast microscopic phenomena are described.

Chapter 2

Femtosecond Beam Generation

2.1 Theory and Operation of Femtosecond Terawatt Lasers

F. FALCOZ

V. MORO

E. MARQUIS

THALES Laser, RD 128, BP 56,

91401 ORSAY, FRANCE

2.1.1 *Ultrashort pulses: theory and generation*

The first issue in the field of terawatt (TW) lasers is the oscillator principle for generating ultrashort pulses.

Conventional pulsed lasers are based on the Q-switching principle. The laser cavity is made of a gain medium and a switching element that triggers pulse emission. The Q switch, which can be either acousto-optical or electro-optical (the Pockels cell principle), is a variable loss element (oscillator Q modulator) and usually exhibits a switching time in the nanosecond range. Such devices do not generate pulses below the nanosecond range.

A specific combination of techniques should be used to generate femtosecond pulses. This combination is detailed below.

2.1.1.1 *Principle of mode locking for short pulse generation*

A laser is an optical oscillator made of an amplifying medium and an optical cavity. Oscillation in the cavity is only possible when the two following conditions are fulfilled:

(1) Conservation of phase after a 2π round trip: only frequencies defined

Femtosecond Beam Science

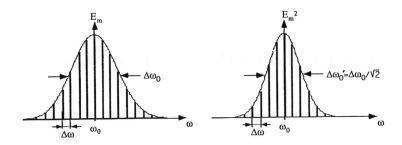

Fig. 2.1   Spectrum of short pulse

by $\nu = \frac{kc}{2L}$ can oscillate in the cavity (L cavity length and k an integer). These frequencies define the longitudinal modes of the cavity.

(2) The gain for the longitudinal modes must be above the loss threshold.

In a laser, three different regimes can be defined:

- Multimode, with random phase
- Single mode
- Mode locked, with an established phase relationship between the modes.

This last regime is commonly used for ultrashort pulse generation. It corresponds to a state where all oscillating modes in the cavity have a constant phase. We will demonstrate that this operating condition leads to ultrashort pulses.

Assuming that the laser emission consists of several modes with a Gaussian amplitude distribution centered on ω_0, we can define the laser beam electric field as

$$E_m = E_0 \mathrm{e}^{-\left(\frac{2m\Delta\omega}{\Delta\omega_0}\right)^2 \ln 2} , \qquad (2.1)$$

with

E_0 : field amplitude at ω_0,

m : mode index; $\omega_m = \omega_0 + m\Delta\omega$,

$\Delta\omega = c/2L$: mode spacing,

$\Delta\omega_0$: full width at half maximum (FWHM) of Gaussian distribution,

as shown in Fig.2.1. If we assume a null phase, we have

$$E(t) = \sum_{m=-\infty}^{m=\infty} E_m e^{i\omega_m t}, \tag{2.2}$$

then

$$E(t) = e^{i\omega_0 t} \sum_{m=-\infty}^{m=\infty} E_m e^{im\Delta\omega t}, \tag{2.3}$$

where $E(t)$ is a pulsating field at ω_0, slowly modulated by the $k(t)$ function

$$k(t) = \sum_{-\infty}^{+\infty} E_m e^{im\Delta\omega t}. \tag{2.4}$$

Being a complex function sum, $k(t)$ has a period T defined by

$$T = \frac{2\pi}{\Delta\omega} = \frac{2L}{c}, \tag{2.5}$$

as shown in Fig. 2.2. The laser emission is then modulated with a period linked to the cavity length. Knowing that $k^2(t)$ has a Gaussian distribution, we can easily demonstrate that

$$k^2(t) = \exp\left\{ -\left(\frac{2t}{\Delta t}\right)^2 \ln(2) \right\}, \tag{2.6}$$

where Δt is the pulse width at half maximum,

$$\Delta t = \frac{2}{\pi \Delta\nu_0'} \ln(2), \tag{2.7}$$

and

$$\Delta\nu_0' = \frac{\Delta\omega_0'}{2\pi} = \frac{\Delta\omega_0}{2 \cdot \sqrt{2} \cdot \pi}. \tag{2.8}$$

If the spectral emission of a laser is Gaussian and all modes are perfectly locked, the emission is an infinite Gaussian pulse train. All pulses are Δt wide with spacing $T = \frac{2L}{c}$. This is equivalent to a short pulse traveling in the cavity and leaking a part of its energy after each round trip through a coupling mirror.

We note that:

- The pulse duration Δt is inversely proportional to the intensity spectral width.
- All calculations assume a Gaussian shape.

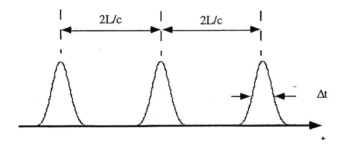

Fig. 2.2　Temporal profile of short pulse

- A pulse of Δt duration is Fourier transform limited if the following relation applies:

$$\Delta t \Delta \nu = K, \qquad (2.9)$$

where K depends on the intensity shape. Examples of K are given in Table 2.1.

Table 2.1　K values for several pulse shapes of femtosecond lasers.

Intensity shape	$k^2(t)$	K
Gaussian	$\propto \exp\left\{-\left(\frac{2t}{\Delta t}\right)^2 \ln 2\right\}$	0.411
Sech^2	$\mathrm{sech}^2\left(\frac{1.766}{\Delta t}\right)$	0.315
Rectagle	1 for $0 < t < \Delta t$	1

Ultrashort pulses are generated in a laser cavity by the combination of several lasing modes, where the pulse duration is inversely proportional to the emission spectral width and the pulse repetition frequency is directly linked to the cavity length.

Commercially available products now operate in the 30–300 fs range with a repetition rate of 20–50 MHz.

2.1.1.2　*Mode-locking techniques*

As explained above, a multimode combination with a specific constant phase relationship is needed for short pulse operation in an oscillator. As only a specific subset of phase values will lead to short pulse emission and as the behavior of the laser medium leads to dispersion which needs to be compensated for, a femtosecond oscillator uses specific techniques to reach an optimized operating point and keep it stable.

The first basic principle is to introduce a modulating element in the cavity, so that phase conditions can be reached for the oscillating modes. The modulating element can be either active (acousto-optical) or passive (saturable absorbers, Kerr lens effect, etc.).

We will only detail the Kerr lens effect, which is the most common non-linear effect used in femtosecond oscillators. In this technique the crystal itself plays the role of a saturable absorber. It is also called *self-mode-locking*, as no extra elements are needed in the cavity.

The mechanism used is a third-order nonlinear effect. When a high intensity pulse propagates through a medium of refraction index n_0, there is an index variation defined by

$$n(I) = n_0 + n_2 I \,, \tag{2.10}$$

n_2 being the nonlinear refractive index. Assuming a Gaussian pulse, we have

$$I(\Omega, Z) = I_\phi(Z)e^{-\frac{2\Omega^2}{\omega^2}} \,, \tag{2.11}$$

where $I_\phi(Z)$ is the peak intensity on the axis and ω is the $1/e^2$ beam dimension. The effective index can then be written as

$$n = n_0 + n_2 I_0(Z)e^{-\frac{2r^2}{\omega^2}} \,. \tag{2.12}$$

As n_2 is positive and the intensity is higher in the center of the beam, a plane wave propagating through the medium will acquire a phase curvature, focusing the beam. The medium behaves as a converging lens, known as the *Kerr lens*, and this phenomenon is called *self-focusing*.

In a laser cavity working in a pulsed regime, the gain medium is equivalent to a converging lens influencing the cavity stability conditions and the beam diameter as shown in Fig. 2.3.

A slit is introduced at a specific location where the beam is small. As long as the laser is working in the pulsed regime, losses will be kept low. However, if the mode-locking conditions are not satisfied and the laser emission switches to continuous wave (CW), the Kerr lens disappears and losses increase. Such a system acts as a discriminating factor between mode locking (the minimum loss point) and the CW regime.

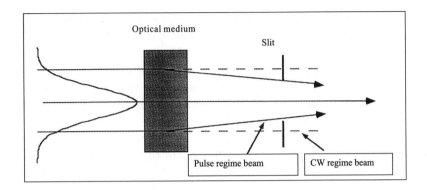

Fig. 2.3   Principle of slit beam diameter filtering.

The combination of slit and Kerr lens acts as a saturable absorber. We note that the time response of such an absorber is in the femtosecond range, whereas the time response of semiconductor absorbers is in the picosecond range. Nevertheless, this condition is not sufficient for short pulse emission, as a wide spectrum is also required.

Pulse intensity is also a time dependent function:

$$n(t) = n_0 + n_2 I(Z, t).$$ (2.13)

A wave propagating through a Kerr lens is phase shifted with the relation

$$\phi(Z, t) = \phi_0(Z, t) + \Delta\phi(Z, t).$$ (2.14)

This phenomenon is called *self-phase-modulation*. The instantaneous wavelength can be written with a Gaussian profile,

$$\omega(Z, t) = \frac{\partial\phi(Z, t)}{\partial t},$$ (2.15)

and then

$$\omega(Z, t) = \omega_0 + \Delta\omega(Z, t).$$ (2.16)

As n_2 is positive, the self-phase-modulation will shift to shorter frequencies at the rising edge of the pulse and to greater frequencies at the trailing edge (see Fig. 2.4). The pulse spectrum is then broadened, keeping the same temporal profile.

Hence, self-phase-modulation generates the wide spectrum required for short pulse operation.

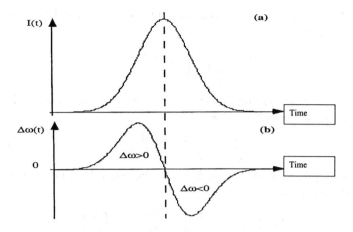

Fig. 2.4 Frequency shifting in self-phase-modulation.

However, in a dispersive medium this effect leads to a temporal broadening of the pulse. Dispersion compensation techniques should be implemented in the cavity to avoid this broadening.

In optical materials, the refractive index is inversely related to the wavelength. As a femtosecond pulse has a wide spectrum, dispersion occurs for propagation through a medium, as the speed of the longer wavelengths is higher than that of the shorter wavelengths. Temporal stretching of the pulse occurs.

In such conditions, the wave vector can be approximated by

$$k(\omega) = k(\omega_0) + \left(\frac{\partial k}{\partial \omega}\right)_{\omega_0} \Delta\omega + \frac{1}{2}\left(\frac{\partial^2 k}{\partial \omega^2}\right)\Delta\omega^2 + o(\Delta_\omega^2), \qquad (2.17)$$

where $k = \frac{2\pi}{\lambda}$. The first term of this equation describes wave propagation at ω_0. The second term describes the pulse group velocity G_V,

$$G_V = \frac{1}{\partial k/\partial \omega} = \frac{\partial \omega}{\partial k}. \qquad (2.18)$$

The third term is the dispersion of the group velocity and can be written for a medium of index $n(\omega)$ as

$$\frac{\partial^2 k}{\partial \omega^2} = \frac{\lambda_0^3}{2\pi c^2} \cdot \frac{\partial^2 n}{\partial \lambda^2}. \qquad (2.19)$$

If $\frac{dn}{d\lambda} < 0$ and $\frac{d^2n}{d\lambda^2} > 0$, low frequencies will travel faster than high frequencies. We have then

$$\frac{t_0}{t_i} = \sqrt{1 + \left(\frac{4\ln 2\phi''}{t_i^2}\right)^2},\tag{2.20}$$

where t_0 and t_i are the output and input pulse durations and ϕ'' is the phase quadratic dispersion. Then

$$\phi'' = \frac{\partial^2\phi}{\partial\omega^2} = \frac{\partial^2 k}{\partial\omega^2}L = \frac{\lambda_0^3 L}{2\pi c^2}\frac{\partial^2 n}{\partial\lambda^2},\tag{2.21}$$

with L being the medium length. It is also important to study third-order phase dispersion, as

$$\phi''' = \frac{\partial^3\phi}{\partial\omega^3} = -\frac{\lambda_0^4 L}{4\pi^2 c^3}\left(3\frac{\partial^2 n}{\partial\lambda^2} + \lambda_0\frac{\partial^2 n}{\partial\lambda^2}\right).\tag{2.22}$$

Table 2.2 gives values of the most commonly used materials.

Table 2.2   Material parameters for mode locking.

	n	$n' =$ $dn/d\lambda$ $[\mu m^{-1}]$	$n'' =$ $d^2n/d\lambda^2$ $[\mu m^{-2}]$	$n''' =$ $d^3n/d\lambda^3$ $[\mu m^{-3}]$	ϕ'' $[fs^2]$	ϕ''' $[fs^3]$
Silica	1.4533	-0.0173	0.0398	-0.2387	361.03L	274.35L
SF10	1.7112	-0.0496	0.1755	-0.9971	1589.04L	1042.27L
LaK31	1.6874	-0.0294	0.0775	-0.4411	701.54L	420.87L
Ti^{3+}:Al$_2$O$_3$	1.7602	-0.0268	0.0641	-0.3773	580.35L	420.87L
Cr^{3+}:LiSAF	1.4078	-0.0107	0.0151	-0.1511	136.54L	290.66L

Prisms or chirped mirrors must be used for compensation, as the operation criteria of a laser is

$$\phi_1'' + \phi_2'' + \phi_3'' = 0,\tag{2.23}$$

where ϕ_1'' is chirped mirror contribution (negative), ϕ_2'' is cavity elements contribution (positive), and ϕ_3'' is self-phase-modulation contribution (positive). This is the key equation driving the design of femtosecond laser sources.

The successful operation of an ultrashort oscillator is possible when:

• A saturable absorber (active or passive) puts the system in a stable phase operating condition, namely mode locking.

- Elements inserted in the cavity (prisms, chirped mirrors) compensate the natural optical elements dispersion in order to satisfy the phase dispersion stability criteria.

2.1.2 *Stretching and compressing laser pulses*

2.1.2.1 *Chirped pulse amplification principle*

As shown above, it is necessary to stretch a short pulse before amplification. This is mainly due to the following reasons:

- The density has to be lower than the damage threshold of the optics to prevent its destruction. A 100 fs pulse with 100 mJ energy leads to power over 1 TW. The interaction surface with the gain medium, lenses, mirrors and other optical elements is usually less than 1 cm^2; at these levels of density, nonlinear effects that occur when passing through materials lead to a distortion of the beam that can sometimes damage the gain medium.
- The variation of the index as a function of intensity ($n = n_0 + n_2 I$) causes nonlinear effects (the Kerr effect mainly) that create self-phase-modulation and wave front distortion that can alter the focusing properties of the pulse after amplification. This effect is characterized by the so-called B integral.

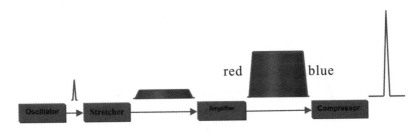

Fig. 2.5 CPA principle.

In 1987, Mourou demonstrated the validity of the chirped pulse amplification (CPA) technique [Ferré *et al.* (2000)], which is now known to be the only technique for amplifying short pulses (< 10 ps) in solid-state materials. As shown in Fig. 2.5, the technique is made of three steps:

(1) Stretching: as the output energy available from the amplifiers is inde-

pendent of the input pulse duration, by stretching the pulse one can reduce its peak power. The pulse power is reduced by the stretching factor (typically 1000–5000).

(2) Amplification: the pulse can then be amplified without risk. The longer the stretched pulse is, the higher the amplified pulse energy that can be reached without damage.

(3) Compression: the pulse is finally compressed back to its original duration, allowing a much higher peak power.

2.1.2.2 *Stretcher/compressor operation*

Most of the difficulties of chirped pulse amplification come from the stretching and compression stages. The stretching must be reversible (spectral filtering to lengthen the pulse may cause problems). The main stretching technique known is based on the dispersion properties of gratings or prisms. The principle of the technique is to make the optical paths for each wavelength different, as shown in Fig. 2.6.

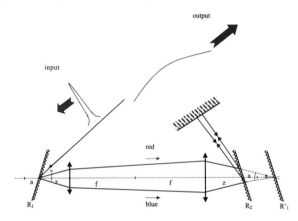

Fig. 2.6   Stretching principle.

The stretching device is made of two gratings R_1 and R_2 between which is set an afocal telescope (L_1 and L_2, which are identical) of magnitude -1. R_2 is set parallel to R'_1, the image of R_1 through the focal device. The distance $D = R_2 R'_1$ is an algebraic value that can be positive, negative or null. In the case of Fig. 2.6, it is negative.

We note that the wavelengths diffracted by R_2 are parallel, but translated from one another. To get rid of this *spectral shift*, one can fold down the beam using two folding mirrors. The reverse is the same, but at different heights. This feedback leads to a chirped pulse in the time domain, but not in the spatial domain.

The following are some important parameters of stretchers:

- Dispersive factor of gratings,

$$\beta = \frac{d\theta}{d\omega} = \frac{N \cdot \lambda^2}{2\pi c \cdot \cos\theta}. \qquad (2.24)$$

- Quadratic term of the velocity group dispersion,

$$\phi'' = \frac{d^2\phi}{d\omega^2} = -2k\beta^2 D. \qquad (2.25)$$

(a) Discussion about the D value

- In the case where $D < 0$, $\phi'' > 0$, the dispersion is normal and we are in the stretcher configuration. This is possible by using a virtual grating. In this case R_2 is the virtual image of the real object R_1.
- In the case where $D > 0$, $\phi'' < 0$, the dispersion is abnormal. It is the compressor configuration.
- In the case where $D = 0$, $\phi'' = 0$, it is necessary to modify the phase and amplitude of the spectrum using masks in the Fourier plan.

2.1.2.3 *The Offner triplet configuration*

In 1996, Cheriaux and Chambaret demonstrated using the Offner triplet configuration in the stretcher to reduce aberrations [Cheriaux *et al.* (1996)].

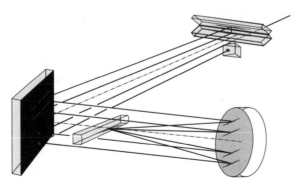

Fig. 2.7 Offner triplet optical configuration.

As shown in Fig. 2.7, the configuration is made of an all-reflective triplet combination: a set of two spherical mirrors, the first concave, the second convex. The main characteristic of this combination is that it is completely symmetrical, so that only symmetrical aberration can appear, such as astigmatism or spherical aberration. The setup of the two mirrors eliminates these aberrations. Hence, the advantages of the Offner triplet are that it has no on-axis coma and no chromatic aberration. Since the first experiments, this combination has been the most common stretching technique used in CPA lasers.

2.1.2.4   *Compressor subsystem*

Usually, this stretcher design is coupled with a classical two-grating compressor design as described on Fig. 2.8:

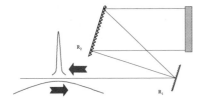   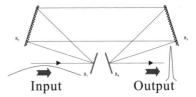

Input   Output

Fig. 2.8   Compressor optical configuration.

Fig. 2.9   Alternative compressor optical configuration.

In some cases, a mismatch of the groove number of the gratings in the stretcher and in the compressor can lead to a better compensation of the velocity group dispersion introduced by both the stretcher and the different amplification stages.

The gratings have to be parallel to get good compression; after two diffractions, the spectral components of the beam are parallel to each other, but spatially shifted. The use of the same dihedral mount will change the height of the beam to allow it to go back onto the gratings and to go out of the compressor with a homogeneous repartition of the spectral components.

Another setup that leads to exactly the same result is shown in Fig. 2.9.

2.1.3   *Amplification process*

The second step of the CPA technique is the amplification stage. Several solutions can be implemented, but in this chapter we will mainly focus on regenerative amplification and multipass amplification.

2.1.3.1 *Regenerative amplification*

The regenerative amplifier is a laser cavity that is seeded by a short pulse coming from the stretcher. The design of this cavity depends on the repetition rate of the laser and of the pumping energy. For a low repetition rate the cavity is stabilized by the gain, because of the gain guiding effect [Cheriaux *et al.* (1996)], and a large mode operation is preferred to maintain the fluence below the damage threshold.

As the repetition rate increases, the size of the mode has to be reduced to maintain the gain level. In this case a concave cavity (Z or confocal) can be used. This configuration generally leads to better stability and optimal overlap between the seeded beam and cavity mode.

The principle of the regenerative amplifier is as follows:

- The seeding of a short pulse in the amplifier is obtained by polarization switching in a Pockels cell. This pulse comes from the oscillator at a MHz repetition rate. The regenerative amplifier is seeded at Hz or kHz as the Pockels cell is also used as a pulse selector.
- The cell generally uses a KD*P crystal. The natural birefringence of this crystal allows the crystal to be set so it is equivalent to a quarter-wave plate. This is accomplished by introducing a small angle between the crystal optical axis and the cavity optical axis.
- The pulses coming from the stretcher are vertically polarized after an optical path selector. The seeding polarizer reflects them and flips their polarization by 90° after a round trip in the Pockels cell. They are then transmitted through the polarizer and travel a round trip in the cavity. After passing through the Pockels cell a second time, the polarization is flipped back to vertical and the pulse exits the cavity by reflection from the polarizer.

In order to trap a pulse in the cavity, a quarter-wave voltage is applied on the Pockels cell when the pulse is between the polarizer and the rear mirror. This makes the Pockels cell equivalent to a half-wave plate and a round trip through the cell does not affect the polarization of the pulse. The pulse is then amplified by successive passes through the Ti:Sapphire crystal (Ti:Sa) (between 20 and 30 round trips). When the pulse reaches its maximum energy it is dumped from the cavity by applying a second quarter-wave voltage step to the cell. This principle is summarized in Fig. 2.10.

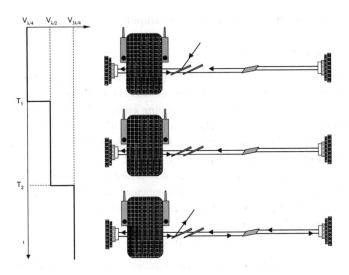

Fig. 2.10   Regenerative amplifier principle.

The main advantages of a resonator compared to a multipass amplifier (described in the following Section) are that it benefits from the intrinsic properties of the cavity:

- The pulse tends to take the spatial mode of the cavity, which is usually TEM_{00} or quasi-TEM_{00}.
- The laser beam inherits the stability of the cavity, which can be very high when using a diode pumped laser as a pump for the Ti:Sa.

2.1.3.2   *Multipass amplification*

The output energy of a regenerative amplifier is generally limited to the mJ range to avoid damage and nonlinear effects. If higher energy is required, the geometrical multipass amplifier is the best compromise between efficiency, reliability and cost. This type of amplifier generally uses a bow-tie configuration where each pass in the gain medium is separated geometrically as shown in Fig. 2.11. The number of passes is limited by the complexity of the design and the difficulty in maintaining the size of the amplified beam

on the Ti:Sa at each round trip. The typical number of passes is from two
to eight depending on the gain and the energy required; Fig. 2.11 shows a
typical four pass amplifier geometry with symmetrical pumping geometry.

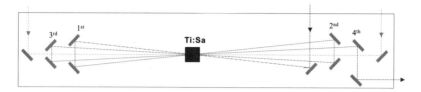

Fig. 2.11   Multipass amplifier principle

Nanosecond pulses, generated by YAG or YLF lasers, pump the amplifier. The YAG laser is commonly used for low repetition rate lasers (few
tens of Hz), with an available energy > 1 J per laser. As the repetition
rate increases (> 1 kHz), the best compromise is the YLF with available
average power > 20 W. Above 10 kHz, YAG is again the best choice as its
fluorescence lifetime is lower that YLF.

The small signal gain of the amplifier is defined by

$$g_0 = \frac{J_{\text{STO}}}{J_{\text{SAT}}}, \qquad (2.26)$$

where J_{STO} is the pump fluence (in J/cm^2) stored in the medium and J_{SAT}
is the saturation fluence (also in J/cm^2). J_{STO} can be generally calculated
using

$$J_{\text{STO}} = \frac{E_p \alpha L}{S} \frac{\lambda_p}{\lambda_L} \qquad (2.27)$$

where E_p is the pump energy on the crystal amplifier (in J) α is the absorption of the pump (in cm^{-1}) L is the length of the amplifying medium (in
cm), S is the cross section of the pump beam (in cm^2) and λ_p and λ_L are
the wavelengths of the pump and of the laser. The value of J_{SAT} depends
on the material. Table 2.3 gives examples for typical crystals.

Generally, we can define two different regimes in the amplifier depending
on the input energy E_{in}. If the pulse which is amplified has a low fluence
(E_{in}/S) compared to the saturation fluence (J_{SAT}), the amplifier is working
in a non-saturated regime. The output energy in this case is simply given

Table 2.3   Saturation energy in gain medium.

Amplifier medium	J_{SAT} (J/cm^2)	Spectral range
Dyes	~ 0.001	Visible
Excimers	~ 0.001	UV
Nd:YAG	0.5	1064 nm
Ti:Al$_2$O$_3$	1	800 nm
Nd:Glass	5	1054 nm
Alexandrite	22	750 nm
Cr:LiSAF	5	830 nm

by

$$E_{\text{out}} = E_{\text{in}} \exp(g_0) \,. \tag{2.28}$$

When the input fluence reaches a level comparable to the saturation fluence we can use the formula derived from Frantz and Nodvick theory,

$$J_{\text{out}} = \frac{E_{\text{out}}}{S} = J_{\text{SAT}} \ln \left[g_0 \left\{ \exp \left(\frac{J_{\text{in}}}{J_{\text{SAT}}} \right) - 1 \right\} + 1 \right] , \tag{2.29}$$

where J_{out} is the output fluence of the pulse after amplification.

For example, a four pass Ti:Sa amplifier working near the saturation regime exhibits an amplification factor $J_{\text{out}}/J_{\text{in}} > 20$. It means that if three cascaded multipass amplifiers (pumped by 8 J in a second harmonic generator (SHG) at 10 Hz) are used, the output amplified beam before compression will achieve more than 3 J in the infrared. After a 70% efficiency compressor and with a 20 fs pulse, one can then achieve greater than 100-TW femtosecond lasers.

2.1.3.3   *20-TW laser system*

An existing 20-TW system consists of a single grating stretcher, a 10 Hz repetition rate regenerative amplifier, a multipass pre-amplifier, two multipass amplifiers and a two-grating compressor.

The stretcher and the compressor use ruled gratings specially designed for short pulse compression. Their optical configuration allows high order dispersion compensation to shorten the output pulse duration as close as possible to the Fourier limit.

The regenerative amplifier linear cavity and the two multipass amplifiers use Ti:Sa as gain medium for their broadband tunability (700–1100 nm). A Pockels cell (MEDOX Twin Peak system) is used to seed and dump the

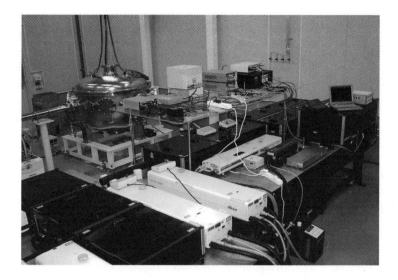

Fig. 2.12   View of 20 TW system in the laboratory

pulse from the regenerative cavity.

In addition to the Nd:YAG pump lasers (one COMP 10/10 SHG and six SAGA 230/10 SHGs), the system can be divided into several modules (see Fig. 2.12):

- The oscillator and its pump laser.
- An Offner stretcher and a synchronization module within the same breadboard.
- The regenerative amplifier and the pre-amplifier within the same breadboard.
- A first multipass amplifier in one breadboard.
- The main multipass amplifier in one breadboard.
- The compressor set in a vacuum chamber.

The output energy achieved after compression is 1.350 J with a total pump energy of 8 J. A special pumping geometry (THALES Laser patented) sets the angles between the four pass beams to a minimum, increasing the overlap inside the crystal and thus increasing the beam quality after amplification.

Special care has been taken to improve the Nd:YAG profiles to produce the smoothest gain profile possible. Wherever needed, an imaging configuration is used to avoid propagation distortion of the pump beam. A special mount for cooling the Ti:Sa has been designed (THALES LASER patented) to reduce very efficiently and at a low cost the aberration introduced by the thermal load. From our experience on high peak power TW lasers, the cryogenic device is very heavy and only needs to be set up for systems over 50 TW.

The results obtained with this laser show outstanding specifications and the contrast is better than 10^{-6}:1 at 20 ps.

2.2 Linear Accelerator

2.2.1 *Photoinjectors*

I. BEN-ZVI

Collider–Accelerator Department Building 817,

Brookhaven National Laboratory,

Upton,

NY 11973, USA

H. IIJIMA

M. UESAKA

Nuclear Engineering Research Laboratory,

University of Tokyo,

22-2 Shirane-shirakata, Tokai, Naka,

IBARAKI 319-1188, JAPAN.

2.2.1.1 *RF cavity and laser*

The device called a *photoinjector* is a fascinating element of linear accelerators (linacs). The number of photoinjectors around the world is large and growing fast, and there is a large body of research on its many aspects. In this Section we will explore the science and technology of photoinjectors. This is not a survey document and does not attempt to fully describe the many variations of photoinjectors, but tries to be complete in the description of the most common photoinjector, the S-band 1.6-cell device developed at Brookhaven National Laboratory (BNL). We will start by explaining why a discussion of photoinjectors is relevant in a book dedicated to ultrashort pulses.

An ultrashort pulse with only a few photons is usually not very useful. To generate ultrashort pulses of electrons, or of photons that are produced by the electrons, we need an electron bunch that has a high electron density in 6-dimensional phase space. Figure 2.13 illustrates the 2-dimensional phase space $(p_x\text{–}x)$. The phase space area due to a broad spread of momentum is larger than that due to a narrow spread of momentum, even if the spatial distributions are the same. Therefore, assuming the same number of electrons, the electron density in a large phase space is smaller than that in a small phase space. Hence, it is not sufficient to have a short bunch if its transverse dimensions are large. Tolerance alone would make such a bunch useless, since it would be impossible to make use of the small longitudinal dimension as long as the transverse size dominates. The relation between electron brightness and photon brightness can be seen in the following example: Eq. (2.30) provides the brightness (in the usual units, photons per second per mm-mrad squared per % bandwidth) of an undulator source, where N_w is the number of periods of the undulator, K is the strength parameter, λ is the wavelength of the photons, q is the electron beam bunch charge, τ is the bunch length and ε_x and ε_y are the (non-normalized) emittances in the two transverse planes:

$$br \cong \frac{2 \cdot 10^{18}}{(\varepsilon_x + \lambda/4\pi)\,(\varepsilon_y + \lambda/4\pi)} \frac{N_w K^2}{1 + K^2/2} \frac{q}{\tau}. \qquad (2.30)$$

The brightness is proportional to the charge divided by the 5-dimensional volume of phase space (if one neglects the photon's "emittance" $\lambda/4\pi$). The expression for brightness does not contain the 6th missing phase space dimension of energy spread only because this expression assumes that the energy spread is much smaller than the undulator natural width.

As seen in Secs. 2.2.2 and 2.2.3, various bunch compression schemes can trade off energy spread and bunch length, therefore there is no need to generate the electron beam with a short bunch; high brightness is all that matters.

This leads to the question: how does one generate the highest brightness electron bunch that still has a useful number of electrons? The answer is to use a photoinjector.

The photoinjector [Fraser *et al.* (1985); Fraser and Sheffield (1986)] is a basic tool for the production of high brightness electron beams. The term photoinjector stands for laser photocathode RF gun. The name indicates the principles of this device. A photocathode is located inside a short microwave resonant cavity, usually followed immediately by one or more additional cavities. A high power source of microwaves energizes the gun's

Photoinjector   Thermionic injector

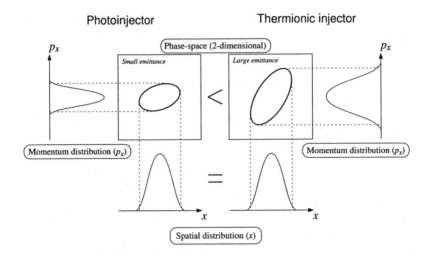

Fig. 2.13   Schematic view of phase space plane. In the case of a photoinjector, the emittance growth is smaller than that of a thermionic gun because the electron beam is accelerated up to relativistic energies in a short distance.

cavities, thus operating as a short linac. The frequency of the microwaves (or radio frequency, RF) can be almost anything. Guns have been built and operated from 144 MHz to 17 GHz, but the two most common frequencies are about 1.3 GHz and about 2.8 GHz. A laser illuminates the photocathode with a short pulse (short relative to the period of the microwave) synchronized with the microwave frequency. Electrons are produced at the cathode surface by the photoelectric effect. The timing (or phase) of the laser relative to the microwave field is such that the electrons are accelerated by the electric field of the cavity. A system of solenoids provides an axial electric field to perform what is called *emittance compensation* (see later). The magnetic field at the cathode is usually set at zero, although there are some specialized applications (flat beam or magnetized beam) that require a particular value of the field at the cathode.

Due to the combination of the high surface field on the cathode and the high yield of electrons possible by photoemission, a very large current density, $J \sim 10^4$–10^5 A/cm^2, is possible. This current density is much larger than that possible by thermionic emission (about 10 A/cm^2). The normalized thermal root mean square (rms) brightness is proportional to the current density. Rapid acceleration also serves to reduce the space charge induced emittance growth. It also makes the photoinjector a very compact accelerator.

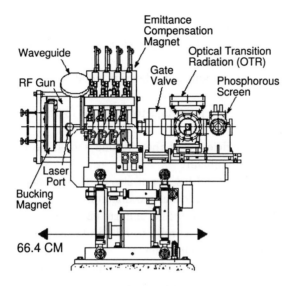

Fig. 2.14   Photoinjector assembly, including solenoid and diagnostics.

Figure 2.14 shows a schematic view of a gun assembly, depicting additional features that are necessary for a working photoinjector. The photoinjector is precision mounted on an emittance compensation solenoid (a technique adopted from BNL Gun III onwards). The gun and solenoid assembly are mounted on a support table with precision alignment adjustments. An in-line vacuum valve is provided to protect the gun if downstream sections are vented to the atmosphere. The gun is pumped through the waveguide connection as well as the beam pipe, usually by titanium sputter pumps. The figure also shows diagnostic chambers with various electron beam diagnostics (and an optional window for laser injection and/or visual inspection and alignment).

The following considerations affect the performance of the photoinjector:

(1) The electric field of the gun, which must be as high as possible.
(2) The geometry of the gun should be such that the high-order terms in the electric field are minimized and the ratio of the cathode field to the peak surface electric field is maximized.
(3) The geometry of the cells and RF power coupler should be such that

departures from cylindrical symmetry are minimized.

(4) The surface of the cavity must be smooth and clean to minimize "dark current" (field emission).

(5) The gun material must be a good electrical and thermal conductor and minimize dark current.

(6) The vacuum in the gun must be very good to prolong the life of the cathode.

(7) The location of the photoinjector relative to the rest of the linac is important.

(8) The quantum efficiency of the cathode should be high enough that lasing is not too difficult.

(9) The cathode size (and laser spot size on the cathode) should be optimized, depending upon other parameters.

(10) The cathode should produce a uniform emission.

(11) The cathode should be as robust and durable as possible (to withstand a rather demanding environment).

(12) The laser illumination should be as uniform as possible, transversely and longitudinally, and should provide a round spot on the cathode.

(13) Short term and long term stability of the various systems are extremely critical, including the RF power, laser pulse energy, laser pointing, laser mode and laser phase.

(14) The solenoid producing the magnetic field must be longitudinal with high precision to avoid breaking the cylindrical geometry.

(15) Good diagnostics must be provided for the laser and electrons.

A photograph of an assembled S-band photoinjector (BNL Gun IV) is shown in Fig. 2.15, showing the massive waveguide input coupling port and its opposing symmetrizing port, the vacuum flange for the electron beam pipe, the water cooling tubes and the stainless vacuum sealing surface between the gun body and the cathode plate.

The mode-locked lasers that drive the photoinjector provide interesting possibilities. The pulses can be made extremely short (to sub-picosecond) and intense (tens of nC at a few ps). The spatial and temporal laser power distributions can be tailored to arbitrary profiles. Particular profiles can lead to the reduction of the emittance of the photoinjector [Gallardo and Palmer (1990a)]. The pulse format is very flexible and pulse trains of arbitrary length and spacing can be generated.

Fig. 2.15   Photograph of BNL Gun IV.

2.2.1.2   *Cathode and quantum efficiency*

The ideal photocathode material would have high emission efficiency (for drive laser cost-containment) and high ruggedness. A study of various materials [Srinivasan-Rao *et al.* (1991a)] for the photocathode has shown that certain metals have a good combination of quantum efficiency, high damage threshold and good mechanical and chemical stability. Copper and yttrium metal cathodes have proved particularly robust. Yttrium has a work function of about 3.1 eV and a quantum efficiency of up to 10^{-3} at 266 nm. Copper's work function is 4.3 eV and it has a quantum efficiency of up to 10^{-4}. Magnesium is widely used in photoinjectors and has demonstrated relatively high quantum efficieny under modest vacuum conditions [Wang *et al.* (1995)]. Magnesium metal can be reliably attached to the copper backplate by friction welding. The magnesium cathode is prepared mechanically by polishing, using three different sizes of diamond polishing compounds progressing from 9 μm to 6 μm and then to 1 μm grain size. The polished surface is rinsed with hexane and then immersed in an ultrasound cleaning hexane bath for 20 minutes, blown with dry nitrogen and finally placed in a high-vacuum chamber for bake-out at 150^{circ}C [Srinivasan-Rao *et al.* (1991b)].

A cathode has to be laser cleaned in order to achieve its peak performance. Laser cleaning is an easy and dependable technique for improving the quantum efficiency of metallic cathodes [Srinivasan-Rao *et al.* (1997);

Wang *et al.* (1998)]. Following the cleaning, the quantum efficiency achieved is about 0.3% with a very high uniformity. This quantum efficiency can remain stable for a few months at a time. The quantum efficiency is highly dependent on the electric field strength, in what is known as the Schottky effect [Schottky (1914); Herring and Nichols (1949)]. For electrons with energy very close to the threshold for emission $(E - E_T \ll E_T)$, the quantum efficiency is given by the following expression:

$$\eta = K \left(h\nu - W + \sqrt{e/4\pi\varepsilon_0}\sqrt{\beta E} \right)^2. \qquad (2.31)$$

Here η is the quantum efficiency, K is a material dependent constant, $h\nu$ is the photon energy, W is the work function, E is the surface electric field at the time of photoemission and β is the field enhancement coefficient. At fields of the order of 100 MV/m the enhancement due to the term with the electric field is appreciable, about 0.38 eV. The dependence of the quantum efficiency on photon energy, work function, electric field strength, field enhancement and intensity of laser cleaning has been studied [Smedly (2001)].

   Another important photocathode material is cesium telluride, which is intermediate between metallic cathodes and cathodes such as CsK_2Sb in terms of quantum efficiency and lifetime. Such cathodes find application in guns that require an intermediate average current. The Tesla Test Facility (TTF) [Schreiber *et al.* (2002)], where beam currents in excess of 1 mA are produced, is one good example. Quantum efficiency better than 0.5% has been reported, but the vacuum required to maintain this performance is 10^{-10} torr, requiring a lock-load system.

   At the University of Tokyo, Nuclear Engineering Research Laboratory (UTNL), Japan, an intense and ultra-short electron bunch from a photoinjector with an S-band accelerating tube and a chicane-type magnetic bunch compressor is utilized for pulse radiolysis experiments. Details of the bunch compressor and the pulse radiolysis experiments are seen in Secs. 2.2.2 and 4.1.2. The photoinjector is same as the type of BNL Gun IV. Copper metal was chosen as the cathode material, which has been changed to magnesium metal recently, and intense charge production was achieved from the copper cathode [Kobayashi *et al.* (2002)]. Figure2.16 shows the produced charge as a function of laser energy for different spot size of the laser on the cathode. Maximum charge of 7 nC was obtained at 250 μJ/pulse laser energy and 5.2 MW RF power (85 MV/m peak gradient). A dash

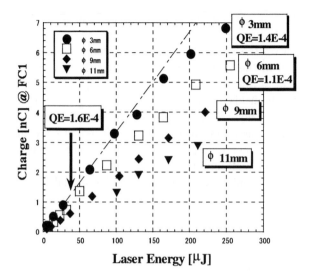

Fig. 2.16   Quantum efficiency for different spot size of the laser on the cathode.

line shown in Fig.2.16 indicates a fitting result up to the energy of 100 μJ which corresponds to 1.6×10^{-4} quantum efficiency, however a decline of QE is found at higher laser energy, for example, QE at 7 nC point is equal to 1.4×10^{-4}. This is caused by the Schottky effect.

A photoinjector constructed by Eindhoven University of Technology (TUE) has a characteristic feature of a part of cathode[De Loos *et al.* (2002)]. The cathode of copper metal is set on outside of a cavity which is a type of 2.5-cell S-band. A wageguide is coupled with the cavity on axis. The laser is driven from front of the cavity on axis, and illuminates the cathode to pass through a whole on the end of cavity whose size is 0.5 mm. Gap length between the cathode and the cavity-end is 2 mm. High pulsed-voltage with amplitude of 2 MV/m and pulse width of 1 ns is applied to the cathode, therefore emitted electron is accelerated up to relativistic energy in the gap. Consequently the emittance grouth due to the space-charge effect in the cavity is reduced. They have estimated the emittance to be less than 1 πmm·mrad for producted charge of 100 pC/bunch.

2.2.1.3   *Emittance control*

An early analytical model for the emittance and energy spread of a photoinjector was developed by Kim [Kim (1989)]. This model provides scaling laws that allow insight into the relationship of some of the design parameters of photoinjectors. The Kim model calculates the effects of space charge and RF fields on the emittance. The space charge component is created by a variation in the space charge force along the electron bunch. The RF component is due to differential focusing applied to various parts of the bunch by the fields at the exit of the cavity. These are linear effects. Kim's model does not account for possible charge distribution changes in the bunch, to thermal emittance or to the process of emittance compensation, which does not take place in the photoinjector, but in the space downstream of the photoinjector.

Using practical units, Kim obtained for the normalized, space charge related emittance

$$\varepsilon_{sc} \approx 3.8 \times 10^3 q \left(2\sigma_x + \sigma_b\right)^{-1} \left(E_0 \sin \phi_0\right)^{-1} \qquad (2.32)$$

and for the normalized, RF related emittance

$$\varepsilon_{rf} \approx 2.7 \times 10^{-5} E_0 f^2 \sigma_x^2 \sigma_b^2 \qquad (2.33)$$

where the emittances are expressed in π mm·mrad, E_0 is the cathode peak electric field in MV/m, f is the gun frequency in GHz, q is the charge in nC, σ_b is the rms bunch length in ps and σ_x is the rms transverse size in mm. ϕ_0 is the launch phase, typically 30–60°.

For a given cathode electric field, charge and beam size, the emittance is optimized by

$$f_{\mathrm{opt}} = 1.2 \times 10^4 \left(\sigma_b \sigma_x E_0 \sin \phi_0\right)^{-1} q^{1/2} \left(\sigma_b + 2\sigma_x\right)^{-1/2} \qquad (2.34)$$

and the optimized total emittance (neglecting correlations as well as thermal emittance) is

$$\varepsilon_{\min} \approx \left[\varepsilon_{rf}^2 + \varepsilon_{sc}^2\right]^{1/2} \approx 5.4 \times 10^3 q \left(E_0 \sin \phi_0\right)^{-1} \left(\sigma_b + 2\sigma_x\right)^{-1} . \qquad (2.35)$$

Since the minimum emittance is proportional to the charge q (and thus to the peak current), the highest brightness is not necessarily associated with the highest charge. Since we have left out the thermal emittance in these expressions, one should not conclude that the brightness is maximized for an extremely small charge.

The minimum emittance (using the given optimized frequency) is inversely proportional to the electric field. Thus, for a given set of beam parameters (charge and bunch size), the highest possible electric field should be applied. As we increase the field with other parameters kept constant, the optimal frequency is lowered. However, for a number of practical reasons the technically achievable field is smaller at lower frequencies. At some fields and frequencies, the photoinjector will operate at the limit of breakdown or available RF power. Once we cross that limit the assumptions of the optimization break down and we cannot apply these results. In addition, the thermal emittance limit may be approached. The thermal emittance of a photocathode with electron excess emission energy Δ and a cathode radius ρ is given by

$$\varepsilon_N = \frac{\rho}{4}\sqrt{\frac{2\Delta}{mc^2}}\,. \qquad (2.36)$$

For excess emission energy of 0.5 eV (which is believed to approximately represent metal cathodes) the thermal normalized emittance is 0.35 μm for a cathode radius of 1 mm. The issue of excess photon energy in calculating the thermal emittance requires more discussion, as the orientation of the excess photon momentum is not strictly random for electrons that have sufficient perpendicular momentum to overcome the work function. The orientation of the photon momentum is random initially, but the requirement that the electrons follow a trajectory that allows escape (step 3 in the three-step model) imposes a selection rule on emitted electrons that forward-peak their energy distribution, and thus reduces the thermal emittance.

The beam of a photoinjector has significant correlations to the longitudinal position and transverse phase space. This is the key to its emittance correction schemes (to be discussed later). When one uses an emittance correction scheme, the space charge emittance is reduced. This invalidates the conditions of the calculation presented above, pushing the optimum towards lower frequencies, lower electric fields and smaller beam sizes. In addition, the high brightness of the photoinjector will be diluted by any one of a large collection of effects: wake fields, beam transport aberrations, dispersion, skew quadrupoles, and others. For example, an ambient magnetic field B on a cathode (from magnetic lenses or ion pumps) produces an emittance increase given by

$$\varepsilon_n \approx \left[\varepsilon_{n0}^2 + e^2 B^2 \sigma_r^4 m^{-2} c^{-2}\right] . \qquad (2.37)$$

Thus fields of the order of 10 G may be detrimental to emittance. Good design practices call for a rapid acceleration of the beam to a few tens of MeV before applying dipole fields. Thus, magnetic pulse-compression is better done above, say, 70 MeV. Pulse compression has always been part of conventional electron gun technology. Although the beam pulses of a photoinjector start out short, the brightness can be further increased by magnetic pulse-compression.

An important point that was not fully appreciated in early work is the significance of the uniformity of the emission from the cathode. A non-uniformity can result from either non-uniform laser illumination or non-uniform quantum efficiency across the area of the cathode. It turns out that the laser cleaning technique described above leads to a very uniform quantum efficiency, which may explain in part the high brightness of the BNL Accelerator Test Facility (BNL-ATF) beam, where this cleaning technique is routinely applied. A recent experimental study [Zhou *et al.* (2002)], characterized emittance as a function of non-uniformity (creating artificial non-uniformity by laser masks). One of the products of this study is shown in Fig. 2.17, in which the emittance is measured for a range of rms non-uniformity values for a checkerboard pattern of non-uniformity. With a highly uniform emission from its high quantum efficiency magne-

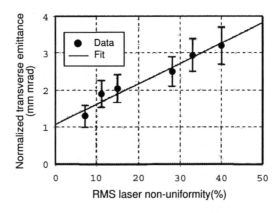

Fig. 2.17   Normalized transverse emittance as a function of the laser non-uniformity.

sium cathode, the BNL-ATF photoinjector exhibited a record brightness, as described by Yakimenko. The result was an rms normalized emittance of 0.8 μm for a bunch charge of 0.5 nC.

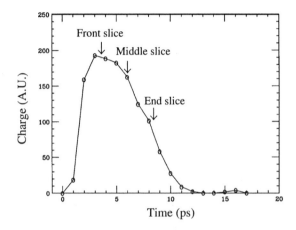

Fig. 2.18   Longitudinal charge distribution of electron beam bunch [BNL website (2003)].

An interesting photoinjector subject is emittance correction. We define the *slice emittance* as the transverse emittance measured for a short longitudinal slice of the bunch. It has been observed computationally [Carlsten (1989)] that the slice emittance is considerably smaller than the total emittance (that is integrated over the full length of the bunch). This effect is due to the variation of the space charge force as a function of longitudinal position in the bunch within a regime of longitudinal laminarity. Indeed, since the electrons of a slice do not mix with electrons of other slices (through lack of synchrotron motion), each slice behaves like an independent beam subject to a space charge field that varies from slice to slice (the amount of variation, in a rms sense, is minimized for a uniform distribution of the current in the bunch). Carlsten [Carlsten (1989)] proposed a simple scheme for reducing the total emittance by using the space charge force to compensate its own effect. The method employs a lens set to produce a beam size waist with no crossover. The electrons "reflect" relative to the beam axis due to space charge forces. This condition, which can restore the effects of the linear space charge force, has been verified in an experiment [Qiu *et al.* (1996)]. In this experiment, the emittance of a picosecond long slice of an electron bunch was measured, as shown in Figs. 2.18 and 2.19. A short

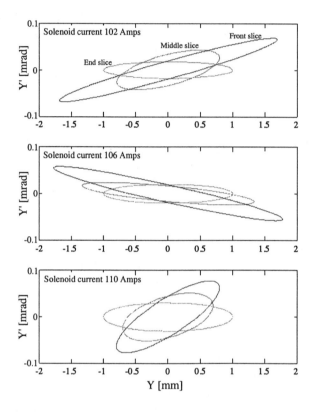

Fig. 2.19   Plots of transverse phase space ellipses for each electron bunch slice.

slice is selected out of an energy chirped beam by a slit in a dispersive region. The emittance is measured using the quadrupole scan technique. The process of emittance compensation of the beam is observed by repeating the measurement for various values of the compensating solenoid and for several slices.

The Carlsten emittance correction technique corrects linear space charge effects. Other correction schemes have been proposed to produce the same correction by laser pulse shaping [Gallardo and Palmer (1990a)], RF quadrupoles [Gallardo and Palmer (1990b)] and asymmetric RF cavities [Serafini *et al.* (1992a)]. However the Carlsten scheme is simple and has been tested experimentally. Other correction schemes have been proposed to correct RF time dependent effects [Serafini *et al.* (1992b)] and nonlinear space charge effects [Serafini *et al.* (1992a);

Gao (1991)]. Finally, a correction scheme for ultrashort, disk-like bunches using an optimized charge distribution has been proposed by Serafini [Serafini (1992)]. Rosenzweig and Colby [Rosenzweig and Colby(1995)] have provided wavelength and charge scaling laws for photoinjectors as a function of various parameters, such as bunch size and electric field. One finds that, when optimum conditions are maintained for a variable wavelength, the gun electric field and solenoid magnetic field must scale inversely with the wavelength; the beam size σ (in any dimension), the charge q and the emittance scale with the wavelength; and the brightness with the wavelength to the inverse second power:

$$E, B \propto \lambda^{-1}, \qquad \sigma, q, \varepsilon \propto \lambda, \qquad br \propto \lambda^{-2}. \tag{2.38}$$

If the charge is constrained to a particular value, the scaling is different [Rosenzweig et al. (1999)]. The beam size then would be proportional to the cube root of the charge:

$$\sigma \sim q^{1/3}. \tag{2.39}$$

Also, the normalized rms emittance is given as a function of the charge as:

$$\varepsilon_N = \sqrt{aq^{4/3} + bq^{8/3}}, \tag{2.40}$$

where a and b are constants that depend on the specific photoinjector. Rosenzweig et al. [Rosenzweig et al. (1999)] provide the values for a plane wave transformer (PWT) photoinjector as $a = 1.34$ and $b = 0.11$ when the emittance is in μm and the charge in nC.

The emittance correction mechanism is fully explained in a beautiful work by Serafini and Rosenzweig [Serafini and Rosenzweig (1997)]. They developed a theory of longitudinal slice-by-slice electron bunch propagation with space charge, electric and magnetic fields that describe the photoinjector and the linac that follows it. In this theory there is an invariant envelope (which is basically a beam envelope for a matched beam that will undergo optimal emittance correction). This predicts an evolution of the beam rms spot size through the booster linac following the gun. The rms beam size scales like the inverse of the accelerating gradient times the square root of the current divided by the beam energy. If the beam is properly matched out of the gun into the booster linac by adjusting a beam waist at the linac entrance and regulating the linac gradient in such a way as to fulfill the invariant envelope condition, the final emittance at the linac exit is minimized. Advanced designs of high brightness injectors for X-ray free electron lasers, like the Lincac Coherent Light Source (LCLS) injector, make full use of these theoretical criteria, achieving the best emittances in simulation [Bolton *et al.* (2002)].

A technique of laser pulse shaping was developed for low-emittance electron beam generation by Sumitomo Heavy Industires[Yang *et al.* (2002)]. The emittance growth due to space-charge and RF effects in the RF gun was experimentally investigated with square and Gaussian temporal laser pulse shapes. The temporal pulse shaper was accomplished through a technique of frequency-domain pulse shaping. The spectrum of the incident femtosecond laser pulse was dispersed in space between a pair of diffraction gratings separated by a pair of lenses (see Fig.2.20). A computer-addressable liquid-

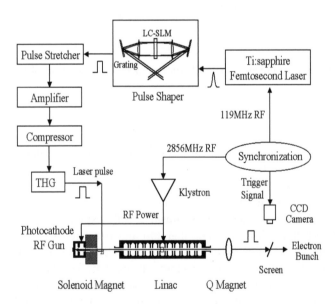

Fig. 2.20   Experimental arrangement.

crystal spatial light modulator (LC-SLM) with 128 pixels was used as the phase mask. The resolution of the phase shift on LC-SLM was near 0.01π.

The pulse shaper was located between the oscillator and the pulse stretcher to reduce the possibility of damage on the optics. The typical Gaussian and square-shaped temporal distributions of the UV laser pulses with a pulse length of 9 ps FWHM are shown in Fig.2.21 (left). The data was measured by an X-ray streak camera with a time resolution of 2 ps, resulting a rise time of 1.5 ps for the square pulse shape. The pulse-to-pulse fluctuation of the shaped pulse length was 7%. The spatial profile of the laser beam on the cathode is shown in Fig.2.21 (right). The beam spot size was 1.2 mm and 0.4 mm FWHM in the horizontal and vertical directions, respectively. The normalized rms horizontal emittance measured

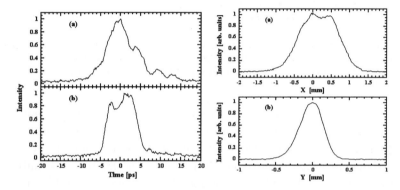

Fig. 2.21 Temporal distribution of the Gaussian laser pulse (a) and square laser pulse 9 ps FWHM (b). Spatial profile of the laser beam in the horizontal (c) and vertical (d) directions.

as a function of the laser pulse length is shown in Fig.2.22 (left) for the Gaussian and square temporal pulse shapes. The electron bunch charge was fixed at 0.6 nC and the solenoid field was set to 1.5 kG which was optimal for compensating the space charge emittance at 0.6 nC. The data shows that the emittance increases at shorter and longer laser pulse length regions for both the Gaussian and square pulse shapes. This is behaved in emittance growth due to space charge and rf effects. The normalized rms horizontal emittance was also measured as a function of the bunch charge for the Gaussian and square temporal pulse shapes with a pulse length of 9 ps FWHM, as shown in Fig.2.22 (right). The measured data was fit as a function of

$$\varepsilon = \sqrt{(a'Q)^2 + b'^2} \tag{2.41}$$

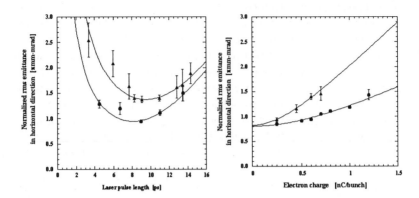

Fig. 2.22   The left figure is the emittance versus laser pulse at 0.6 nC for the Gaussian (triangle) and square (dot) pulse shapes. The right figure is the emittance versus bunch charge for the Gaussian (triangle) and square (dot) pulse shapes at a pulse length 9 ps FWHM.

where a' is a fitting parameter referred to space charge force, and b' in $\pi mm \cdot mrad$ is a zero charge emittance. It is found that the square pulse shape reduced the space charge force of about 50% comparing with the Gaussian pulse shape. Consequently, the optimal normalized rms emittance of $1.2\pi mm \cdot mrad$ at 1 nC was obtained by a square temporal laser pulse shape with a pulse length of 9 ps FWHM.

2.2.2   *Magnetic bunch compression*

M. UESAKA
K. KINOSHITA
Nuclear Engineering Research Laboratory
University of Tokyo,
22-2 Shirane-shirakata, Tokai, Naka,
IBARAKI 319-1188, JAPAN.

2.2.2.1   *Analogy with chirped pulse amplification (CPA) for femtosecond lasers*

Magnetic bunch compression, used to generate femtosecond electron pulses, is analogous to chirped pulse amplification (CPA) in femtosecond lasers, as shown in Fig. 2.23. CPA consists of chirping (longitudinal energy modulation), stretching, amplification and compression. Generally a set of gratings is used for chirping and compression. Higher/lower energy photons are located in an earlier/later part of laser pulses for downward/upward chirping. If we replace photons with electrons, chirping and compression become magnetic bunch compression. Downward and upward chirping are respectively done by putting the electron beam on an increasing or decreasing RF phase of the accelerating traveling wave in an accelerating tube. Then the beam is passed through a magnet assembly consisting of bending and focusing magnets. The chirping is transformed to path length modulation because the bending radius is inversely linearly proportional to the electron energy. Therefore, the later electrons can catch up with the earlier electrons and the beam is compressed.

2.2.2.2 *Theory*

Magnetic bunch compression can be considered to be the rotation of an electron bunch in longitudinal phase space, as shown in Fig. 2.24 [Emma (2002)]. In order to perform the compression effectively, the electron distribution should be thin enough initially. In other words, the longitudinal emittance of the bunch should be low enough. For this purpose, a photoinjector as explained in Section 2.2.1, should be used as an electron source. An electron distribution as straight as possible along the vertical axis is preferred. We perform energy modulation by using the curvature of the sinusoidal RF wave in an accelerating tube. Thus, the energy dependence on the longitudinal coordinate has not only a linear component but also higher order components. This nonlinearity contributes to deformation of the final electron distribution, so that the compression is not even. We can cancel the distortion in the compression by the nonlinearity of the transformation of the energy modulation to path length difference at a magnetic bunch compressor. For a linear transformation, the matching between the energy modulation and path length modulation induces the over-/under-compression shown in Fig. 2.24. The effect of the higher order components gives the deformation of the electron distribution. Matching of the higher order components suppresses the deformation. The relevant theory of beam dynamics [Lee (1999)], a typical design and an experimental verification are introduced here. The electron coordinates in an accelerator can be charac-

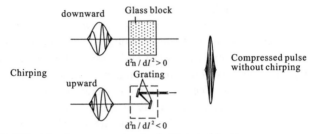

(i) Chirped pulse compression for femtosecond laser

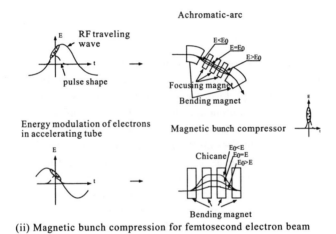

(ii) Magnetic bunch compression for femtosecond electron beam

Fig. 2.23   (i) Chirped pulse amplification for laser pulse and (ii) magnetic compression for electron pulse.

terized by a state vector:

$$
W = \begin{pmatrix} W_1 \\ W_2 \\ W_3 \\ W_4 \\ W_5 \\ W_6 \end{pmatrix} = \begin{pmatrix} x \\ x' \\ z \\ z' \\ \beta c \Delta t \\ \delta \end{pmatrix}, \tag{2.42}
$$

where βc is the speed of the electron, $\beta c \Delta t$ is the path length difference with respect to the reference orbit and $\delta = \Delta p / p_0$ is the fractional energy

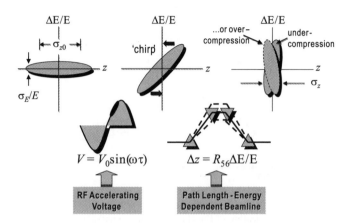

Fig. 2.24 Magnetic bunch compression with over-/under-compression [Emma (2002)].

deviation of an electron. The transport of the state vector in a linear approximation is given by

$$W_i\left(s_2\right) = \sum_{j=1}^{6} R_{ij}\left(s_2|s_1\right) W_j\left(s_1\right), \quad \left(i, j = 1, \cdots, 6\right). \tag{2.43}$$

Note that the 2×2 diagonal matrices for the indices 1,2 and 3,4 are respectively the horizontal and the vertical M matrices. The R_{13}, R_{23}, R_{14} and R_{24} elements describe the linear betatron coupling. The R_{16} and R_{26} elements represent the dispersion. Without synchrotron motion, we have $R_{55} = R_{66} = 1$. All other elements of the R matrix are zero.

In general, the nonlinear dependence of the state vector can be expanded as

$$W_i\left(s_2\right) = \sum_{j=1}^{6} R_{ij}W_j\left(s_1\right) + \sum_{j=1}^{6}\sum_{k=1}^{6} T_{ijk}W_j\left(s_1\right) W_k\left(s_1\right)$$

$$+ \sum_{j=1}^{6}\sum_{k=1}^{6}\sum_{l=1}^{6} U_{ijkl}W_j\left(s_1\right) W_k\left(s_1\right) W_l\left(s_1\right) + \cdots. \tag{2.44}$$

If we neglect the coupling between the longitudinal and transverse motions for simplicity, we get

$$W_5 = R_{56}\delta + T_{566}\delta^2 + U_{5666}\delta^3 + \cdots. \tag{2.45}$$

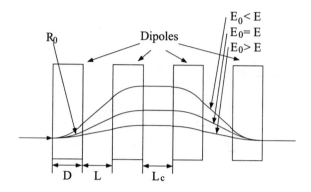

Fig. 2.25   Configuration of the chicane-type magnetic bunch compressor.

Here, we try to obtain R_{56} and T_{566} in the realistic chicane-type bunch compressor. Its configuration is depicted in Fig. 2.25. The path length of a nominal electron having energy E_0 can be formulated as

$$S = 4R \arcsin\left(\frac{D}{R}\right) + \frac{2LR}{\sqrt{R^2 - D^2}} + L_c , \qquad (2.46)$$

where R is the bending radius in the bending magnet of the chicane, D is the width of the magnets and L and L_c are the gaps between the magnet (see Fig. 2.25). R is written as

$$R = \frac{\gamma m_0 v}{eB} ,$$

where γ is the Lorentz factor, m_0 is the electron mass, e is the unit charge and B is the magnetic field. Assuming that the electron is relativistic with $v \approx c$, R is linear with γ and we have

$$R = R_0 \frac{\gamma}{\gamma_0} .$$

When we substitute this into Eq. (2.46) we have,

$$S(\gamma) = 4R_0 \frac{\gamma}{\gamma_0} \arcsin\left(\frac{\gamma_0 D}{\gamma R_0}\right) + \frac{2LR_0 \frac{\gamma}{\gamma_0}}{\sqrt{R_0^2 \frac{\gamma^2}{\gamma_0^2} - D^2}} + L_c . \qquad (2.47)$$

Next, we calculate the movement of the electron in the bunch, which has an energy distribution, when the bunch passes through the chicane. We expand Eq. (2.47) using the energy deviation $\delta = (\gamma - \gamma_0)/\gamma_0$. Expressing Eq. (2.47) using δ we get

$$S(\delta) = 4R_0(1 + \delta) \arcsin\left(\frac{D}{R_0(1 + \delta)}\right) + \frac{2LR_0(1 + \delta)}{\sqrt{R_0^2(1 + \delta)^2 - D^2}} + L_c. \quad (2.48)$$

Since δ is small, $S(\delta)$ can be expanded up to the third-order term as

$$S(\delta) = S(0) + R_{56}\delta + T_{566}\delta^2 + U_{5666}\delta^3, \quad (2.49)$$

where

$$R_{56} = 4R_0 \arcsin\left(\frac{D}{R_0}\right) - \frac{4DR_0}{\sqrt{R_0^2 - D^2}} - \frac{2LD^2R_0}{(R_0^2 - D^2)^{3/2}}, \quad (2.50)$$

$$T_{566} = \frac{2D^3R_0}{(R_0^2 - D^2)^{3/2}} + \frac{3LD^2R_0^3}{(R_0^2 - D^2)^{5/2}} > 0, \quad (2.51)$$

$$U_{5666} = \frac{-8D^3R_0^3 + 2D^5R_0}{3(R_0^2 - D^2)^{5/2}} - \frac{4LD^2R_0^5 + LD^4R_0^3}{(R_0^2 - D^2)^{7/2}}. \quad (2.52)$$

We write the normalized longitudinal coordinate ζ of the electron in a bunch as

$$\zeta = \frac{\omega}{c}(z_0 - z), \quad (2.53)$$

where z is the longitudinal coordinate of the electron, z_0 is the center of the bunch, ω is the RF frequency and c is the speed of light.

The energy of the electron can be expressed by ζ in the following:

$$E = C_0 + C_1\zeta + C_2\zeta^2 + C_3\zeta^3, \quad (2.54)$$

and

$$\frac{\gamma - \gamma_0}{\gamma_0} = \frac{C_1}{C_0}\zeta + \frac{C_2}{C_0}\zeta^2 + \frac{C_3}{C_0}\zeta^3. \quad (2.55)$$

Eqs. (2.54) and (2.55) express the nonlinear energy modulation of the acceleration RF wave in an acceleration tube. Substituting Eq. (2.55) into Eq. (2.46) gives

$$S(\zeta) = S(0) + \frac{C_1}{C_0}b_1\zeta + \left(\frac{C_2}{C_0}b_1 + \frac{C_1^2}{C_0^2}b_2\right)\zeta^2 + \left(\frac{C_3}{C_0}b_1 + \frac{2C_1C_2}{C_0^2}b_2 + \frac{C_1^3}{C_0^3}b_3\right)\zeta^3. \quad (2.56)$$

The normalized coordinate after the chicane becomes

$$\zeta' = \zeta + S(\zeta) - S(0) \tag{2.57}$$

$$= \zeta_1 + \zeta_2 + \zeta_3, \tag{2.58}$$

where

$$\zeta_1 = \left(1 + \frac{C_1}{C_0}R_{56}\right)\zeta, \tag{2.59}$$

$$\zeta_2 = \left(\frac{C_2}{C_0}R_{56} + \frac{C_1^2}{C_0^2}T_{566}\right)\zeta^2, \tag{2.60}$$

$$\zeta_3 = \left(\frac{C_3}{C_0}R_{56} + \frac{2C_1C_2}{C_0^2}T_{566} + \frac{C_1^3}{C_0^3}U_{5666}\right)\zeta^3. \tag{2.61}$$

Making ζ' zero gives perfect compression. Since ζ is small, ζ_1 mainly contributes to ζ'. We choose the parameters of the chicane (D, R_0, L, L_C) and of the RF wave (C_0, C_1, C_2, C_3) so as to minimize ζ'. We show now an example of the parameters. If we put the bunch at the RF phase and amplitude as shown Fig. 2.26, C_0, C_1, C_2 and C_3 become 16.0, 14.7, -25.3 and 80.2, respectively, and we have

$$R_{56} = -\frac{C_1}{C_0} = -0.92. \tag{2.62}$$

C_1 is negative for the FODO cell type or achromatic arc type and therefore $R_{56} > 0$ in those cases. By choosing D, L and R_0 to satisfy Eq. (2.62), we try to minimize ζ_2 by changing L as shown in Fig. 2.27. Finally, we determined D, L and R_0 to be 115 mm, 191 mm and 82 mm. The numerical result of compression for the original 7 ps FWHM bunch using the PARMELA software package is shown in Fig. 2.28. We can compress it down to 200 fs FWHM, but both tails remain due to the higher order terms.

2.2.2.3 *Experiment*

Here we introduce the Femtosecond Quantum Phenomena Research Facility at the Nuclear Engineering Research Laboratory, University of Tokyo, [Uesaka *et al.* (2001)], as shown in Fig. 2.29. At the 35-MeV linac we have an achromatic-arc-type compressor for downward chirping, and at the 18-MeV linac we have a chicane-type compressor for upward chirping. Bunch shapes obtained with and without bunch compression are shown in Fig. 2.30. The 13-ps bunch is compressed to 440 fs FWHM [Uesaka *et al.* (2000)]. The

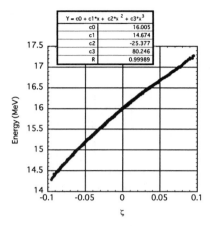

Fig. 2.26   Energy distribution of the electron bunch on the sinusoidal RF wave.

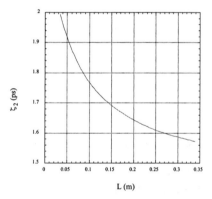

Fig. 2.27   Second order longitudinal deviation vs. chicane magnet gap.

best pulse compression was done around the phase of the best emittance. After this measurement, we calibrated the streak camera by using an 86-fs FWHM Ti:Sapphire laser (Spectra-Physics, Tsunami). The 86-fs laser pulse was elongated to 390 fs. Thus, according to the error propagation law, the error at the camera was estimated to be 370 fs FWHM. We found that this error was mainly due to the degradation of the microchannel plate (MCP) in the streak tube. If we assume independent Gaussian distributions

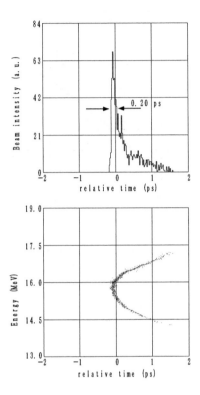

Fig. 2.28   Numerical result of bunch compression by Parmela.

for both the electron bunch and the error, the error propagation law gives a bunch length of 240 fs FWHM after the error reduction [Uesaka *et al.* (2000)]. When we do not assume that the distributions are Gaussian, and use the error propagation function evaluated from all measured data, the bunch length becomes 290 fs FWHM. After this experiment, we replaced the degraded MCP with a new one, the error of which had been evaluated by the same procedure to be 240 fs. A 700 fs FWHM electron single bunch was obtained at the achromatic-arc compressor [Uesaka *et al.* (1994); Uesaka *et al.* (1998)].

Sophisticated higher order compensation was analyzed and proposed for the design of an X-ray FEL [LCLS (2002)], such as a two-staged compressor system [Emma (1998)], X-band RF harmonic compensation [Emma (2001)],

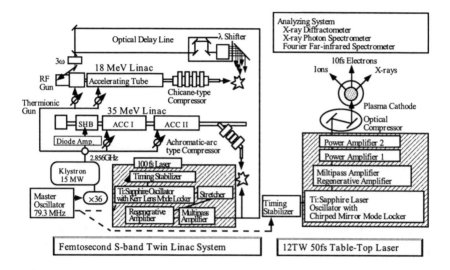

Fig. 2.29 Femtosecond Ultrafast Quantum Beam Research Facility of the Nuclear Engineering Research Laboratory, University of Tokyo.

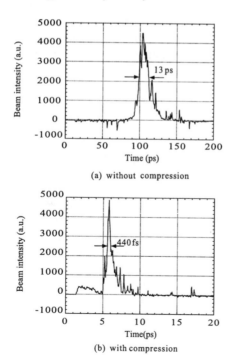

Fig. 2.30 Pulse shapes (a) without and (b) with compression, measured by femtosecond streak camera.

and 433 MHz–1300 MHz linac system [Dowell *et al.* (1995)].

2.2.2.4   *Other subjects*

Several works are under way in order to improve the precision of compression by taking account of the following subjects:

- Geometric wake field
  A compensation scheme has been numerically proposed and analyzed using the point charge wake field function developed by Bane [LCLS (2002); Bane *et al.* (1998)].
- Diagnosis
  Bunch compression must proceed along with the progress of diagnostic methods of such short electron bunches in the longitudinal and transverse directions. The details are described in Chapter 3.
- Stability
  Due to the RF phase and amplitude jitters, the best compression is found for several shots. Therefore, it is crucial to have a sophisticated diagnostic tool, such the femtosecond streak camera, to watch this instability. We have to stabilize the RF by using a very high quality RF modulator with voltage jitter of less than 0.2%. Also, a feedback control system should be implemented in the near future.

2.2.2.5   *Effect of CSR force*

R. HAJIMA
Advanced Photon Research Center,
Japan Atomic Energy Research Institute (JAERI)
Tokai, Naka,
IBARAKI 319-1195, JAPAN

A magnetic bunch compressor is an intrinsic device for the generation of femtosecond electron bunches in electron linacs. One of the important issues in the design of magnetic bunch compressors is emittance growth by coherent synchrotron radiation [Derbenev *et al.* (1995); Saldin *et al.* (1997); Li (1999)].

When a relativistic electron bunch travels in a circular path, such as that induced by a bending magnet, the electrons emit synchrotron radiation. For radiation wavelengths longer than the electron bunch length, the synchrotron radiation from each electron builds up with the same radiation phase. This is known as *coherent synchrotron radiation* (CSR). Owing to this coherent build-up of radiation, the power of the coherent radiation is proportional to the square of the number of electrons in the bunch, $P_{coh} \propto N^2$, while incoherent synchrotron radiation with wavelengths shorter than the bunch length obeys a scaling law, $P_{incoh} \propto N$.

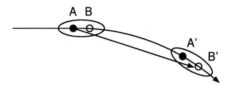

Fig. 2.31 Electromagnetic self-interactions inside an electron bunch with a bending path. Radiation from an electron in the bunch tail overtakes another electron in the bunch head.

Coherent synchrotron radiation can be also be explained by electromagnetic self-interactions inside an electron bunch with a bending path. Figure 2.31 shows that an electromagnetic field is generated by an electron located at the bunch tail overtaking another electron in the bunch head. This overtaking field induces a non-uniform energy spread in the bunch and results in a bending angle error beyond the bending magnet. Since the CSR-induced energy loss is a function of local electron density in the bunch, each bunch slice has a different deviation from the designed trajectory, as shown in Fig. 2.32. As a result of the CSR kick, transverse emittance in the bending plane dilutes (Fig. 2.32). Note that the emittance growth by coherent synchrotron radiation is mainly a growth of the projection emittance; the growth of the slice emittance is not so large.

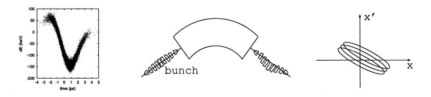

Fig. 2.32 (left) CSR-induced energy spread in an electron bunch for parameters $\sigma_s = 0.3$ mm, $\rho = 2$ m, $\theta = 20°$, $Q = 1$ nC. (center) Deviation of bunch slices after bending due to CSR. (right) Emittance growth by CSR kick.

As we have seen in Fig. 2.31, the power of coherent synchrotron radiation does not depend on the electron energy and is determined by geometrical conditions, such as curvature radius, bunch length and electron density. For a bunch with Gaussian distribution, the CSR power is given by [Schiff (1946)]

$$P_{coh} \simeq \frac{Q^2 c}{\rho^{2/3} \sigma_s^{4/3}} \frac{3^{1/6} \Gamma^2(2/3)}{2\pi}.$$ (2.63)

If we assume the parameters, bunch length $\sigma_s = 0.3$ mm (1 ps), bending radius $\rho = 2$ m, bending angle $\theta = 20°$, bunch charge $Q = 1$ nC and electron energy $E = 100$ MeV, the total radiation energy along the bending path becomes ~ 70 μJ, which corresponds to the average energy loss of the electrons, $\Delta E/E = 70$ keV/100 MeV= 0.07%.

For a femtosecond electron beam application that requires small emittance, such as X-ray FEL, the effect of coherent synchrotron radiation is a critical source of emittance dilution. In the design study of LCLS (SLAC XFEL), it was found that the CSR-induced emittance growth can be partially compensated by a tandem bunch compressor, as shown in Fig. 2.33 [LCLS (2002)]. The emittance compensation is performed by canceling two CSR kicks which have opposite directions from two bunch compressors.

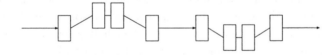

Fig. 2.33   Tandem bunch compressor to reduce CSR-induced emittance growth.

2.2.3 Velocity bunching

L. SERAFINI
Department of Physics,
INFN-Milan and University of Milan
Via Celoria 16,
20133 MILAN, ITALY

In this Section we describe a new method to reduce the bunch length of high brightness electron beams, which enhances the present capabilities of photoinjectors towards attaining the beam quality needed by X-ray FELs. Since the present technology of photoinjectors is capable of bunches no shorter than a few ps for bunch charges of a few nC (at the necessary emittance for driving SASE-FELs), while the pulse length requirements of FELs are in the 10–100-fs range, a bunch compression mechanism is required. Furthermore, there are many other applications of femtosecond long electron bunches that well justify the effort of reducing the bunch length of high-brightness electron beams.

In the previous section a technique based on dispersive devices, such as magnetic chicanes, was described. In this section we present a technique called *velocity bunching* [Serafini and Ferrario (2001)], based on a rectilinear compression process performed in a booster linac following a gun. This technique has been extensively explored by numerical simulation studies of the whole photoinjector system which generates ultrashort electron bunches up to 150 MeV [Serafini *et al.* (2001); Boscolo *et al.* (2002)]. However, recently the first experimental confirmations of simulations and theoretical predictions have been observed at a number of laboratories [Piot *et al.* (2002); Musumeci (2002); Musumeci and Rosenzweig (2003)]. Other experiments in the past showed some evidence of this process [Wang *et al.* (1996)]. The promise is to attain 10–100-fs long bunches with charges ranging from 10 pC up to a few nC.

Before discussing this technique in detail, let us remember that rectilinear bunching techniques have been applied extensively in the past, both in the field of thermionic injectors and in the field of photoinjectors. These are based on inducing an energy chirp on a moderately relativistic beam and letting it drift in free space to turn the energy correlated chirp into a rotation of the longitudinal phase space distribution, hence leading to a compression of the bunch. In this way the peak current is increased, leading to strong space charge effects caused by an increase of the phase space density occurring at a constant energy. As has been discussed in [Boscolo et al. (2002)] and mentioned later, this unavoidably causes a transverse emittance deterioration. We will call this well-known technique *ballistic bunching*, mainly referring to the ballistic behavior of the applied phase space rotation in free space.

The reason for the name *velocity bunching* is the following: although the phase space rotation in this process is still based on a correlated velocity chirp in the electron bunch in such a way that electrons in the tail of the bunch are faster than electrons in the bunch head, this rotation does not happen in free space but inside the longitudinal potential of a traveling RF wave, accelerating the beam, for instance inside a long multi-cell traveling wave RF structure. This is possible if the injected beam is slightly slower than the phase velocity of the RF wave, so that, when injected at the crossing field phase (no field applied), it will slip back to a phase where the field is accelerating, but at the same time it will be chirped and compressed. The key point is that compression and acceleration take place at the same time throughout the photoinjector, from a few MeV (> 4) up to about 30 MeV.

Therefore, the name *velocity bunching* refers to a correlation between particle velocity and position or phase within the bunch, as opposed to the magnetic compression technique, which relies on a correlation between the particle path length through the device and the particle phase. Since the chicane performing magnetic compression is often called the *magnetic compressor*, the linac section performing *velocity bunching* is often called the *RF compressor* or *bunch compressor*. Of course, this is not to be confused with compression of RF pulses such as that accomplished in a SLED device.

2.2.3.1 Theory

(a) Longitudinal dynamics
The theoretical model explaining velocity bunching is as follows. The interaction of a beam with an RF wave is given by

$$E_z = -E_0 \sin(\omega t - kz + \psi_0)\,, \qquad (2.64)$$

which is described by a Hamiltonian

$$H = \gamma - \beta_r \sqrt{\gamma^2 - 1} - \alpha \cos \xi\,, \qquad (2.65)$$

where the normalized electron energy is $\gamma = 1 + T/mc^2$ and $\xi = kz - \omega t - \psi_0$ is its phase with respect to the wave, while $\alpha \equiv eE_0/mc^2 k$ is the dimensionless vector potential amplitude and $\gamma_r = 1/\sqrt{1 - \beta_r^2}$ is the resonant gamma of the wave (conventional traveling wave structures operate at $\beta_r = 1$ and $k = \omega/\beta_r c$. A wave with phase velocity slightly smaller than c, so that $k = \omega/c + \Delta k$, is characterized by $\beta_r = 1 - c\Delta k/\omega$ and $\gamma_r = \sqrt{\omega/2c\Delta k}$ $(c\Delta k/\omega \ll 1)$.

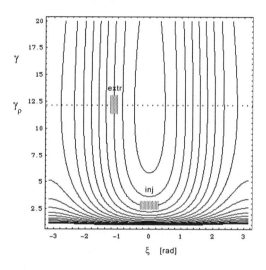

Fig. 2.34   Phase space plots of a slow RF wave (γ_r finite) showing the basics of phase compression in a linac.

The basic mechanism underlying the rectilinear compression effect is illustrated in Fig. 2.34, where the contour lines of the Hamiltonian associated with a slow RF wave having $\gamma_r = 12$ (i.e. $\beta_r = 0.9965$ and $\Delta k = 0.0035 \, \omega/c$) and $\alpha = 0.2$ are plotted. If the bunch is injected into the wave at zero phase (i.e. when the field is not accelerating) at an energy lower than the synchronous energy (which corresponds to γ_r), it will slip back in phase and increase in energy. By extracting the beam from the wave at the time it reaches the resonant γ_r (i.e. when it becomes synchronous with the wave), we make the bunch undergo one quarter of a synchrotron oscillation. In doing this, the beam is compressed in phase, as depicted in the figure.

The achievable compression ratio has been calculated in L. Serafini and M. Ferrario (2001) to be

$$C = \frac{2\delta\psi_0 \left|\sin\bar{\xi}_{ex}\right|}{\sqrt{\delta\psi_0^4 + \left(\frac{1}{\alpha\gamma_0}\frac{\delta\gamma_0}{\gamma_0}\right)^2}} , \qquad (2.66)$$

where $\delta\psi_0$ and $\delta\gamma_0/\bar{\gamma}_0$ are the initial phase spread and energy spread of the bunch and $\bar{\xi}_{ex}$ is the average beam exit phase at $\gamma = \gamma_r$. This expression gives a good first order estimate of the compression for an uncorrelated longitudinal phase space distribution at injection.

We will consider here the layout of the SPARC photinjector at INFN-Milan [Serafini (2002)], depicted in Fig. 2.35. This is one of the most

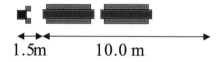

1.5m   10.0 m

Fig. 2.35   Layout of SPARC photoinjector, consisting of an RF gun, focusing solenoid and two traveling wave accelerating sections.

advanced photoinjectors recently proposed, based on the use of two S-band traveling wave structures boosting a 6.5 MeV beam produced by an RF gun. The paradigm for this beam is: 1 nC bunch charge, 10 ps laser pulse length on the photocathode and 140 MV/m peak field on the photocathode, resulting in a 100 A peak current at the gun exit with 0.8% energy spread and 0.5 μm rms normalized transverse emittance (as predicted by simulations). A solenoid lens located at the gun exit focuses the beam in order to match its envelope onto the invariant envelope (see below) at the injection in the first traveling wave structure, located 1.5 m away from the cathode. We take this layout and substitute the first traveling wave structure with one that is properly designed to support a slow wave with wave number $\Delta k = 7 \times 10^{-4}c/\omega$ (hence $\beta_r = 1 - 7 \times 10^{-4}$). This structure will have 86 cells over a length of about 3 m, operating at 2.856 GHz.

Since the compression factor C for a wave of amplitude $\alpha = 0.75$ (corresponding to an accelerating gradient of 23 MV/m at an S-band) is a decreasing function of γ_0 for a given γ_r and increases slightly with γ_r for a given γ_0, for the case under consideration we chose the following optimum parameters for the slow wave structure: $\gamma_r = 27$, corresponding to the previously mentioned 0.07% increase in wave number and $\alpha = 0.75$. For this set of parameters the predicted compression factor is $C = 6.5$.

By optimizing the injection phase, we ran the Homdyn code [Ferrario *et al.* (1996)] from the cathode surface all the way through the gun, drift, slow wave structure, up to the exit of the second 3-m long traveling wave structure (operating at $\beta_r = 1$). At the optimum phase (slightly before $0°$) we found a very nice final peak current of 870 A at an energy of 120 MeV, as shown in Fig. 2.36.

The effect of the RF compressor, extending from $z = 1.5$ m up to $z = 4.5$ m, is clearly illustrated. The energy at its exit is 18.5 MeV ($\gamma = 37$, somewhat larger than γ_r) and the compression takes place almost entirely during the acceleration through the compressor. The current and the energy increase at almost the same rate. We will discuss how this is the best condition for minimizing emittance growth due to space charge effects. The compression factor achieved is $C = 8.1$, a bit larger than the analytical prediction.

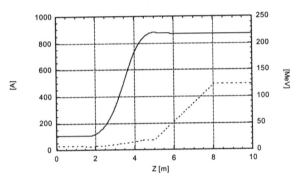

Fig. 2.36   Beam current (solid line, left scale) and energy (dashed line, right scale) along SPARC injector with velocity bunching.

(b)   Transverse dynamics

Space charge dominated beams passing through the photoinjector are described by the invariant envelope model, whose beam equilibrium mode is given by

$$\sigma_{\text{INV}} = \frac{1}{\gamma'}\sqrt{\frac{2I}{I_A\gamma}}, \tag{2.67}$$

where the normalized beam kinetic energy is $\gamma = 1 + T/mc^2$. The normalized accelerating gradient is defined by

$$\gamma = \gamma_0 + \gamma' z, \qquad (2.68)$$

and

$$\gamma' \equiv E_{acc}/mc^2. \qquad (2.69)$$

I is the beam peak current in the bunch. σ_{INV} is an exact analytical solution of the rms envelope equation for laminar beams, typical of photoinjectors up to energies of about 200 MeV. It corresponds to an equilibrium beam condition that ensures emittance correction, i.e. a control of emittance oscillations associated with envelope oscillations so that the final emittance at the photoinjector exit is reduced to an absolute minimum [Serafini and Rosenzweig (1997)]. In order to create this condition, it is necessary to match two types of flow along the photoinjector: the invariant envelope in accelerating sections and Brillouin flow, given by

$$\sigma_{\text{BRI}} = \frac{mc}{eB_{sol}} \sqrt{\frac{I}{2I_A \gamma}} \qquad (2.70)$$

in intermediate drift spaces.

This analysis is valid only for beams carrying constant peak current I, usual in photoinjectors when no compression mechanism is applied (or when space charge debunching is negligible). In order to extend the model to the case of RF compression, we assume a current that increases at the same rate as energy, i.e. $I = I_0 \gamma/\gamma_0$, where I_0 and γ_0 are the initial values of the current and energy at injection into the compressor. This assumption is derived from observations in several simulations, indicating that the best results for beam brightness are achieved with this condition, which represents a new kind of beam equilibrium.

The new exact analytical solution is

$$\sigma_{\text{RFC}} = \frac{1}{\Omega \gamma'} \sqrt{\frac{I_0}{2I_A \gamma_0}}, \qquad (2.71)$$

i.e. a beam flow at constant envelope (instead of $1/\sqrt{\gamma}$ as for the invariant envelope). This is dictated by a new equilibrium between the space charge defocussing term (decreasing now as $1/\gamma^2$) and the focusing and acceleration terms (imparting restoring forces to the beam). Invariant envelope equilibrium is achieved even in the absence of external focusing (see [Serafini and Rosenzweig (1997)]), however, in this case we need to provide external focusing in order to control the envelope. Additional external focusing provided by solenoids around the first accelerating sections is crucial for the control of the final emittance, as has been shown by simulations and confirmed by the first experiments (see below).

For the sake of comparison we note that the solution for Brillouin flow (i.e. a drifting beam at a constant energy and constant current undergoing a rigid rotation in the solenoid field B_{sol}) becomes

$$\sigma_{\mathrm{BRI}}^{\mathrm{BAC}} = \frac{mc}{eB_0} \sqrt{\frac{I_0}{2I_A\gamma_0}} \tag{2.72}$$

in the case of a current increasing linearly along the drift ($I = (\mu z)I_0$) for a corresponding growing solenoid field of the type $B_{sol} = \sqrt{\mu z}B_0$ (also, in this case we obtain a constant envelope matched beam through the system, as in the case of RF compression). $\sigma_{\mathrm{BRI}}^{\mathrm{BAC}}$ describes what typically happens to the beam envelope in ballistic bunching, which needs to be controlled by providing a ramped solenoid field to avoid envelope instability.

Relevant for the emittance correction process is the behavior of the envelope and associated emittance oscillations due to envelope mismatches at injection. These envelope mismatches produce emittance oscillations in laminar beams because of the spread in initial mismatches due to different slice currents [Serafini and Rosenzweig (1997)]. The emittance behaviors for the three flow conditions (invariant envelope at constant current (no compression), Brillouin flow at increasing current (ballistic bunching) and generalized invariant envelope during velocity bunching) are as follows: the emittance oscillates and adiabatically damps as $1/\sqrt{\gamma}$ in the invariant envelope case, it oscillates at constant amplitude along the RF compressor with a frequency scaling like the invariant envelope case for Brillouin flow, while in the case of ballistic bunching the emittance exhibits a completely different scaling, with constant amplitude but frequency increasing as $eB_0\sqrt{\mu z}/mc\gamma$. The basis for successful correction of the transverse emittance in RF compressors is that by carefully matching the two types of flow, we can make the emittance oscillate at a constant amplitude in the RF compressor. Thus we connect these oscillations adiabatically to a damped oscillatory behavior in the accelerating sections following the RF compressor, where the beam is propagated under invariant envelope conditions. This is possible because of the similar frequency behavior of the two flows. It appears to not be achievable in ballistic bunching, where the increase of the emittance oscillation frequency prevents good matching to the invariant envelope regime, and induces the onset of nonlinear space charge effects that prevent the emittance oscillations being fully reversible [Anderson and Rosenzweig (2000)]. Hence, ballistic bunching tends to produce uncontrollable emittance growth.

2.2.3.2   *Application to the SPARC project*

In this section, we discuss the results of a multi-particle simulation study which show that with a proper setting of the focusing provided by the solenoids it is possible to increase the peak current while preserving the beam transverse emittance, as predicted by theory. The accelerating gradient was set at 15 MV/m in the first section and 25 MV/m in the second, while the peak magnetic field produced by the solenoids was 1.1 kG in the first section and 1.4 kG in the second.

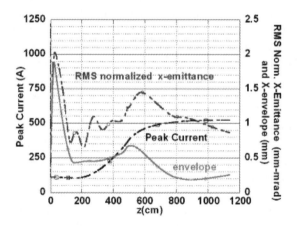

Fig. 2.37   RMS normalized emittance, beam envelope and peak current vs. distance from the cathode.

The plots in Fig. 2.37 represent the behavior of the peak current and the transverse normalized rms emittance as a function of the distance from the cathode, as computed by PARMELA. It shows that a peak current of 510 A can be reached with a transverse rms normalized emittance of 0.9 μm at a final beam energy of 120 MeV. The bunch current profile, initially uniform, tends to a triangular shape. The peak current shown in Fig. 2.37 is actually an average current over the bunch, while the maximum value in the bunch current profile is almost twice the average. Although the beam envelope in the first section (from 1.5 to 4.5 m) is almost parallel, according to the theoretical prescription for the generalized invariant envelope,

$$\sigma_{\mathrm{RFC}} = \frac{\sqrt{I_0/2I_A\gamma_0}}{\Omega\gamma'}, \qquad (2.73)$$

the solenoid focusing is still not fully optimized, as shown by the envelope bump located at $z = 5$ m, within the gap between the first and second linac sections. Associated with this bump we observe a similar bump in the emittance, typical of laminar beams. A more complete optimization of the solenoid magnetic field will definitely lead to a smaller emittance at the exit of the system.

Using the same injector layout as that of the SPARC project we also carried out simulation studies for ultrashort bunches, obtaining a final bunch length of 25 fs for a bunch charge of 25 pC. This requires illuminating the photocathode with a 300-fs rms laser pulse. The final rms normalized transverse emittance was 0.2 μm [Ferrario (2002)]. This shows the potential of velocity bunching for attaining fs long electron bunches at very low transverse emittance.

2.2.3.3 *Experiments*

The first experimental evidence of RF compression was reported by the BNL-ATF group [Wang *et al.* (1996)], which observed a relevant reduction of the electron bunch length with respect to the laser pulse length illuminating the photocathode. This occurred whenever launching phases at the cathode close to 0 (low but increasing electric field) were used. Extracting 40 pC with a 4 ps rms laser pulse, they measured 370-fs electron bunches at the exit of the second linac section (52 MeV) with an rms normalized emittance of 0.5 μm. Comparing this result to simulations, they inferred a bunch length of 1.6 ps at the gun exit, 0.63 ps at the end of the drift space between the gun and the first linac and 0.37 ps at the exit of the whole system. Clearly the type of bunching process applied was a mixed one: velocity bunching in the gun, ballistic bunching in the drift and velocity bunching again in the linac. The last was the smallest contribution and explains the rather large emittance measured for that bunch charge.

A recent paper [Wang and Chang (2002)] proposed repeating the experiment but decreasing the amount of bunching achieved in the drift and increasing the velocity bunching applied in the linac. Simulations reported

in this reference show that a 25-pC bunch can be extracted by launching a
4-ps laser pulse onto the cathode, getting a 1-ps electron bunch at the gun
exit and 15-fs bunch length at the exit of the second linac (almost no compression performed in the drift space). The final value for the emittance
is 1 μm in the case where there is no additional focusing applied around
the linac sections, while 0.5 μm can be achieved with a solenoid located in
the first linac section, where most of the velocity bunching is performed, in
order to control the beam envelope.

The first experiment to verify the velocity bunching concept was performed at the University of Califonia, Los Angeles, Particle Beam Physics
Lab (UCLA-PBPL) [Musumeci (2002)]. Unfortunately, this was performed
in a photoinjector layout not optimized to perform RF compression. Indeed,
the first linac section after the gun was a short (0.6 m) plane wave transformer (PWT) linac, not long enough to perform substantial acceleration
together with bunching, i.e. to allow the bunch to slip back in phase with
respect to the RF wave. However, they measured a 0.39-ps bunch length
after the linac by using a special filter model analysis of the autocorrelator
signal, and they reconstructed by simulations the beam dynamics to show
that the bunching was still a mixed one, i.e. velocity bunching in the PWT
linac and ballistic in the following drift [Musumeci (2002)].

An experiment was performed recently with a photoinjector system operating long TW linac sections after the gun at the BNL deep ultraviolet
FEL (BNL-DUVFEL) laboratory [Piot *et al.* (2002)]. Here they used 1.15-
ps long laser pulses extracting 200 pC of bunch charge, reaching a minimum 0.5-ps bunch length at the exit of the photoinjector system, at a beam
energy of 55 MeV (no compression was performed in the gun or the drift
space). No further compression could be applied to go below 0.5 ps because
of a lack of additional focusing; the beam envelope could not be controlled.
The excellent agreement found in this experiment between simulations and
experimental measurements once again provides a nice confirmation that
this technique is very promising for attaining femtosecond electron bunches
at high brightness.

Most recently, an ongoing experiment at the Livermore National Laboratory (LNL) Pleiades Lab, is producing outstanding results both in terms
of minimum bunch length and emittance achieved. Work is in progress,
with quite good prospects [Musumeci and Rosenzweig (2003)].

A fully optimized dedicated photoinjector for application of the velocity
bunching technique still does not exist. One of the objectives of the SPARC
project is to design and commission such a system. We still haven't explored

the full potentials of such a technique. Femtosecond long bunches will certainly be able to be produced after a full exploitation of its characteristics.

2.2.4 *Microbunching*

W. D. KIMURA

Vice President of Research and Development

STI Optronics, Inc.

2755 Northup Way, Bellevue,

WA 98004-1495, USA

2.2.4.1 *Staged electron laser acceleration (STELLA)*

STELLA [Kimura *et al.* (2001a)] is an example of a velocity or energy modulation technique in which a laser beam electric field modulates the energy of electrons within an electron beam (e-beam) pulse. This process eventually creates a series or train of femtosecond microbunches separated by the laser wavelength. Up to 50% of the electrons in the e-beam pulse can theoretically be bunched together assuming a simple sinusoidal time-varying electric field.

The laser-slicing technique [Schoenlein *et al.* (2000)], developed by the Lawrence Berkeley National Laboratory (LBNL), is a fundamentally different technique because it uses a very short laser pulse to extract a femtosecond-long portion of the electrons from the main e-beam pulse. In STELLA, the laser pulse length is much longer than the e-beam pulse; whereas, in the LBNL technique, the e-beam pulse is much longer than the laser pulse. This difference is summarized in Fig. 2.38.

A technique, analogous but different to the LBNL technique, has been proposed for the laser-driven cyclotron autoresonance accelerator (LACARA) experiment [Marshall *et al.* (2002)]. In the LACARA technique, a spiralling (helical) trajectory is imparted to the e-beam, thereby creating an annular e-beam. A beam stop can be placed downstream of the accelerator, which has a narrow slit that intersects only a small sector of the annulus. This transmits a short electron pulse whose duration is determined by the width of the slit.

The STELLA experiment was the first to demonstrate staging between two laser accelerators. The experiment is currently being conducted at the Brookhaven National Laboratory Accelerator Test Facility (BNL-ATF). This work was motivated by the great progress being made in laser acceler-

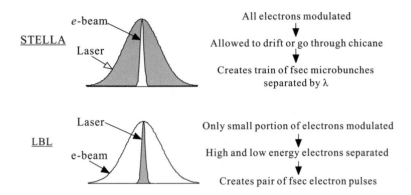

Fig. 2.38   Two techniques for generating femtosecond microbunches. In STELLA (top figure), the e-beam pulse is much shorter in time duration than the laser pulse. In LBNL laser-slicing (bottom figure), the laser pulse is much shorter than the e-beam pulse.

ation, where gradients of > 100 GeV/m have been demonstrated [Leemans *et al.* (1998)]. However, thus far these experiments have involved single passes of the laser beam with the e-beam over limited interaction lengths of a few millimeters or less. A practical accelerator will need to stage the acceleration process in which the laser beam repeatedly interacts with the e-beam, resulting in high net energy gain.

STELLA uses a pair of inverse free electron lasers (IFEL) [Palmer (1972)] for the laser accelerators. The first IFEL (IFEL1) acts as an energy modulator of the e-beam (i.e. a buncher), which results in the creation of femtosecond microbunches. The second IFEL (IFEL2) then accelerates these microbunches. A key issue is achieving proper phase synchronization between the microbunches formed by the buncher and the laser field in the accelerator (IFEL2).

In an IFEL, the e-beam co-propagates with a laser beam inside a periodic magnetic array called a wiggler or undulator (here we shall simply refer to these devices as undulators). The undulator causes the electron trajectory to oscillate in the plane of the laser beam electric field, as illustrated in Fig. 2.39, thereby projecting a component of this field in the direction of the electron motion. Depending on the sign of the electric field (i.e. its phase relative to the electrons), the field can accelerate or decelerate the electrons.

In order to achieve net energy exchange along the length of the undulator, the following resonance condition must be satisfied [van Steenbergen

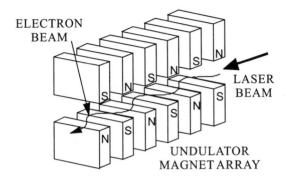

ELECTRON
BEAM

LASER
BEAM

UNDULATOR
MAGNET ARRAY

Fig. 2.39   Illustration of an inverse free electron laser (IFEL) based upon a planar undulator.

et al. (1996)]:

$$\gamma^2 = \frac{\lambda_w}{2\lambda_L} \left(1 + \frac{K^2}{2} \right) , \qquad (2.74)$$

where λ_L is the laser wavelength, λ_w is the undulator wavelength, γ is the Lorentz factor, $K = eB_0\lambda_w/2\pi mc$, e is the electron charge, B_0 is the peak magnetic field, m is the mass of electron and c is the speed of light. Higher energy exchange is possible if the undulator is also tapered, whereby either the magnetic field or magnet period varies along the undulator length [Kroll *et al.* (1981)]. The sign of the taper (whether the resonant energy increases or decreases) determines whether a device is an IFEL or free electron laser (FEL).

This tapering and the resulting difference between an IFEL and FEL is illustrated in Fig. 2.40, following the analysis given in van Steenbergen *et al.* (1996). Plotted schematically is the ponderomotive potential as a function of the phase of the electrons relative to the electromagnetic radiation inside the undulator. Fig. 2.40(a) shows the case for a tapered IFEL where the slope of the potential curve is determined by the amount of taper. The shaded portions on the plot designate the phases corresponding to stable trapping of the electrons in the potential well. As the trapped electrons travel along the undulator they gain energy from the field and the electrons in each of the buckets move upward in energy, i.e. they experience acceleration.

In a tapered FEL, as depicted in Fig. 2.40(b), the taper is reverse and

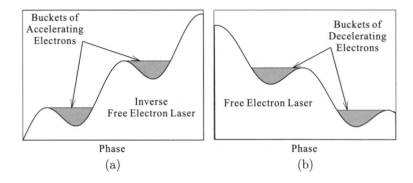

Fig. 2.40   Illustration of ponderomotive potentials for (a) IFEL and (b) FEL

thus the slope is negative. At the proper phase, indicated by shading, the
electrons are trapped and give energy to the field, thereby decelerating. In
both cases, the electrons within the buckets circulate in phase and energy
within the buckets as they are collectively either accelerated (IFEL) or
decelerated (FEL). Electrons on the ponderomotive potential curves that
are outside of the buckets are not in stable phase positions. In general, they
will not experience net energy gain or loss.

A schematic of the STELLA experiment is depicted in Fig. 2.41. The
ATF CO_2 laser beam is split into two beams with approximately 24 MW
sent to the buncher (IFEL1) and up to 300 MW sent to the accelerator
(IFEL2). Both undulators are uniform-gap, i.e. non-tapered devices. Axi-
con lenses convert the Gaussian-profile laser beam into an annular one.
Focusing telescopes focus the beams at the center of each undulator. An
adjustable optical delay stage permits changing of the phase of the laser
beam entering IFEL2 relative to the laser beam entering IFEL1. Each laser
beam enters the beam line vacuum pipe through windows and is directed to
the undulators using in-vacuum mirrors with central holes for transmission
of the e-beam. The separation distance between the exit of IFEL1 and the
entrance to IFEL2 is 2 m. At the end of the beam line is an energy spectrometer featuring a wide energy acceptance ($\pm 20\%$) capable of measuring
the entire electron spectrum in a single shot.

In STELLA, the 45.6-MeV e-beam pulse length is ≈ 3 ps whereas, the
laser pulse length is ≈ 180 ps. This means the electrons enter the STELLA
buncher distributed uniformly over all phases of the laser field inside the
undulator. This is depicted in Fig. 2.42(a), which is a model simulation of
the STELLA buncher. Because the buncher undulator is untapered, there

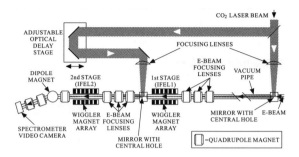

Fig. 2.41   Schematic of the STELLA experiment.

is equal probability for acceleration and deceleration of electrons injected on resonance. Hence, the laser imparts a sinusoidal energy modulation on the e-beam [see Fig. 2.42(b)] with amplitude $\approx \pm 0.5\%$ for 24-MW laser power into IFEL1. This amount of modulation is chosen so that after drifting 2 m to IFEL2, the fast electrons catch up with the slow ones resulting in longitudinal density bunching of the electrons into microbunches [see Fig. 2.42(c)]. Because the laser field induces this modulation, these microbunches have bunch lengths a fraction of the laser wavelength. And, since the laser wavelength (10.6 μm) is much shorter than the e-beam pulse length, a train of 3-fs microbunches is formed with the microbunches spaced apart by the laser wavelength ($\sim$ 30 fs). It is by this means that STELLA generates femtosecond microbunches. These microbunches are then sent into IFEL2 where they are accelerated.

Figure 2.43 compares the model [Kimura *et al.* (2001b)] predictions with the experimental data for the phase delay set at near maximum acceleration and 200 MW laser power delivered to the accelerator. The line profile of the measured electron energy spectrum is plotted in Fig. 2.43(c) along with the energy histogram predicted by the model. We see there is good agreement between the model spectrum and the line profile. Fig. 2.43(a) shows the electron phase distribution that gave rise to the model energy spectrum in Fig. 2.43(c). A concentration of electrons representing the microbunch can be clearly seen. These electrons projected onto the phase axis (see Fig. 2.43(b)) indicate that the microbunch length is $\sim$ 0.8-μm long FWHM corresponding to $\sim$ 2.7 fs in duration. Thus, the microbunch length is inferred from the energy spectrum.

A more compact system is possible by replacing the drift space between IFEL1 and IFEL2 with a magnetic chicane. This is being done for the

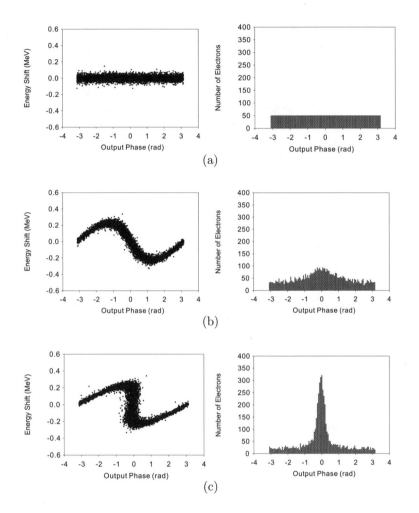

Fig. 2.42   Model simulations for the buncher showing the electron Energy phase distribution and longitudinal density distribution within one optical wavelength (2π of the
optical phase). (a) At entrance to buncher. (b) At exit of buncher. (c) After drifting
2 m to the accelerator.

STELLA-II experiments [Kimura (2002)], where a much higher laser power
will be used to achieve staged monoenergetic laser acceleration. This experiment will also use a highly tapered undulator for IFEL2. The combination of higher laser power and a tapered undulator will enable the trapped
electrons to be accelerated in energy away from the background electrons,

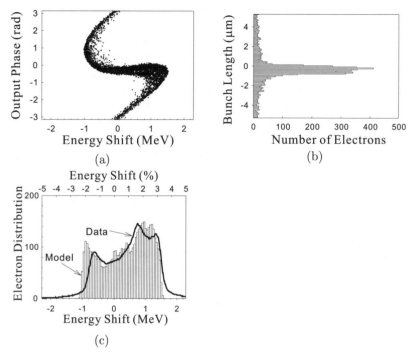

Fig. 2.43 Comparison of STELLA data with model for the case of near maximum acceleration of the microbunch. (a) Model prediction of output phase vs. energy shift from mean e-beam energy. (b) Model prediction for longitudinal density distribution of electrons. (c) Experimentally measured electron energy spectrum overlaid on model prediction.

i.e. the electrons outside of the buckets in Fig. 2.40. These background electrons are still evident in the energy histogram shown in Fig. 2.43(c). As a femtosecond source of electrons, these background electrons are undesirable. By separating in energy the microbunch electrons from the background electrons it will be possible to further accelerate only the trapped microbunch electrons. Maintaining a narrow energy spread during this process is also important for many applications. Thus, the STELLA-II experiment represents another step towards realizing practical laser accelerators and useful femtosecond e-beam sources.

2.2.4.2   Future potential issues

One of the challenges of generating femtosecond microbunches is accurately measuring their length. Some schemes, such as the LBNL technique, have a convenient means for measuring the length using conventional correlation techniques. Nevertheless, it is not clear how effective correlation techniques are when the pulse length is < 10 fs. On the other hand, the STELLA accelerator can be viewed as a bunch length analyzer in that the shape of the energy spectrum produced by the accelerator gives indirect, but real time information about the electron microbunches. By sweeping the phase delay, it is possible to observe the movement across the spectrum of the energy peak associated with the microbunches. The energy width of this peak is directly related to the bunch length. This indirect technique can provide information on microbunches < 1 fs in duration and is limited primarily by the resolution of the energy spectrometer.

Other techniques for inferring the bunch length have been considered, such as observing enhanced coherent transition radiation (CTR) [Rosen-zweig et al. (1995)] when the microbunches pass through a foil. The characteristics of the CTR are directly related to the bunch length; however, it is also affected by other parameters such as the diameter of the e-beam on the foil. Thus, the use of CTR as a bunch length diagnostic is limited by the ability to measure these other parameters with high precision.

There is always the desire to maximize the charge in the microbunches since this increases the amount of, say, X-ray emission from them. One way to increase the amount of charge in an energy modulation technique, such as STELLA, is to alter the laser electric field temporal shape inside the undulator from a sinusoidal one to a sawtooth shape. A sawtooth electric field can in theory yield near 100% bunching of the electrons. Of course, making a perfect sawtooth field is not easily done, but an approximation can be made by sending two laser beams through the undulator, one at the fundamental frequency and the other at the second harmonic frequency. By adjusting the magnitude and phase of these two beams relative to each other, the superposition of their fields can begin to approximate a sawtooth field (this process is equivalent to using the first two terms of a series expansion for a sawtooth waveform).

As methods for increasing the charge in the microbunches are implemented, another issue will arise. Space charge effects will eventually become a serious limitation to the maximum charge because of the small dimensions of the microbunches [Steinhauer and Kimura (1999)]. In microwave accelerators, the aspect ratio of the width-to-length of the e-beam pulse is typically small, e.g. a submillimeter diameter pulse is several millimeters in length. For femtosecond microbunches, this ratio becomes very large because the bunch length is now much shorter than the diameter of the e-beam pulse. Essentially the microbunches have a pancake shape. In microwave accelerators, space charge tends to widen the diameter of the e-beam before it affects the pulse length. For femtosecònd microbunches, space charge will first spread out the length of the microbunch before it affects the beam diameter. Since the beam diameters for either situation are essentially the same, this implies for the same energy e-beams that the femtosecond microbunches will be more sensitive to space charge effects than conventional microwave accelerator bunches.

2.3 Synchrotron

2.3.1 *Synchrotron*

A. ANDO
Laboratory of Advanced Science and Technology for Industry,
Himeji Institute of Technology,
NewSUBARU/Spring-8,
Kamigoori, Ako,
HYOGO 678-1205, JAPAN

There are two approaches to obtain very short bunches; in a synchrotron or in a storage ring. The essential difficulties in maintaining a short bunch for a long time are the particle density and the life time. This is because of the Coulomb repulsion effects on the particle distribution in a bunch and the many instabilities that can easily grow through the interaction of the bunch with the environment, such as the vacuum chamber for very high peak currents.

The normal approach is to control the parameters of synchrotron oscillation, that is, to make the accelerating RF voltage very large and/or to make the momentum slippage factor η very small. η describes the time change of revolution due to the momentum shift of a beam. An rms bunch length of a few millimeters or a few tens of picoseconds is now more easily obtained and the corresponding coherent radiation can also be observed [Abo-Bakr *et al.* (2002)]. However, much effort must still be made to get FWHM bunch lengths of less than picoseconds. This method has the fundamental problem of having a weak beam current. Another method is to produce a modulation in the longitudinal density distribution by introducing a special interaction with an external laser beam or by controlling longitudinal instabilities. This method is effective for electron beams, but it is very difficult to keep the modulation stable. Coherent radiation is observed at some storage rings as a burst due to somewhat accidental microwave instability [Carr *et al.* (2001); Podobedov *et al.* (2001); Kramer and Podobedov (2002)]. In this section, these approaches are briefly summarized and some of the trials made to overcome the difficulties are also explained.

2.3.1.1   *Bunch length in synchrotron and storage ring*

Each particle in a bunch oscillates in energy–time phase space, called *synchrotron oscillation*. This oscillation is described as follows when the oscillation amplitude is small [Bruck (1968)]:

$$\epsilon = \sqrt{A/k} \, \sin \Omega t \, , \quad \tau = \sqrt{kA} \, \cos \Omega t \, ,$$

where ϵ and τ are the energy and time difference of synchronous particles (or from the center of a bunch), πA is the phase space area surrounded by the ellipse of motion (longitudinal emittance, normally measured in eV), Ω is the angular frequency of synchrotron oscillation and

$$k = \frac{|\eta|}{\beta_s^2 E_s \Omega} \, , \quad \Omega = \frac{\omega_s}{\beta_s} \sqrt{\frac{-h\eta eV \cos \phi_s}{2\pi E_s}} \, .$$

E_s, ω_s and β_s are the energy, angular revolution frequency and Lorentz kinematical factor of a synchronous particle, h is the harmonic number of the RF frequency, e is the charge of a particle, V is the RF peak voltage and $eV \sin \phi_s$ equals the energy gain per turn of an ion beam for an accelerating synchrotron or the energy loss per turn for an electron storage ring.

The above equations are obtained for a single particle, but are also applicable for statistically treating values over a distribution in a bunch. k is given by $k_0 \sqrt{|\eta| / (V |\cos \phi_s|)}$ and therefore the bunch length τ becomes shorter as V increases or η decreases. η is defined as

$$\Delta T/T = \eta \, \Delta p/p = \eta \beta_s^{-2} \Delta E/E \, , \quad T = L/v \, ,$$

where T is the revolution period, L is the circumference of the ring and v is the velocity of the beam. η is approximated by α_p for very high energy beams such as most electron beams where $\Delta L/L = \alpha_p \Delta p/p$ for momentum p, because

$$\Delta T/T = \Delta L/L - \Delta v/v = (\alpha_p - G)\Delta p/p,$$

with the kinematical term G vanishing for a relativistic beam. α_p is determined by the ring optics and depends on $\Delta p/p$. This coefficient is called the *momentum compaction factor* and can be treated as a constant when $\Delta p/p$ is small and α_p is not too small.

In an electron storage ring, synchrotron radiation and RF acceleration, to compensate for the energy loss due to this radiation, determine the electron distribution, which becomes Gaussian. The standard deviations of $\epsilon(\sigma_\epsilon)$ and $\tau(\sigma_\tau)$ are usually used. The corresponding emittance is well approximated by [Sands (1970)]

$$A\,[\text{eVs}] \simeq 2.2 \times 10^2 E_s^2\,[\text{GeV}]\,B\,[\text{T}]\,\frac{|\eta|}{\Omega},$$

where B is the strength of the bending magnets measured in T. Here the factor $|\eta|/\Omega \propto \sqrt{|\eta|/(V|cos\phi_s|)}$ also appears. We can also obtain the explicit equation for σ_τ as

$$\sigma_\tau^2 = C_q \gamma^2 \alpha_p^2 I_3/(\Omega_s^2 I_2 J_E),\quad C_q = \frac{55\hbar}{32\sqrt{3}mc}, \tag{2.75}$$

where I_2 and I_3 are synchrotron integrals, J_E is the longitudinal damping partition function, $\hbar$ is the Planck constant, m is the electron mass and c is the speed of light. This equation can be approximated as

$$\sigma_\tau^2\,[\text{ps}^2] = 1.3 \times 10^6 \frac{\alpha_p E_s^3\,[\text{GeV}^3]\,L^2[\text{m}^2]}{h\rho\,[\text{m}]\,|V\cos\phi_s|\,[\text{kV}]}. \tag{2.76}$$

As an example, typical values of the 1.0 GeV storage ring of NewSUB-ARU [Ando *et al.* (1998)] are

$$\sigma_\tau \simeq 17\,[\text{ps}],\quad A \simeq 7.1 \times 10^{-6}\,[\text{eVs}], \tag{2.77}$$

where $V \simeq 100$ kV, $\eta = \alpha_p \simeq 1.0 \times 10^{-3}$, $\rho \simeq 3.2$ m, $L \simeq 118.7$ m and $\Omega \simeq 3.1 \times 10^4$. There is a technical limit in increasing V due to discharge and space for RF cavities, and hence a better way to reduce τ is to reduce η. If V is increased to 1 MV and η is reduced to 10^{-6}, τ is reduced by a factor of 100 and σ_τ reachs 170 fs. The NewSUBARU storage ring is designed to obtain a 1-mm FWHM bunch length, i.e. $\sigma_\tau \simeq 1.4$ ps, at the first stage and to realize coherent radiation in the mm-wave region, which has been demonstrated at the linac of Tohoku University, Japan [Nakazato *et al.* (1989)]. In preliminary results, a bunch length of 5.7 ps FWHM has already been realized, but the intensity was $\sim 10^{-4}$ of that in normal operations, as seen in Fig. 2.44. In this case there are still $\sim 3 \times 10^6$ electrons

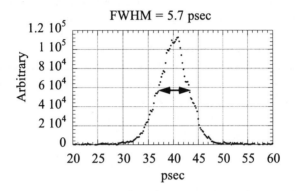

Fig. 2.44   Preliminary data of streak camera

in the bunch, because there are $\sim 3 \times 10^{10}$ electrons in normal operations. Therefore, the mm wavelength coherent radiation becomes stronger than the normal incoherent radiation by ~ 100 times.

2.3.1.2   *Small α_p and intrinsic problems*

A simple way to obtain a very small α_p is as follows.

- Take a double bend achromat (DBA) lattice structure.
- Insert an inverse bending magnet (BI) between two bending magnets (BM), as shown in Fig. 2.45.
- Adjust the momentum dispersion at the BI by two quadrupoles (Q1 and Q2).

First, without the BI we have a suitable lattice with circumference L, $\alpha_p = u$ and momentum dispersion D at the point where the BI would be. Then we introduce the BI with a radius of curvature ρ and total inverse bending angle in a ring A, so as to satisfy the equation $Lu = \oint D ds/\rho = DA$. In the case of NewSUBARU, we have $A \simeq 0.7$ radian (40.1°) for $L \simeq 100$ m, $u \simeq 0.01$ and $D \simeq 1.5$ m. Since the NewSUBARU lattice has six DBA cells, each bending angle of the BI must be near 6.7°. Finally, we achieved a bending angle of 8° with precise calculations.

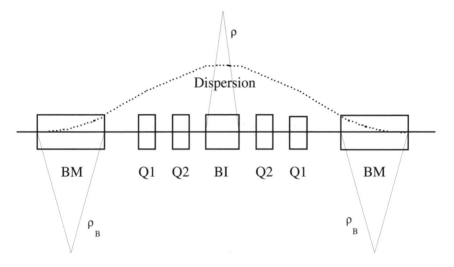

Fig. 2.45   Modified DBA lattice. Q1 and Q2 are adjusted to give the desired dispersion with zero derivative at the center of BI.

When η decreases to near zero, the contribution of the higher order terms cannot be negligible because η is not constant and depends on the momentum deviation $\delta = \Delta p/p$ [Ando and Takayama (1983)]. One contribution is the kinematical effect $\Delta v/v = G\Delta p/p$, with $G = (dv/dp)/(v/p)$. Another is the effect of the sextupole field on $\alpha_p = (\Delta L/L)/(\Delta p/p)$, which is necessary for the chromaticity correction to get a long enough lifetime, but is avoidable in the error field of bending magnets. Taking the new variables $\phi = h\omega_s\tau + \phi_s$, $\delta = \beta_s^{-2}\epsilon/E_s$ and the independent variable $s = h\omega_s t$, the synchrotron oscillation is described by the Hamiltonian

$$H(\phi, \delta : s) = \Sigma_{n=1} \frac{\eta_n}{(n+1)} \delta^{n+1} + \lambda(\cos\phi + \phi\sin\phi_s),\qquad(2.78)$$

where $\eta = \Sigma_{n=1}\eta_n\,\delta^{n-1}$. There appear new fixed points at $\delta \neq 0$ from $\partial H/\partial\delta = \Sigma_{n=1}\eta_n\,\delta^n = 0$, which are clearly observed [Murphy and Kramer (2000)]. An example is shown in Fig. 2.46. The shape and energy of bunches are very sensitive to the higher order terms of η. Therefore, many families of correction sextupole magnets must be installed to keep the bunch in the desired shape and distribution. There are five families of magnets in the

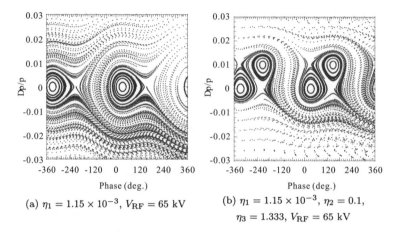

(a) $\eta_1 = 1.15 \times 10^{-3}$, $V_{RF} = 65$ kV

(b) $\eta_1 = 1.15 \times 10^{-3}$, $\eta_2 = 0.1$,
$\eta_3 = 1.333$, $V_{RF} = 65$ kV

Fig. 2.46   Phase space plot of synchrotron oscillation at NewSUBARU. (a) Normal "fish" shape for $\eta = \eta_1$ (constant). (b) η polynomial up to δ^2.

NewSUBARU storage ring; two are for chromaticity correction and three, called *harmonic sextupoles*, minimize the higher order effects. Numerical tracking programs and optical calculation code, such as SYNCH or MAD, can be used to calculate the off-momentum orbit, though the analytical equations can be obtained by expanding D in polynomials of δ [Wiedemann (1976)]. It is important that the entrance and exit angles of the magnets are different from those at the on-momentum orbit in this calculation. These effects are very sensitive in calculating circumference. Only SYNCH treats these correctively, as far as the author knows.

The most significant problems are the Touschek effect in electron storage rings and beam instabilities. The Touschek effect is the result of momentum exchange between the longitudinal and transverse directions through Coulomb scattering of electrons in a bunch. In the center-of-mass frame (CM), the transverse momentum is the same as that of the laboratory frame (Lab). Consider a head-on collision between two electrons moving transversely in opposite directions with the same momentum $p_\perp$ and recoil in the longitudinal direction in the CM. The longitudinal momentum in the Lab now becomes $\gamma p_\perp$, where $\gamma = (1 - \beta_s^2)^{-1/2}$, which can be much larger than the longitudinal momentum acceptance of a ring. This collision frequency is proportional to the particle density, which increases as the bunch length becomes shorter for the same beam current. Therefore, the beam intensity becomes so small that this effect becomes negligible.

The second problem is beam instability. There inevitably exist many electromagnetic fields called *wake fields*. Charges moving with very high velocity, radiate an electromagnetic field. This field is very complicated because it depends on its environment, such as a vacuum chamber. Some fields not only disturb the charge distribution but also enhance some oscillations if bunches have any trace of oscillation, which causes beam loss. These traces are inevitable, particularly for electron beams, because of the synchrotron radiation. The wake field becomes stronger as the peak current of the beam is higher, i.e. the bunch is shorter. If the amplitude of the longitudinal momentum oscillation is large enough, an unstable oscillation can be smeared out because the oscillation parameters depend strongly on the longitudinal momentum. For example, the magnetic focusing force becomes weak for a higher momentum. η is the scale of this momentum oscillation. Also, the bunch length and the energy spread of a beam increase and the peak current becomes so small that the effect of wake fields can become negligible.

2.3.1.3 *Intrinsic bunch length in an electron storage ring*

There are some assumptions made in deriving the earlier formula of bunch length in an electron storage ring:

- Every electron radiates photons at the same location in each revolution.
- The total effect is obtained by averaging over locations and photons in a ring.

In reality the energy loss through radiation is completely statistical and the radiant location is always changing. Suppose a ring is isochronous; if the radiation of every electron occurs at the same location, the revolution period or the arrival time at the RF cavities are the same for all electrons. However, if the location of radiation differs, the revolution period of a beam has a spread. This effect was first pointed out by Shoji of NewSUBARU [Shoji *et al.* (1996)] who worked out the intrinsic bunch length of an electron storage ring. The rms energy spread is given by

$$\sigma_{\epsilon 1}^2 = \frac{1 + (\Omega T)^2 J/\eta^2}{1 - (\Omega T/2)^2} cdot\sigma_{\epsilon}^2 \,, \tag{2.79}$$

$$J = <(B- <B>)^2>\,, \quad B = \frac{1}{L}\int_{s_j}^{L}\frac{D(s)}{\rho(s)}ds \,, \tag{2.80}$$

where $D(s)$ and $\rho(s)$ are the momentum dispersion and radius of curvature at location s. The brackets $< \ldots >$ denote an average over the radiation locations s_j and number of photons emitted in one turn. The rms bunch length is now written [Shoji *et al.* (1997); Soutome *et al.* (1999)]

$$\sigma_{\tau 1} = \{[\alpha_p(\delta = 0)/\Omega]^2 + T^2 J\}^{1/2}(\sigma_\epsilon/E). \qquad (2.81)$$

The second terms in the above equation give the intrinsic energy spread and bunch length due to the fluctuation of quantum radiation. J can be written for an isochronous DBA lattice $(\alpha_p(\delta = 0) = 0)$, given in Fig. 2.45, as

$$J = \frac{(\rho/L)^2}{\theta_B + \theta_I} I, \qquad (2.82)$$

$$I = \sum_{k=B,I} \left[\frac{\theta_k^3}{3} + 2(1 + D_k)(\theta_k \cos\theta_k - \sin\theta_k) \right.$$
$$\left. + \frac{(1 + D_k)^2(2\theta_k - \sin 2\theta_k)}{4} \right], \qquad (2.83)$$

where θ_B and $2\theta_I$ are the deflecting angles of each of the BM and the BI, and D_I is the dispersion at the center of the BI $(D_B = 0$ and $\rho = \rho_B$ are assumed). The intrinsic bunch length at 1 GeV in NewSUBARU is about 75 fs for $T \simeq 400$ ns, $\sigma_\epsilon/E \simeq 5 \times 10^{-4}$, $\rho = 3.22$ m and $L = 117.83$ m. The bending angle per magnet should be small to achieve a bunch length of a few fs. This value is estimated to be 1 fs for SPring-8 [Soutome *et al.* (1999)]. Therefore, a storage ring obtains a very short bunch by increasing in size for almost the same reasons it obtains a very small transverse emittance.

2.3.1.4   *Microwave instability*

A novel way to obtain a quasi-short bunch is by controlling the microwave instability or turbulence and maintaining a density modulation in time (or a longitudinal density modulation). However, this method is so far only under discussion and has yet to be tried. The source accelerating force is proportional to the peak bunch current and the coefficient is described by the longitudinal impedance, which has a frequency dependence, and its strength at any frequency, n/T, where n is an integer, is important for this instability. For a proton synchrotron, the kinematical contribution in η or G in the previous section changes drastically in acceleration and there is a transition energy where $\eta = 0$, and after crossing this transition there is usually microwave instability, as seen in Fig. 2.47. The microwave structure, i.e. less than 1-ns structure in a few tens of ns bunch length, keeps its modulation (or structure) over a duration of the order of s. Once

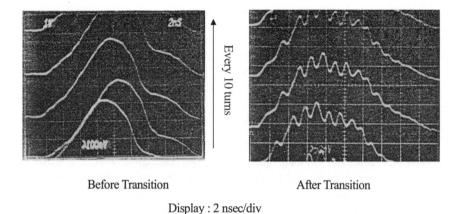

Before Transition   After Transition

Display : 2 nsec/div

Fig. 2.47   Bunch shape of High Energy Accelerator Research Organization (KEK) 12-GeV proton synchrotron near the transition energy.

this instability occurs, the energy spread of a beam widens, satisfying the equation [Ruggiero and Vaccaro(1968)]

$$\delta^2 = F \frac{I}{|\eta|(E/e)} |\frac{Z}{n}|, \tag{2.84}$$

where F is a form factor of about 1 and I is the peak current. If the rise time of this instability is equal to the damping time in an electron storage ring, it may in principle be possible to keep the instability and the modulation for a long time. The modulation scale or quasi-bunch-length is determined by the frequency characteristics of Z. It is very difficult to visualize this modulation for electron beams with a few tens of ps bunch length because of the short structure. The resulting bunch lengthening or energy widening is simply evidence of this. For transverse instability, such as coupled bunch instability [Sacherer (1974)], the "stable" oscillation of the instability can be kept to the order of ten minutes at NewSUBARU. Microwave instability is also going to be tested.

2.3.2 Femtosecond e-beams in storage rings

V. N. Litvinenko,
Department of Physics and FEL laboratory,
Duke University, Durham,
NC 27708, USA

2.3.2.1 Strong longitudinal focusing

It is the traditional belief that it is impossible to achieve both low transverse emittance and low longitudinal emittance (i.e. sub-picosecond electron bunches) in a storage ring because of strong coherent synchrotron radiation and the resulting microwave instability [Kim *et al.* (1985)]. The peak current threshold of the microwave instability for an electron beam with energy E is given by Boussard's criterion [Boussard (1975)],

$$I_{peak} = \frac{2\pi \alpha_c E}{e(Z_n/n)} \left(\frac{\sigma_E}{E}\right)^2 , \qquad (2.85)$$

where Z_n is the longitudinal coupling impedance at the nth harmonic of the revolution frequency $f_0 = v_e/C$, σ_E is the rms energy spread and α_c is the momentum compaction factor. The most striking feature of Boussard's criterion is that it does not depend on such parameters of the longitudinal motion as the synchrotron tune Q_s or on the RF voltage V_{RF} or its harmonic h_{RF} and it is practically identical to the Keil–Schnell criterion derived for a coasted beam [Keil and Schnell (1969)]. The absence of dependence on the longitudinal focusing, which has been confirmed by numerous experimental results, is a clear indication that it does not play any significant role in the development and saturation of the microwave instability. This effect can be explained by the weakness of longitudinal focusing in existing storage rings where the synchrotron tune $Q_s \ll 1$ is smaller than the increment of the microwave instability ξ_{MW} or the decrement of the Landau damping ξ_L:

$$Q_s \ll \max\{\xi_{MW}, \xi_L\} . \qquad (2.86)$$

It was recognized that with weak longitudinal focusing there is no chance to suppress the microwave instability and to achieve sub-picosecond duration electron bunches in storage rings [Blum *et al.* (1996)]. In 1996 we suggested [Litvinenko (1996)] that the use of strong longitudinal focusing with $Q_s \sim 1$ will modify the criterion of Eq. (2.85) and allow the compression of electron bunches to femtosecond duration. We also suggested the use of an inverse mm or sub-mm FEL mechanism for strong longitudinal focusing [Litvinenko (1996); Litvinenko *et al.* (2002)] (see following Sections). In the case of strong longitudinal focusing, the synchrotron motion should be treated in a similar way as betatron oscillations, i.e. averaging over one period is no longer a valid technique. Thus,

$$\frac{ds}{dz} = -\frac{\eta \cdot \delta}{\rho}, \qquad \frac{d\delta}{dz} = \frac{eV_{RF}}{E}\sin(k_{RF}l + \phi_s) - \frac{eV_{loss}}{E}, \qquad (2.87)$$

where $s = -v_e(t_0 - t)$ is the position of the electron with respect to the bunch center, ρ is the radius of curvature, η is the transverse dispersion function (η-function), $\delta = (E - E_0)/E_0$ is the relative energy deviation, $k_{RF} = 2\pi h_{RF}/C$ is the RF wavenumber and z is the azimuth along the beam orbit. For a storage ring with a single RF cavity, the linearized synchrotron oscillations are described by Courant–Snyder parameterization [Courant and Snider (1958)],

$$s = \sqrt{\beta_s}\cos\Psi_s, \qquad \delta = -\sqrt{\beta_s}(\sin\Psi_s + \alpha_s\cos\Psi_s),$$
$$\frac{d\Psi_s}{dz} = \frac{1}{\beta_s(z)}, \qquad \alpha_s = -\frac{1}{2}\frac{d\beta_s}{dz}. \qquad (2.88)$$

It is easy to show that the synchrotron tune and the initial value of the β-function at the location of the RF cavity are

$$\mu_s \equiv 2\pi Q_s = \arccos\left(1 - 2\pi\alpha_c\frac{eV_{RF}h_{RF}\cos\phi_s}{E}\right), \qquad \beta_{s0} = \frac{C}{\sin\mu_s}. \qquad (2.89)$$

Thus, in order to attain significant synchrotron tunes in a modern storage ring with $E \sim 1$ GeV and $\alpha_c \sim 10^{-3}$, one should either use impractical GV levels of RF voltage in a conventional RF system with $h_{RF} \sim 100$, or use a very short RF wavelength of the order of mm. Because an RF cavity for a mm wavelength cannot be built, one can use an inverse FEL as a novel RF "cavity". In this system electrons co-propagate with an intense FEL wave through a wiggler, which is tuned at resonance. For a helical wiggler with period λ_w and dimensionless wiggler parameter K_w the resonant condition is well known:

$$\lambda_{RF} \equiv \frac{C}{h_{RF}} = \frac{\lambda_w}{2\gamma^2}(1 + K_w^2), \qquad (2.90)$$

where $\gamma = E/mc^2$ is the relativistic factor of the electron. As shown in [Litvinenko *et al.* (1997)], optimal focusing of the FEL beam with power p

inside a wiggler with N_w periods provides an accelerating voltage of

$$V\,[\text{MW}] \cong e^{-1/(2\pi N_w)}\sqrt{3N_w \cdot p}\,[\text{GW}]\,. \qquad (2.91)$$

Using mm or sub-mm FEL radiation with GW levels of intra-cavity peak power and a wiggler with few periods ($N_w \sim 2\text{--}4$) is sufficient to provide $Q_s \sim 0.5$ in most modern storage rings.

In addition to strong longitudinal focusing, such an unusual RF system provides very high values of dV_{RF}/dt as well as significant dependence of the synchrotron tune on the amplitude of the synchrotron oscillations. As we will see in the following Section, these features are critically important for the stability of femtosecond electron bunches in storage rings. Let us note that it is advantageous [Litvinenko (1996)] to combine a traditional (MHz) RF system with a short wavelength (THz) RF system as shown in Fig. 2.48. A standard "500-MHz" RF system compensates for radiation losses and provides a large energy acceptance, i.e. a long life time. In contrast, the inverse FEL RF system provides only strong longitudinal focusing and is responsible only for micro-bunching. In this case, the inverse FEL RF system power interacts reactively with the electron beam providing strong longitudinal focusing, and there is no net energy transfer between the e-beam and the mm-wave FEL. In contrast, the conventional RF system compensates for both incoherent and coherent energy losses of electrons in the ring, but its contribution to the longitudinal focusing is practically negligible. A sketch of the resulting separatrix for the e-beam is shown in Fig. 2.49.

Our computer simulation testing of this concept proved that Bousard's stability criterion is not applicable to the case of strong longitudinal focusing. Furthermore, we demonstrated that the microwave instability caused by broadband impedance of the vacuum chamber can be suppressed using strong-focusing RF systems, especially those with very short RF wavelengths in the mm and sub-mm range. The main limiting factor, as expected, is the coherent synchrotron radiation (CSR). CSR and its effects on the dynamics of very short electron bunches has been at the center of attention of the accelerator physics community for at least a decade [Murphy and Krinsky (1994)].

2.3.2.2   *Coherent synchrotron radiation and stability criteria*

For very short electron bunches, the screening of the vacuum chamber can be neglected and the longitudinal electric field of the CSR can be calculated

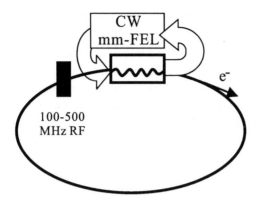

Fig. 2.48 Storage ring with conventional RF cavity and strong focusing inverse FEL RF system driven by a continuous wave (CW) mm-wave FEL. Electrons interact with the FEL wave in a wiggler, which is tuned to the FEL resonance.

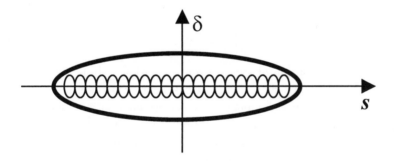

Fig. 2.49 Sketch of the separatrix for the combined RF system in Fig. 2.48. Low frequency RF provides a large size separatrix filled with small separatrices caused by the inverse mm-wave FEL system.

using textbook formulae [Landau and Liftshitz (1975)]. Subtraction of the trivial, and well-studied, static electric field gives an expression for the CSR longitudinal field, which is finite at all distances:

$$E_s = \mp \frac{e\beta^2(\cos\phi \mp \beta)}{2\rho^2(1 \mp \beta\cos\phi)^3}$$
$$\pm \frac{e}{4\rho^2\gamma^2} \left\{ \left(\frac{\cos\phi}{\sin^2\phi(1 \mp \beta\cos\phi)^2} - \frac{1}{(\phi \mp \sin\phi^2)} \right) + \frac{\beta}{(1 \mp \beta\cos\phi)^3} \right\},$$
$$(2.92)$$

where $t - t'$ is the delay between two electrons, $c(t - t') = 2\rho |\sin \phi|$ and ρ is the radius of curvature in a bending magnet. In contrast with the wake field, the CSR field from the tail of the bunch acts on the particles at the head of the bunch. In addition, for femtosecond e-bunches, the integrated "wake potential" of CSR exceeds that of the vacuum chamber by many orders of magnitude. The presence of such a strong wake potential will inevitability cause longitudinal instability for femtosecond electron bunches with high charge. We have derived two stability criteria for femtosecond electron bunches in a storage ring. The first is based on an approximate solution of the Vlasov equation. This criterion comes from comparing the growth rate of the CSR instability and corresponding density modulation with period λ_m with the de-coherence time

$$\zeta_D \approx \frac{2\pi^5 Q_s \sigma_l^2}{\lambda_{RF}^2} \left(\frac{\sigma_l}{\lambda_m} \right)^2 , \qquad \tau_D = \frac{1}{\zeta_D} , \qquad (2.93)$$

caused by the anharmonicity of the synchrotron oscillations (some authors call it Landau damping)

$$\frac{dQ_s}{d(l^2)} \cong -\frac{\pi^2 Q_s}{4\lambda_{RF}^2} \qquad (2.94)$$

and the finite amplitude of synchrotron oscillations. It is obvious from Eqs. (2.93) and (2.94), that having high synchrotron tune and short RF wavelength provides fast damping of the density modulation, and therefore provides a higher stability threshold. Because of space limits, we present the details elsewhere [Litvinenko and Shevchenko(2003)].

For the case that we had studied, the threshold turned out to be higher. This simple criterion is based on the fundamental principle of stability of synchrotron motion: fragmentation of the electron beam will not occur provided that dV/dt of the RF system is larger than dV/dt induced by the beam wake field:

$$\left| \frac{dV_{wake}}{dt} + \frac{dV_{RF}}{dt} \right| \neq 0 . \qquad (2.95)$$

This process is similar to potential well distortion for self-consistent distributions of electrons. The stability criterion is provided by an equation similar to that derived by Haïssinski [Haissinski(1973)]. When the CSR wake potential changes the sign of the total dV/dt, the beam is fragmented and nonlinear motion causes an increase in the longitudinal emittance. A

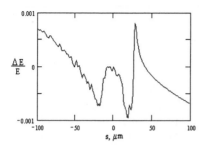
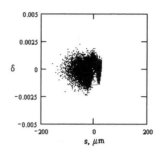

Fig. 2.50 Typical fragmentation of electron beam in (s, δ) phase space caused by CSR (right-hand plot) after 400 turns and effective longitudinal voltage (RF voltage plus wake fields) seen by the beam (left-hand plot). The charge is 10 pC, $E = 750$ MeV, $\alpha c = 10^{-4}$ with uniform η-function, $V_{(178 \text{ MHz})} = 700$ kV, $V_{(1 \text{ mm})} = 1$ MV, natural energy spread is 0.0435% and the natural bunch length is 1.5×10^{-3} cm (50 fs). The bunch length and energy spread grow by a factor of 2.2 due to well developed CSR–microwave instability. This is a clear indication that with this set of parameters a bunch with 10 pC is not stable and cannot be compressed to 50 fs.

typical picture of such fragmentation is shown in Fig. 2.50. It is easy to see that dV/dt changes sign along the bunch and causes it to fragment. It is easy to write this criterion for a femtosecond Gaussian electron beam with constant bunch length σ_s much longer than the critical wavelength of the synchrotron radiation $\lambda_c = 2\rho/3\gamma^3$. In this case the CSR impedance has a rather simple form [Bane *et al.* (1994)] (where Z_0 is the impedance of vacuum and $\omega_0 = 2\pi c/C$):

$$Z(\omega) = Z_0 \frac{\Gamma(2/3)}{3^{1/3}} \left[\frac{\sqrt{3}}{2} + \frac{i}{2} \right] \left(\frac{\omega}{\omega_0} \right)^{1/3}, \qquad (2.96)$$

which is easily integrated with the Gaussian beam. It is worth mentioning that due to the asymmetry of dV_{CSR}/dt, the stability criterion of a femtosecond bunch in Eq. (2.95) depends on the sign of the orbit compaction factor α_c. The maximum positive amplitude of $e(dV_{CSR}/dt)$ is 1.383 times smaller than the maximum negative amplitude. This implies that a stable femtosecond e-bunch (with the same duration) can have ~ 1.4 times higher charge in a storage ring with negative α_c. Because the value of the momentum compaction factor must be very close to zero for attainment of femtosecond e-bunches, it is practical to use negative compaction factors.

In practical units this simple stability criterion is

$$Q\,[\text{pC}] \leq \begin{Bmatrix} 1.5 \\ 2.1 \end{Bmatrix} \cdot 10^{-4} \cdot f_{RF}\,[\text{THz}] \cdot V_{RF}\,[\text{MV}] \cdot \rho\,[\text{m}] \cdot \sigma_l^{7/3}\,[\text{fs}]\,, \quad (2.97)$$

where Q is the charge in a single femtosecond electron bunch, the top number is for $\alpha_c > 0$ and the bottom number is for $\alpha_c < 0$. This criterion is the obvious explanation for the choice for a ultrahigh frequency inverse FEL RF system (1 THz corresponds to an FEL wavelength of 0.3 mm). For a storage ring with $\rho \sim 5$ m, 1 THz, ~ 1 MV RF system, one can expect to maintain a stable 1 pC electron bunch with rms duration ~ 20 fs. Our numerical simulations (see next paragraph) did show that this criterion works reasonably well (well within a factor of 2) for exact self-consistent distributions of electrons.

In practice, low values of the compaction factor $|\alpha_c| \leq 10^{-3}$ are attained by changing the sign of the η-function in the arcs of the ring [Abo-Bakr *et al.* (2002)]. In this case, in the presence of strong longitudinal focusing, the longitudinal β-function and the bunch length strongly oscillate [Litvinenko (2002)] along the ring. As a result, the electron bunch stays short only in a few locations around the ring (for example in straight sections) and the stability criterion Eq. (2.97) should be modified to

$$Q\,[\text{pC}] \leq \begin{Bmatrix} 1.5 \\ 2.1 \end{Bmatrix} \cdot 10^{-4} \cdot f_{RF}\,[\text{THz}] \cdot V_{RF}\,[\text{MV}] \cdot \rho\,[\text{m}]\langle\sigma_l^{7/3}\,[\text{fs}]\rangle_c\,, \quad (2.98)$$

where the average $\langle \ldots \rangle$ is taken only in the bending magnets. We have simulated the case of an alternating sign of the η-function for the Duke Storage Ring [Litvinenko (2002)], and found that beam charge can be increased by about a factor of 10 in magnitude compared with the case when the sign of the η-function is constant.

We consider a lattice with alternating sign of the η-function is a natural choice for a low compaction factor storage ring with femtosecond electron bunches and strong longitudinal focusing.

2.3.2.3 *Selected results of computer simulations*

In addition to studies of the theoretical issues related to the dynamics of femtosecond e-bunches in storage rings, we have developed a numerical code based on a macro-particle model to investigate this concept and its applications [Litvinenko (2002)]. The parameters used for tracking of the longitudinal dynamics are written in Table 2.4. An electron bunch with N_e electrons is split into a large number of macro-particles $N_{mp} \ll 1$ comprising

$n_e = N_e/N_{mp}$ electrons. In order to reduce shot noise, each macro-particle is assumed to have a Gaussian distribution of electrons with rms duration of 1 μm. Each arc is split into a large number of sections ($i = 1, \cdots, m$) with individual longitudinal dispersions $d_i = \int_{z_i}^{z_{i+1}} dz \cdot \eta(z)\rho$. At each position, we calculate the total wake field (from both CSR and the vacuum chamber) and apply it to each individual macro-particle using the following equations:

$$s_n^i = s_n^i + d_i \cdot \delta_n^i, \qquad W_{total}(s) = \frac{Q}{N} \sum_{n=1}^{N} W_0(s, S_n^{i+t}), \qquad (2.99)$$

$$\delta_n^{i+1} = \delta_n^i(1 - \xi l/m) - \frac{eW_{total}(s_n^{i+1})}{E_0} + \Delta\delta_{SR}/\sqrt{m}, \qquad \{i = 1, m\}, \qquad (2.100)$$

where ξ_d is the radiation damping and $\Delta\delta_{SR}$ is the fluctuation of the spontaneous radiation. Finally, at each turn we apply the interaction with the RF system as

$$\delta_{n+1}^l = \delta_n^m - \frac{eV_{RF}(S_n^m)}{E_0}. \qquad (2.101)$$

After running the macro-particles for a large number of damping times, we obtain self-consistent distributions of the electron beam below the instability threshold (see Fig. 2.51) or a typical microwave type bunch lengthening and energy spread growth above the threshold (see Fig. 2.50). The parameters attained via computer simulation are in good agreement with the stability criteria above.

2.3.2.4 *Proposed applications of femtosecond electron bunches in storage rings*

We propose to use the above scheme to compress and sustain femtosecond electron bunches in a storage ring. These electron bunches can generate femtosecond pulses of various types of radiation, from coherent far IR [Litvinenko (2002)] to incoherent X-rays [Litvinenko et al. (2002)]. Even though the charge per bunch is rather small, the number of electron microbunches can be very large, $\sim 20\,000$ in a compact storage ring [Litvinenko (2002)]. These beams will generate coherent far IR radiation at unprecedented kW levels of average power and can be considered a new generation of broadband IR sources.

Use of bending magnets and insertion devices is a natural way of generating intense incoherent femtosecond soft X-ray and X-ray radiation from

Table 2.4   Parameters used for tracking longitudinal dynamics

Standard parameters	
Electron beam energy	0.75 GeV
Circumference	107.46 m
Radius of curvature	2.1 m
Revolution frequency	2.78 MHz
Frequency of standard RF	178 MHz
Voltage, standard RF	700 kV
Natural energy spread	0.0435%
Losses per turn (incoherent)	12.4 kV

Variable parameters	
Focusing ($\lambda = 0.25$–1 mm) RF	300–1200 GHz
Voltage, focusing RF	0–2 MV
Orbit compaction factor, α_c	8.6×10^{-3}–1×10^{-5}
Max number of macro particles, N_{mp}	20 000
Number of damping times	0–100
Interactions with wake per arc, m	1–40

such an electron beam. We also propose developing a hard X-ray source at a low energy storage ring using Compton back scattering of femtosecond electron beams from intense intra-cavity mm-wave FEL radiation, similar to that used to generate high intensity γ-ray beams at Duke [Litvinenko *et al.* (1997)]. Figure 2.52 shows one of the possible schemes [Litvinenko *et al.* (2001)] for generating hard X-ray femtosecond beams using the same mm-wave FEL for both strong longitudinal focusing and Compton back scattering. The process of Compton back scattering shortens the wavelength of radiation by a factor of $4\gamma^2$, which produces hard X-ray photons from mm-wave photons.

Such a source installed at the medium energy Duke Storage Ring will provide [Litvinenko *et al.* (2002); Litvinenko *et al.* (2001)] X-ray energy tunable from 1.4 to 75 keV, with a flux of up to 1.2×10^{12} photons/s and average spectral brightness of 0.4–5×10^{14} photons/s/mm²/mrad²/0.1%bandwidth.

It seems feasible to generate and maintain electron bunches with femtosecond duration in a storage ring using a mm wave inverse FEL as a novel RF system. Our theoretical studies and direct computer simulations support this hypothesis and provide a reasonable estimation of the limits attainable by such a scheme.

Overall, a femtosecond storage ring light source has many complementary features to linac-based systems and should eventually be a part of the femtosecond light source family.

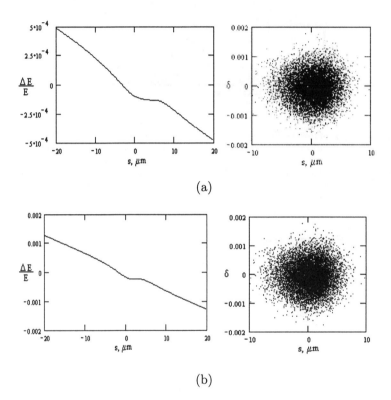

Fig. 2.51   Results of computer simulations with stable femtosecond electron bunches with parameters (energy deviation per turn in the left-hand plots, Poincaré plots in the right-hand plots): (a) $Q_{th} = 1.0$ pC, $V_{(1/4\,mm)} = 2$ MV, $\alpha_c = 0.2510^{-4}$, $\sigma_t = 11.2$ fs; (b) $Q_{th} = 1.3$ pC, $V_{(1/3\,mm)} = 1$ MV, $\alpha_c = 0.3310^{-4}$, $\sigma_t = 14.7$ fs. In all the above cases the energy spread remains natural and unchanged, while the longitudinal distribution obtains a self-consistent non-Gaussian shape caused by the CSR wake field.

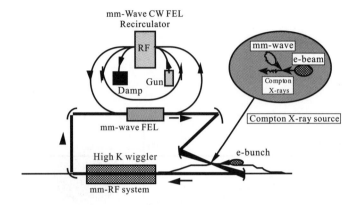

Fig. 2.52　A CW mm-wave FEL based on an energy recovery linac generates intense mm-waves with intra-cavity power ~ 1 GW and an average power of a few kW. The feasibility of such a system was demonstrated by the FEL laboratory in the Thomas Jefferson National Accelerator Facility. The mm waves co-propagate with the electron beam through a high K wiggler to provide compression of the electron bunches to femtosecond duration. The mm wave power is also focused to a collision point, where electrons generate hard X-rays via Compton back scattering.

2.4　Laser Plasma Acceleration

2.4.1　*Electron*

2.4.1.1　*Laser plasma wake field acceleration*

D. UMSTADTER
University of Michigan,
ANN ARBOR, USA

M. UESAKA
T. HOSOKAI
Nuclear Engineering Research Laboratory, University of Tokyo
2-22 Shirane-shirakata, Tokai, Naka,
IBARAKI 319-1188, JAPAN

The technique of chirped pulse amplification has enabled the construction of compact (table-top size) solid state lasers that produce ultrashort pulses with intensities three to four orders of magnitude higher than was previously possible. These pulses have multi-terawatt peak power, and when focused can produce the highest electromagnetic intensities ever achieved, exceeding 10^{20} W/cm^2. The electric field at the laser focus for 1-μm light at this intensity is approximately 3×10^{11} V/cm. These pulses also have extremely short time durations, in the range of 20 fs to 1 ps, which can be shorter than the timescale of significant ion motion or even the plasma period, thus permitting the isolation of plasma instabilities.

These recent technological developments in the design of high peak power lasers, and novel ideas about how to use them to accelerate electrons, may soon revolutionize accelerators and high energy photon sources. Ever since the development of the chirped pulse amplification (CPA) technique in 1987 [Maine et al. (1988)], the size of high power lasers has been decreasing. Linear accelerators, on the other hand, in terms of field gradient have not changed much since their inception. The reason for this is that the dielectric breakdown of the radio frequency electric fields on the cavity walls limits the maximum field gradients to ≤ 1 MV/cm. Lasers may be used to accelerate electrons via the electrostatic fields of large amplitude plasma waves [Tajima and Dawson (1979)], which, because breakdown cannot occur, have a maximum axial electric field measured to be three orders of magnitude higher (2.5 GV/cm) [Umstadter et al. (1996)]. Consequently, just as the size of high power lasers has recently been reduced by several orders of magnitude, a similar reduction may soon occur in the size of accelerators and the high energy photon sources that employ them.

Laser wake field accelerators are at an exciting stage of development. Besides the huge jump in laser power that has accompanied the CPA technique, there have recently been significant improvements in the theoretical analysis of the interaction physics, and advances in computer power have made it possible to numerically simulate the physics in three dimensions. Experimentally, electrons have been accelerated by lasers up to an energy of 40 MeV [Modena et al. (1995)] with large acceleration gradients (> 2 GeV/cm) [Le Blanc et al. (1996)]. Several techniques for optical guiding have also been demonstrated, although they have yet to be incorporated into a resonant wake field accelerator. Even without guiding, it is now possible with available laser power to envisage a practical table-top GeV energy accelerator with superior electron beam qualities.

A major technological component that is missing, however, is a suitable means to jitterlessly inject electrons into the extremely short duration (femtosecond) acceleration buckets that accompany high field gradient accelerators. After presenting the underlying physical concepts and the current status, we discuss a method to remedy this situation.

(a) Laser wake field acceleration (LWFA)

When an intense short laser pulse goes through a plasma, the ponderomotive force of the laser pulse pushes plasma electrons [Tajima and Dawson (1979); Nakajima (1996)]. As a result of the electron density oscillations, a large amplitude plasma wave, i.e. a wake field is excited behind the laser pulse. For a cold plasma of electron fluid with density n_e and stationary ions, the momentum, the continuity and Poisson's equations are linearized, resulting in a simple harmonic oscillator equation for the wake potential Φ of the electron fluid driven by the ponderomotive force $\nabla\Phi_{NL}$, defined by averaging the nonlinear force over $2\pi/\omega_0$ exerted on the plasma electrons by a laser field with frequency ω_0. Thus, the ponderomotive potential is $\Phi_{NL} = -(m_e c^2/2)|\mathbf{a}^2(r,z,t)|$ for circular polarization, where $\mathbf{a}(r,z,t) = e\mathbf{A}(r,z,t)/(m_e c^2)$ is the normalized vector potential of the laser field. Assuming that all of the axial and time dependencies can be expressed as a function of a single value, $\zeta = z - v_\phi t$, with a phase velocity v_ϕ of the plasma wave, the linearized equation is

$$\frac{\partial^2 \Phi}{\partial \zeta^2} + k_p^2 \Phi = \frac{1}{2} k_p^2 m_e c^2 a^2(r,\zeta)\,, \qquad (2.102)$$

where $k_p = \omega_p/v_\phi$ and $\omega_p = \sqrt{4\pi e^2 n_e/m_e}$ is the plasma frequency. The phase velocity of the plasma wave v_ϕ, which is equal to the group velocity of the laser pulse in a plasma, is given by $v_\phi = c\sqrt{1 - \omega_p^2/\omega_0^2}$. An analytical solution for the wake potential is obtained for a bi-Gaussian pulse given by $|\mathbf{a}(r,\zeta)| = a_0 \exp(-r^2/\sigma_r^2 - \zeta^2/2\sigma_z^2)$, where σ_z is the rms pulse length and σ_r is the rms spot size. The axial and radial wake fields for $\zeta \ll -\sigma_z$ are obtained as follows,

$$eE_z(r,\zeta) = \frac{\sqrt{\pi}}{2} m_e c^2 a_0^2 k_p^2 \sigma_z \exp\left(-\frac{2r^2}{\sigma_r^2} - \frac{k_p^2 \sigma_z^2}{4}\right) \times \cos k_p \zeta\,, \qquad (2.103)$$

$$eE_r(r,\zeta) = -2\sqrt{\pi} m_e c^2 a_0^2 k_p \sigma_z \frac{r}{\sigma_r^2} \exp\left(-\frac{2r^2}{\sigma_r^2} - \frac{k_p^2 \sigma_z^2}{4}\right) \times \sin k_p \zeta\,. \qquad (2.104)$$

The maximum gradient is achieved at a plasma wavelength $\lambda_p = \pi\sigma_z$, $(eE_z)_{max} = 2\sqrt{\pi}e^{-1}m_e c^2 a_0^2/\sigma_z$, where $a_0^2 = 0.73 \times 10^{-18} I\lambda_0^2$, I is the laser intensity in units of W/cm^2 and λ_0 is in units of μm. Figure 2.53 shows axial and radial wake fields excited by a laser pulse with $\lambda_p = \pi\sigma_z$. The wake fields go in the z-direction with the group velocity of the laser pulse.

Electrons should be injected and synchronized with the wake field, and then accelerated in the wave.

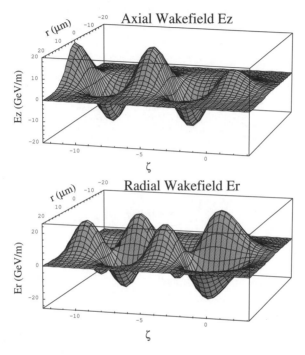

Fig. 2.53   Axial and radial wake fields excited by a laser pulse propagating along the z-axis with $a_0 = 0.5$, $\omega_0 = 10$ μm and $\sigma_z = 18$ μm.

(b)   Injection into wake fields

The injection of energetic electrons into wake fields for their further acceleration is a crucial part of LWFA. Usually the injection of a high quality electron beam from an RF accelerator is assumed [Nakajima *et al.* (1995)]. However, this requires highly precise synchronization between the wake field and the injection. The pulse duration of the electron bunches from an RF accelerator is too long, larger than a wake field bucket. Another method of electron injection exploits injection produced by the laser pulse itself, the so-called *self-injection*. However, because self-injected electrons have random phase with respect to the plasma wave, self-injection results in an electron beam with a large energy spread. There are three major self-injection techniques of LWFA which employ not only single, but two or more laser pulses. These techniques are called *transverse optical injection* [Umstadter *et al.* (1996)], *colliding pulse injection* [Esarey *et al.* (1997)], and *injection by wave-breaking of plasma waves* [Bulanov *et al.* (1999a)]. Figure 2.54(a)–(c) shows the major Injection schemes of LWFA. In trans-

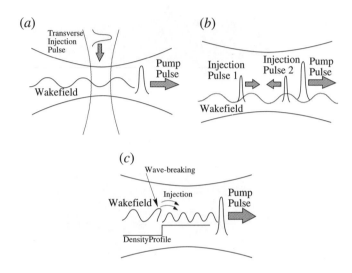

Fig. 2.54   Injection scheme of LWFA. (a) Transverse optical injection. (b) Colliding optical injection. (c) Injection by wave-breaking of plasma waves

verse optical injection, shown in (a), a transverse pulse injects energetic electrons into a longitudinal wake field excited by the pump pulse. In colliding optical injection, shown in (b), the collision of two counter-propagating laser pulses produce a beat wave that traps background electrons and injects them into wake fields excited by the pump laser pulse. In injection by wave-breaking of plasma waves, shown in (c), a plasma wave produced by a single intense laser pulse breaks at a steep change of plasma density with ejection of plasma electrons, which are injected into further wake fields. These injection schemes are described in detail in the next sections.

2.4.1.2 *Laser injected laser accelerator concept (LILAC)*

D. UMSTADTER
University of Michigan,
ANN ARBOR, USA

High field gradient laser wake field accelerators have extremely short duration (femtosecond) acceleration buckets, making acceleration of electrons by them challenging. Extremely short duration electrons need to be injected into the plasma wave in order to produce monoenergetic beams. The injection of electrons can occur uncontrollably by trapping of hot background electrons, which are preheated by other processes such as Raman back scattering and side scattering instabilities, or by wave-breaking (longitudinal or transverse). It can also be achieved by specific injection schemes [Umstadter *et al.* (1996); Esarey *et al.* (1997); Rau *et al.* (1997); Hemker *et al.* (1998); Bulanov *et al.* (1999b); Moore *et al.* (1999)] to control the characteristics of the generated electron beam. Normally, electrons oscillating in a plasma wave cannot be accelerated by a wake field since they are out of phase with it. Electrons that are not part of the plasma wave, however, can become trapped, i.e. continuously accelerated, by the plasma wave, provided they are moving in the correct phase at nearly the phase velocity of the wave. Since this velocity is close to the speed of light, it is generally thought that the required pre-acceleration can only be accomplished with a conventional linac. However, the lower field gradient ($< 100\ \mathrm{MeV/m}$) of a first-stage conventional linac prolongs the time during which beam emittance can grow before the beam becomes relativistic; after this point, self-generated magnetic fields can balance the effects of space charge. Even with state-of-the-art electron guns, the pulse width of the electron bunch can be considerably longer than the plasma wave period of a second-stage laser plasma accelerator. It will thus fill multiple acceleration buckets uniformly in phase space, resulting in a large energy spread. Also, it is difficult to position and focus the electron beam in the plasma channel with μm accuracy and to synchronize the electron beam with the plasma wave acceleration phase. We have shown theoretically that it can simply be done with an additional laser pulse [Umstadter *et al.* (1996)].

The basic idea is that once a laser wake field is excited by the longitudinal ponderomotive force of one laser pulse (the pump pulse), the momentum kick due to a second, orthogonally directed, laser pulse (the injection pulse) can then be used to locally alter the trajectories of some of the plasma wave electrons such that they become in phase with the wave's electric field and thus are accelerated to the trapping velocity [Umstadter *et al.* (1996)]. In the original concept, it was the transverse ponderomotive force of the injection pulse that delivered the required kick; we have recently begun to explore another idea in which the interference of two like-polarized waves locally heat electrons to give them a kick. In either case, there are several different means to drive the plasma wave: (1) a resonant wake field and (2) a self-modulated wake field. In the former case, the two pulses are both short. In the latter case, the injection pulse is much shorter than the pump pulse.

By permitting the use of laser wake fields in the injection stage, this concept not only dispenses with the need to merge two dissimilar technologies, it also significantly increases the capabilities and applications of plasma-based accelerators. It is easier to synchronize in phase and overlap in space the processes of electron injection and acceleration by employing the same basic mechanism for both. Improved electron beam emittance may result from an increased field gradient in the first acceleration stage, by minimizing the time during which electrons are non-relativistic and thus most susceptible to space charge effects. A device based on this concept can be used either as a stand-alone accelerator system or as an injector for either conventional or plasma-based high energy accelerators. Both the energetic electrons and the high energy photons into which they can be converted have numerous industrial, medical and scientific applications. For instance, this technique produces a single ultrashort-duration (fs) electron or X-ray pulse (without the need for pulse selection or beam compression), which is synchronized with an ultrashort laser pulse and thus suitable for the study of ultrafast dynamics on femtosecond timescales. This is two orders of magnitude shorter duration for a single pulse of electrons than has ever been produced by a conventional linac.

After the initial theoretical paper on the LILAC concept, in which numerical simulations of the injection process were performed and trapping conditions in 1-D were obtained [Umstadter *et al.* (1996)], 2-D numerical simulations were performed and trapping equations in 2-D were obtained [Dodd *et al.* (1999)].

The LILAC concept consists mainly of three different stages. First a large amplitude wake field is generated; second, electrons are dephased; and then third, electrons are trapped in the pump's wake field due to dephasing. The process of dephasing and then trapping electrons with secondary laser pulses is quite general. Just as there are myriad ways to drive plasma waves, by means of either a single pulse, multiple pulses or beat waves, there are many ways to use lasers to inject electrons. Several orientations of the laser pulses are also possible: collinear, counter-propagating or orthogonal. Besides combining laser pulse characteristics, ionization or density gradients are other possible means to dephase electrons. We will study the specific case of an orthogonal geometry, as shown in Fig. 2.55. The idea is to move oscillating electrons from the background across the

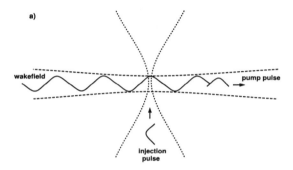

Fig. 2.55   Schematic of the crossed laser plasma wave accelerator concept.

separatrix and thus accelerate them to any desired energy (see Fig. 2.56). The oscillations of electrons in large amplitude plasma waves take them very close to the separatrix without becoming trapped, such that only a small impulse is needed to make them cross the separatrix. Dephasing may arise from a number of sources, such as density variations, ionization, interactions between multiple waves and the ponderomotive force of additional laser pulses. In ionization, newly freed electrons may appear at velocities different from older electrons at the same phase, i.e. dephased. Also, the wave may interact with other wake fields and the ponderomotive potentials of other laser pulses, causing complicated orbits.

We assume here that electrons interact with the injection pulse mainly through its transverse ponderomotive force. The pulse will dephase electrons via this potential, distorting the plasma wave and separatrix, causing some electrons to enter the separatrix and become trapped, as shown in Fig. 2.56. The phase space area with trapped electrons defines the emittance of the beam, and is shaded in the figure. Those electrons that acquire a ponderomotive drift velocity in the same direction as the pump pulse are velocity-dephased with respect to, and injected into, the wake field, and

Femtosecond Beam Science

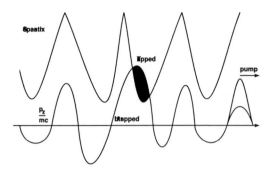

Fig. 2.56   Basic profile of optical injection. The injection pulse dephases electrons in the plasma wave, distorting it and the separatrix. Electrons cross the separatrix, into the accelerating bucket.

will be trapped if inside the separatrix. The local perturbation of the injection pulse can also disrupt the phase of the wake field, disabling LILAC operation in the third stage. However, the propagation velocity of this disturbance is much less than the phase velocity of the plasma wave. This means that once electrons are injected into the plasma wave, the disturbance will not affect the ongoing acceleration.

It is found that with $\gamma_\phi = 10$ and a pump pulse normalized vector potential of $a_0 = 1$, which corresponds to $\epsilon = 0.7$, the analysis predicts that the threshold injection pulse normalized vector potential should be $b_{th} \approx 0.8$.

To determine the importance of 2-D effects and to study geometries that do not require the 1-D approximation, 2-D simulations were run [Dodd et al. (1997); Kim et al. (1997)]. The pump pulse was assumed to be resonant at the plasma frequency. The specific pump pulse parameters used were $a_0 = 1.6$, $r_0 = 8$ μm and $\tau_p = 10\lambda_l/c$, where λ_l is the laser's wavelength. Given the spot size, the Rayleigh range was 180 μm for 2-D. For the injection pulse we used $\tau = 2\tau_p$ and $b_0 = 2.0$ with a spot size of $r_0 = 5$ μm.

The action of the injection pulse on the wake is shown in Fig. 2.57, which shows the longitudinal phase space $p_z/m_e c$ by z. Two sets of particles appear: those in the background and those trapped in the wave. For analysis, trapping was defined by two characteristics, first if the particles had the necessary forward velocity, and secondly we artificially picked only those particles in the bucket of interest for analysis, which allowed us to calculate the properties of a single micro-bunch. In the particular simulation plotted, the injection pulse filled only buckets after it passed. A cross marks where the peak of the injection pulse crossed the pump's axis. As the injection pulse was scanned through the different phases of the wave, electrons were injected into every bucket within the simulation domain, shown in Fig. 2.57. The first three buckets were analyzed for every run. A correlation between momentum and position is visible, a characteristic observed in the previous 1-D simulations mentioned. This chirp in the bunch comes from the electrons having been injected over a finite period of time. A large area of phase space was covered as the wave advanced through different phases, subjecting the particles to remarkably different acceleration gradients. Therefore after some acceleration, the correlation between momentum and position appeared. This chirp opens the possibility of compressing the already short electron bunch by use of conventional electron bunch compression techniques.

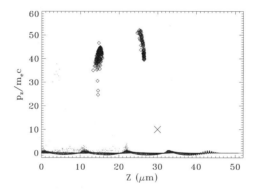

Fig. 2.57 Laser wake field accelerator (LWFA) with an injection pulse. Electrons trapped and accelerated in the wake can be seen. Note that buckets after the one intended for injection are filled due to partially dephased electrons bouncing in the wake, and falling behind. The cross shows the relative position of the injection pulse.

(a) Preliminary injection experiments
Although the short duration laser pulses required for monoenergetic beams are unavailable, a series of experiments with relatively long duration (400 fs) pulses were begun. In the process, we demonstrated for the first time the proof-of-principle of the LILAC concept: one laser pulse was used to

control the energy and emittance of an electron beam driven by a separate laser pulse [Umstadter et al. (2001); Zhang et al. (2003a)]. We have also preliminarily explored another LILAC mechanism, namely heating in the electrostatic field of a density modulation driven by the interference between two pulses.

When two beams, each of duration 400 fs and wavelength 1 μm, were overlapped in space and time in a plasma, the following was observed: (1) enhanced Thomson scattered light from a stationary density modulation, (2) a decrease in beam emittance in the direction of the lower power beam and (3) an increase in the temperature of electrons accelerated in the direction of the low power beam.

Injection in this case is likely to be caused by the combination of the ponderomotive force due to the transverse gradient of the injection pulse in combination with the heating of electrons in the potential well of the ponderomotive beat pattern created along the bisector of the two beams. The heating of the electrons allows them to be caught in the plasma wave created by the low power beam. This is the first time that one laser beam has been used to optically modify the acceleration of electrons by another laser beam [Umstadter et al. (2001); Zhang et al. (2003a)], thus proving the LILAC principle. If the heating pulse was shorter, less than a plasma period, and the overlap between the lasers less than a plasma wavelength, then the simulations predict that a monoenergetic pulse would result.

The experiment was done with a Ti:Sapphire/Nd:glass laser system. It delivered short pulses (400 fs, 1.053 μm) in a single shot with high peak power ($\sim$ 10 TW). As shown in Fig. 2.58, the laser beam was split into two beams by a beam splitter, one contained 80% of the original laser energy and the other 20%. The 20% beam was used as the pump pulse in the experiment and the 80% beam as the injection pulse. These two beams were focused with two $f/3$ parabolic mirrors and the spot sizes in vacuum were about 12 μm FWHM, containing 60% of the total laser beam energy. In the experiment, the laser intensities were about 3×10^{17}–5×10^{17} W/cm^2 for the pump beam and 1.2×10^{17}–2×10^{18} W/cm^2 for the injection beam (the corresponding laser powers were about 1.0–1.5 TW for the pump beam and 4.0–6.0 TW for the injection pump). These two laser beams were perpendicularly overlapped onto the edge of a supersonic helium gas jet. The pressure of the helium gas was 800 psi and the gas was fully ionized by the laser beams (all above 10^{17} W/cm^2) and the corresponding plasma density was about 4×10^{19} cm^3. The background air pressure in the experiment was less than 100 mTorr.

The powers of the pump and injection pulses were both above the critical value for relativistic self-focusing (two times above for the pump and eight times for the injection), and thus the laser pulses underwent relativistic ponderomotive self-channeling. Observed from above the gas jet, the image of these crossed plasma channels was monitored by the side imaging of Thomson scattering of the laser pulse propagating in the plasma and recorded by a CCD camera. A schematic of the experiment is shown in Fig. 2.58.

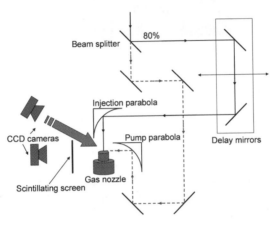

Fig. 2.58 Schematic drawing of the experimental arrangement used for the LILAC proof-of-principle experiments.

The electron energy spectrometer (with range 260 keV–5 MeV) was composed of a dipole permanent magnet with a Kodak LANEX scintillating screen downstream, imaged by a CCD camera. There was an aluminum foil of 40-μm thickness in front of the screen in order to block the laser light. A 1-mm wide slit was placed in front of the magnet in order to collimate the beam, reduce its beam divergence and fix the angle with respect to the spectrometer. The distance between the slit and the gas jet was 30 mm. The profiles of the pump and injection electron beams were observed by LANEX screens and CCD cameras, without the magnetic field.

The spatial overlap of the two crossed laser beams was preliminarily realized by passing these two beams through a 40-mm diameter pinhole at the low laser power output. Then, with high laser power, a good spatial overlap was obtained by adjusting the injection parabolic mirror vertically

Femtosecond Beam Science

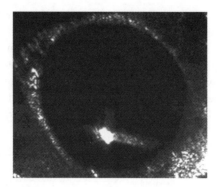

Fig. 2.59   Image of the Thomson scattered light as viewed from the top looking down the throat of the nozzle. A lower power beam propagates from the bottom up, and a higher power beam from right to left.

until we observed from the top view picture an order-of-magnitude brighter spot in the intersection of the two plasma channels (see Fig. 2.59). The temporal overlap of these two beams was checked by laser interference. The delay time between the pump and injection laser beam could be adjusted by moving the position of the delay-line mirrors (see Fig. 2.58).

Figure 2.59 shows the top view of the perpendicularly crossed plasma channels. The bright spot at the intersection of the two plasma channels is quite vivid; generally the intensity of this bright spot was more than 10 times brighter than the other parts of the injection plasma channel. A comparison of top view pictures shows no increase of pump channel length with injection.

Figure 2.60 shows the electron spectrum of the pump electron beam with nearly zero time delay, ±30 fs, between the pump and injection pulses, with and without injection, at a pump laser power of 1.0 TW. It is clear from Fig. 2.61 that the electron number increased and the divergence angle (∼ 10°) decreased with injection, indicating an improvement in transverse geometrical emittance with injection. From Fig. 2.60, it can be seen that with injection there was an increase in the number of electrons with energies greater than 400 keV. Multi-temperature components were observed in the electron spectra. For the low temperature part, the temperature of the electrons increased from 240 keV (without injection) to 390 keV (with injection), a > 60% increase. The increase of the electron temperatures was 30% to 70%. Good correlation was found between the temperature en-

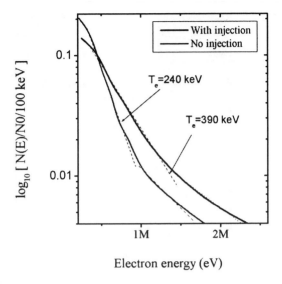

Electron energy (eV)

Fig. 2.60   Longitudinal electron energy spectrum in the direction of the low power beam with and without the injection beam. A shift to higher electron energies is observed with injection.

hancement, the beam overlap and the increase in the Thomson scattering.

Interference is created in the beam intersection of temporally overlapping laser pulses of the same frequency, with, or partially with, the same parallel polarization, which cross each other perpendicularly. The interference will spatially modulate the laser intensity resulting in the creation of an intensity grid directed along the bisector. The effect of the ponderomotive force from such a laser intensity grid on the electrons will modulate the plasma density along the bisector. This modulation is clearly seen in the Thomson scattering top view picture of Fig. 2.59. The intensity of coherent Thomson scattering light is given by $P_s/P_0 \propto \delta n^2$, and for incoherent scattering by $P_s/P_0 \propto \delta n$, where P_s is the scattering light, P_0 is the background scattering and δn is the disturbance of the electron density. Thus, the experimentally measured scattered power $P_s/P_0 > 10$ indicates that the electron density had increased over the background, $\delta n/n \sim 10$. This is a factor of ten to fifty times greater than had been observed previously from plasma density modulation. Analytical calculations indicate that for an intensity enhanced by interference, $I = 1.6 \times 10^{18} \text{ W/cm}^2$

Femtosecond Beam Science

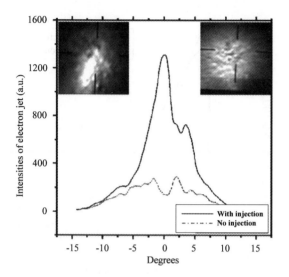

Fig. 2.61   Transverse spatial profile of an electron beam with and without injection. The electron number increases and the divergence angle decreases with injection, indicating an improvement in transverse geometrical emittance with injection.

(four times the pump laser intensity), as in experiment, the ponderomotive force $F_p = m_0 c^2 \nabla (1 + a^2/2)^{1/2}$ should create a plasma density modulation of $\delta n/n > 4$, in good agreement with the experimentally measured value [Zhang *et al.* (2003b)].

Besides energy transfer between the two laser beams, enhancement of the electron beam may also originate from optical injection. It was observed in simulation that laser injection apparently changed the initial velocities or kinetic energies of the background electrons. The increase in initial kinetic energy could result in an increase in the number of background electrons that could be trapped in the plasma wave resulting in the observed increase in the relative number of electrons in the high temperature part of the electron spectrum. The initial energy increase contributed to the observed shift of the electrons in the beam to higher energies. Stochastic heating and ponderomotive acceleration of the electrons were responsible for the enhancement of the initial electron energy by the injection pulse. The results suggest that stochastic heating might be also used to serve as the basis for optical injection, which may produce a monoenergetic electron beam if the two laser pulses have different pulse durations.

2.4.1.3 *Plasma cathode: colliding pulse optical injection*

E. ESAREY
C. B. SCHROEDER
W. P. LEEMANS
Accelerator and Fusion Research Division
Lawrence Berkeley National Laboratory (LBNL)
1 Cyclotron Road, Berkeley,
CA 94720, USA

Perhaps the most basic and simplest form of laser plasma injectors is the self-modulated regime of the laser wake field accelerator (LWFA) [Esarey *et al.* (1996); Leemans *et al.* (2002)], in which a single laser pulse results in self-trapping and the generation of a sub-picosecond electron bunch, with a large energy spread. Typically the self-trapped bunch has a high charge (up to 10 nC), with an energy distribution characterized by a Boltzmann distribution with temperature in the few-MeV range. One possible mechanism for self-trapping is the direct wave-breaking of the plasma wake field [Gordon *et al.* (1998)]. Since the phase velocity of the wake field is near the speed of light, it is difficult to trap the background fluid electrons, which undergo fluid oscillation that sustains the wake field. The wake will trap the background electrons when the separatrix of the wake overlaps the plasma fluid orbit, which is the definition of wave-breaking. Wave-breaking of a cold 1-D plasma wave occurs at $E_{\mathrm{WB}} = [2(\gamma_p - 1)]^{1/2} E_0 \gg E_0$, where $v_p = c\beta_p = c(1 - \gamma_p^{-2})^{1/2}$ is the phase velocity of the plasma wave and $E_0 = cm_e\omega_p/e \simeq 96 \left(n_0 \left[\mathrm{cm}^{-3}\right]\right)^{1/2}$ V/m is the cold 1-D nonrelativistic wave-breaking field, with $\omega_p = (4\pi n_0 e^2/m_e)^{1/2}$ the electron plasma frequency, n_0 the ambient electron density, m_e and e the electron rest mass and charge, respectively, and c the speed of light in vacuum. Thermal and 3-D effects can lower this value, but typically wave-breaking requires non-linear plasma waves with $E_z > E_0$. The observed wake field amplitude, however, as measured in several experiments [Moore *et al.* (1997)], appears to be in the range $E_z/E_0 \sim 10\text{--}30\%$, well below wave-breaking. This suggests that additional laser plasma instabilities may play a role in lowering the effective wave-breaking amplitude.

Alternatively, self-trapping and acceleration can result from the coupling of Raman back scatter (RBS) and Raman side scatter (RSS) to the wake field [Esarey *et al.* (1998)]. As the pump laser self-modulates, it also undergoes RBS, which is the fastest growing laser plasma instability. RBS is observed in intense short pulse experiments, with reflectivities as high as 10–30% [Moore *et al.* (1997)]. RBS generates red-shifted backward light of frequency $\omega_0 - \omega_p$ and wavenumber $-k_0$, which beats with the pump laser (ω_0, k_0) to drive a ponderomotive wave $(\omega_p, 2k_0)$. As the instability grows, the RBS beat wave, which has a slow phase velocity $v_p \simeq \omega_p/2k_0 \ll c$, can trap and heat background plasma electrons [Bertrand *et al.* (1995)]. These electrons can gain sufficient energy and/or be displaced in phase by the beat wave such that they are trapped and accelerated to high energies in the wake field. Simulations [Esarey *et al.* (1998)] indicate that coupling to RBS can lead to self-trapping at modest wake field amplitudes, $E_z/E_0 \simeq 25\%$, much lower than the cold 1-D threshold for direct wave-breaking. In 3-D, this process can be enhanced by coupling to RSS.

When electrons become trapped in the fast wake field, they are accelerated to high energies as they circulate inside the separatrix of the wake. A large energy spread for the trapped electrons results because (a) some fraction of the background electrons are continually being swept up and trapped in the wake field as the laser pulse propagates into fresh plasma, and (b) typically the self-guided propagation distance of the laser pulse is much greater than the detuning length for trapped electrons. This implies that deeply trapped electrons will circulate many revolutions within the separatrix, again resulting in a large energy spread.

For many applications, a small energy spread is desired. This can be achieved by using a standard LWFA (with $L \sim \lambda_p$, where L is the laser pulse length and $\lambda_p = 2\pi c/\omega_p$ is the plasma wavelength), in which the wake field is produced in a controlled manner at an amplitude below the wave-breaking or self-trapping threshold [Esarey *et al.* (1996)]. In principle, if a small energy spread electron bunch of small duration compared to λ_p is injected into the wake field at the proper phase, then the bunch can be accelerated while maintaining a small energy spread. This becomes problematic in the LWFA, since the wavelength of the accelerating field is small, e.g. $\lambda_p \simeq 30$ μm for $n_0 \simeq 10^{18}$ cm^{-3}. Hence, a low energy spread requires an ultrashort bunch duration $\tau_b < \lambda_p/c$ that is injected at the optimal plasma wave phase with femtosecond timing accuracy. These requirements are beyond that of conventional electron beam injector technology (e.g. photo-injectors). On the other hand, the production of ultrashort laser pulses and the femtosecond timing of multiple pulses is routine with compact chirped-pulse amplification (CPA) laser technology. As discussed in this section, ultrashort, high intensity laser pulses can be used to inject electrons into a single bucket (plasma wave period) of a standard LWFA [Umstadter *et al.* (1996);

Esarey *et al.* (1997); Hemker *et al.* (1998); Schroeder *et al.* (1999); Esarey *et al.* (1999)].

Umstadter *et al.* [Umstadter *et al.* (1996)] first proposed using an additional laser pulse to inject background plasma electrons into the wake for acceleration to high energies. To generate ultrashort electron bunches with low energy spreads, the original laser injection method proposed by Umstadter *et al.* [Umstadter *et al.* (1996)] uses two laser pulses that propagate perpendicular to one another. The first pulse (the pump pulse) generates the wake field via the standard LWFA mechanism, and the second pulse (the injection pulse) intersects the wake field some distance behind the pump pulse. The ponderomotive force $F \simeq -(m_e c^2/\gamma)\nabla a^2/2$ of the injection pulse can accelerate a fraction of the plasma electrons such that they become trapped in the wake field. Here $a_0^2 \simeq 3.6 \times 10^{-19}(\lambda \ [\mu m])^2 I \ [W/cm^2]$, for a circularly polarized laser field, with λ the laser wavelength and I the laser intensity. Specifically, the axial (direction of propagation of the pump pulse along the z-axis) ponderomotive force of the injection pulse (propagating along the x-axis) scales as

$$F_z = -(m_e c^2/\gamma)(\partial/\partial z)a_1^2/2 \sim (m_e c^2/\gamma)a_1^2/r_1 , \qquad (2.105)$$

where a_1^2 and r_1 are the normalized intensity and spot size of the injection pulse, respectively. A simple estimate of the change of momentum that an electron will experience due to the ponderomotive force of the injection pulse is $\Delta p_z \simeq F_z \tau_1 \sim (mc^2/\gamma)a_1^2 \tau_1/r_1$, where τ_1 is the injection pulse duration. It is possible for Δp_z to be sufficiently large that electrons are injected into the separatrix of the wake field such that they become trapped and accelerated to high energies. To inject into a single plasma wake bucket, it is necessary for both the injection pulse spot size and pulse length to be small compared to the plasma wavelength, i.e. $r_1^2 \ll \lambda_p^2$ and $c^2\tau_1^2 \ll \lambda_p^2$. Simulations [Umstadter *et al.* (1996)] performed for ultrashort pulses at high densities ($\lambda_p/\lambda = 10$ and $E_z/E_0 = 0.7$) indicated the production of a 10 fs, 21 MeV electron bunch with a 6% energy spread. However, high intensities ($I > 10^{18}$ W/cm^2) are required in both the pump and injection pulses ($a_0 \simeq a_1 \simeq 2$).

Simulations reported in [Hemker *et al.* (1998)] point out that additional electron injection into one or more wake buckets can result due to the influence of the wake associated with the injection pulse, which can be significant due to the high intensity of the injection pulse ($a_1 \gtrsim 1$). Umstadter *et al.* [Umstadter *et al.* (1996)] also discuss the possibility of injection using an injection pulse that propagates parallel, but some distance behind, the pump pulse. The injection pulse would have a tighter focus (and hence smaller Rayleigh length) than the pump pulse, and would be phased appropriately such that it locally drives the wake field to an amplitude that exceeds wave-breaking, thus resulting in local trapping of electrons.

Esarey *et al.* [Esarey *et al.* (1997)] proposed and analyzed the colliding pulse injection method that relies on the beat wave produced by the collision of two counter-propagating laser pulses. Beat wave injection differs intrinsically from the method of ponderomotive injection discussed above in that the source and form of the ponderomotive force differs in these two methods. In ponderomotive injection, injection is the result of the ponderomotive force associated with the *envelope* (time-averaged intensity profile) of a single pulse. In beat wave injection, injection is the result of the ponderomotive force associated with the *slow beat wave* of two intersecting pulses.

Colliding pulse injection [Esarey *et al.* (1997); Schroeder *et al.* (1999); Esarey *et al.* (1999)] uses three short laser pulses: an intense ($a_0^2 \simeq 1$) pump pulse (denoted by subscript 0) for plasma wake generation, a forward going injection pulse (subscript 1) and a backward-going injection pulse (subscript 2), as shown in Fig. 2.62. The frequency, wavenumber and normalized

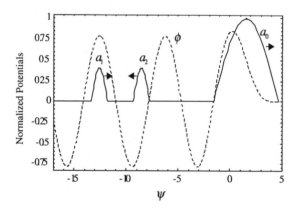

Fig. 2.62 Profiles of the pump laser pulse a_0, the wake ϕ (dashed line) and the forward a_1 injection pulse, all of which are stationary in the $\psi = k_p(z - v_p t)$ frame, and the backward injection pulse a_2, which moves to the left at $\simeq 2c$.

intensity are denoted by ω_i, k_i and a_i ($i = 0, 1, 2$). Furthermore, it is assumed that $k_1 \simeq k_0$, $k_2 \simeq -k_0$ and $\omega_1 - \omega_2 = \Delta\omega \gg \omega_p$. The pump pulse generates a plasma wake with phase velocity near the speed of light ($v_{p0} \simeq c$). The forward injection pulse travels at a fixed distance behind the pump pulse, which determines the position (i.e. phase) of the injected electrons. The injection pulses are orthogonally polarized to the pump laser pulse, such that the pump pulse and backward going injection pulse do not beat. When the injection pulses collide some distance behind the pump, they generate a slow ponderomotive beat wave of the form $a_1 a_2 \cos(\Delta k z - \Delta \omega t)$ (here $\Delta k = k_1 - k_2 \simeq 2 k_0$) with a phase velocity $v_{pb} \simeq |\Delta \omega|/2 k_0 \ll c$. The axial force associated with this beat wave scales as

$$F_z = -(m_e c^2/\gamma)(\partial/\partial z) a_1 a_2 \cos(2 k_0 z - \Delta \omega t) \sim (m_e c^2/\gamma) 2 k_0 a_1 a_2 \,. \quad (2.106)$$

During the time in which the two injection pulses overlap, a two-stage acceleration process can occur, i.e. the slow beat traps and heats background plasma electrons, which, as a result of shifts in their momentum and phase, can be injected into the fast wake field for acceleration to high energies. The ratio of the axial force of the beat wave to that of a single pulse in the ponderomotive injection scheme scales as

$$\frac{F_{z,\text{beat}}}{F_{z,\text{pond}}} \sim \frac{2 k_0 a_1 a_2}{a_p^2/r_p} \,, \quad (2.107)$$

where the subscript p refers to the single ponderomotive injection pulse and the contribution of the relativistic factor γ (which is different for the two cases) is neglected. For comparable injection pulse intensities ($a_1 \simeq a_2 \simeq a_p$), the ratio scales as $4\pi r_p/\lambda_0 \gg 1$, i.e. the axial force of the beat wave is much greater than the ponderomotive force of a single pulse. Consequently, colliding pulses can result in electron injection at relatively low intensities ($a_1 \sim a_2 \sim 0.2$), as well as at relatively low densities ($\lambda_p/\lambda \sim 100$), thus allowing for high single-stage energy gains. Furthermore, the colliding pulse concept offers detailed control of the injection process: the injection phase can be controlled via the position of the forward injection pulse, the beat phase velocity via $\Delta \omega$, the injection energy via the pulse amplitudes, and the injection time (number of trapped electrons) via the backward pulse duration.

To help understand the injection mechanism, it is insightful to consider the electron motion in the wake field and in the colliding laser fields individually. In the absence of the injection pulses, electron motion in a 1-D wake field is described by the Hamiltonian $H_w = \gamma - \beta_p (\gamma^2 - 1)^{1/2} - \phi(\psi)$, where $\phi = \phi_0 \cos \psi$, $v_p = c \beta_p$ is the phase velocity of the plasma wave, $\gamma_p = (1 - \beta_p^2)^{-1/2}$ and $\psi = k_p(z - v_p t)$. The electron orbits in phase space

(u_z, ψ) are given by $H_w(u_z, \psi) = H_0$, where H_0 is a constant, $\gamma^2 = 1 + u_z^2$ and $u_z = \gamma\beta_z$ is the normalized axial momentum given by

$$u_z = \beta_p\gamma_p^2 [H_0 + \phi(\psi)] \pm \gamma_p \left\{ [\gamma_p^2 [H_0 + \phi(\psi)]^2 - 1 \right\}^{1/2}. \qquad (2.108)$$

The 1-D separatrix (the boundary between trapped and untrapped orbits) is given by $H_w(\beta_z, \psi) = H_w(\beta_p, \pi)$, i.e. $H_0 = H_{1D} = 1/\gamma_p - \phi(\pi)$. The maximum and minimum electron momenta on the 1-D separatrix occur at $\psi = 0$ and are (in the limits $2\phi_0\gamma_p \gg 1$ and $\gamma_p \gg 1$) $u_{w,\text{max}} \simeq 4\gamma_p^2\phi_0$ and $u_{w,\text{min}} \simeq 1/4\phi_0 - \phi_0$.

The 1-D theory neglects the effects of transverse focusing. Associated with a 3-D wake is a periodic radial field which is $\pi/2$ out of phase with the accelerating field, i.e. there exists a phase region of $\lambda_p/4$ for which the wake is both accelerating and focusing (as opposed to the $\lambda_p/2$ accelerating region in 1-D). If an electron is to remain in this phase region, it must lie within the "3-D separatrix" defined by $H_w(\beta_z, \psi) = H_w(\beta_p, \pi/2)$, i.e. Eq. (2.108) with $H_0 = H_{3D} = 1/\gamma_p - \phi(\pi/2)$. The extrema on the 3-D separatrix are given by $u_{w,\text{max}} \simeq 2\gamma_p^2\phi_0$ and $u_{w,\text{min}} \simeq 1/2\phi_0 - \phi_0/2$. This value of $u_{w,\text{max}} \simeq 2\gamma_p^2\phi_0$ gives the usual maximum energy gain due to linear dephasing in a 3-D wake.

The background plasma electrons lie on an untrapped orbit (below the separatrix) u_{zf} given by $H_w(u_{zf}, \psi) = 1$, i.e. Eq. (2.108) with $H_0 = 1$. At wave-breaking, the bottom of the separatrix $u_{w,\text{min}}$ coalesces with the plasma fluid orbit, $u_{zf} = u_{w,\text{min}}$. This occurs at the well-known wave-breaking field of $E_{\text{WB}}/E_0 = [2(\gamma_p - 1)]^{1/2}$, assuming a nonlinear wake in 1-D.

We consider the motion of electrons in the colliding laser fields in the absence of the wake field. The beat wave leads to the formation of phase space buckets (separatrices) of width $2\pi/\Delta k \simeq \lambda_0/2$, which are much shorter than those of the wake field (λ_p). In the colliding laser fields, the electron motion is described by the Hamiltonian $H_b = \gamma - \beta_b[\gamma^2 - \gamma_\perp^2(\psi_b)]^{1/2}$, where the space charge potential is neglected. Circular polarization is assumed such that $\gamma_\perp^2 = 1 + a_0^2 + a_1^2 + 2a_0a_1\cos\psi_b$, where $\psi_b = (k_1 - k_2)(z - v_bt)$ and $v_b = c\beta_b = \Delta\omega/(k_1 - k_2) \simeq \Delta\omega/2k_0$ is the beat phase velocity, assuming $\omega_p^2/\omega_0^2 \ll 1$. The beat separatrix is given by $H_b(\beta_z, \psi_b) = H_b(\beta_b, 0)$ with maximum and minimum axial momenta of

$$u_{b,m} = \gamma_b\beta_b \left[1 + (a_0 + a_1)^2 \right]^{1/2} \pm 2\gamma_b(a_0a_1)^{1/2}. \qquad (2.109)$$

An estimate of the threshold for injection into the wake field can be obtained by a simple phase space island overlap criterion. This is done by considering the effects of the wake field and the beat wave individually, as done above, and by requiring that the beat wave separatrix overlap both the wake field separatrix and the plasma fluid oscillation (illustrated in

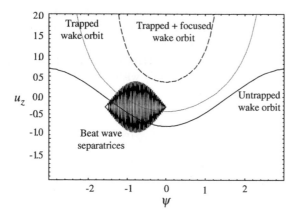

Fig. 2.63 Longitudinal phase space (ψ, u_z) showing beat wave separatrices, an untrapped plasma wave orbit (solid line), a trapped plasma wave orbit (dotted line) and a trapped and focused plasma wave orbit (dashed line).

Fig. 2.63), such that (i) the maximum momentum of the beat wave separatrix $u_{b,\mathrm{max}}$ exceeds the minimum momentum of the wake field separatrix $u_{w,\mathrm{min}}$, i.e. $u_{b,\mathrm{max}} \geq u_{w,\mathrm{min}}$, and (ii) the minimum momentum of the beat wave separatrix $u_{b,\mathrm{min}}$ is less than the plasma electron fluid momentum u_{zf}, i.e. $u_{b,\mathrm{min}} \leq u_{zf}$. Conditions (i) and (ii) imply a beat wave threshold

$$(a_1 a_2)_{\mathrm{th}}^{1/2} = \frac{(1 - H_0)}{4\gamma_b(\beta_p - \beta_b)} \qquad (2.110)$$

and an optimal wake phase for injection (location of the forward injection pulse)

$$\cos \psi_{\mathrm{opt}} = \phi_0^{-1} \left[(1 - \beta_b \beta_p)\gamma_b \gamma_\perp(0) - (1 + H_0)/2 \right], \qquad (2.111)$$

where $H_0 = H_{1D} = 1/\gamma_p + \phi_0$ for the 1-D wake separatrix and $H_0 = H_{3D} = 1/\gamma_p$ for the 3-D wake separatrix (trapped and focused). In the limits $\gamma_p^2 \gg 1$, $\beta_b^2 \ll 1$ and $a_i^2 \ll 1$, Eqs. (2.110) and (2.111) become $4(a_1 a_2)_{\mathrm{th}}^{1/2} \simeq (1 - H_0)(1 + \beta_b)$ and $2\phi_0 \cos \psi_{\mathrm{opt}} \simeq 1 - H_0 - 2\beta_b$ with $H_{1D} \simeq \phi_0$ and $H_{3D} \simeq 0$. As an example, $\phi_0 = 0.7$, $\beta_b = -0.02$ and $\gamma_p = 50$ imply a threshold of $(a_1 a_2)_{th}^{1/2} \simeq 0.25$ and an optimal injection phase of $\psi_{opt} \simeq 0$ for injection onto a trapped and focused orbit.

To further evaluate the colliding laser injection method, the motion of test particles in the combined wake and laser fields was simulated in 3-D [Schroeder *et al.* (1999)]. In the numerical studies, the laser pulse axial profiles were half-period sine waves (linearly polarized with Gaussian radial profiles) with peak amplitude $\sqrt{2}a_i$, such that $\langle a^2 \rangle = a_i^2$, and length L_i. The wake field is assumed to be nonzero for $\psi \le 3\pi/4$ (see Fig. 2.62) and the test particles are loaded uniformly with $\psi > 3\pi/4$ (initially at rest).

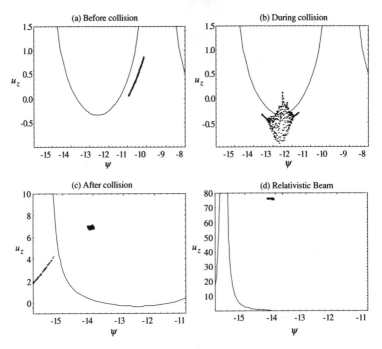

Fig. 2.64   Electron distribution in longitudinal (u_z, ψ) phase space (a) before injection pulse collision ($\omega_p \Delta t = 0$), (b) during collision ($\omega_p \Delta t = 3$), (c) just after collision ($\omega_p \Delta t = 14$) and (d) at $\omega_p \Delta t = 114$ (38 MeV electron bunch with 1 fs duration, 0.2% energy spread and 0.9 mm·mrad normalized emittance). The separatrix between trapped and untrapped wake orbits (solid line) is shown.

An example of the injection process is given in Fig. 2.64, which shows the evolution in longitudinal phase space (u_z, ψ) of the test electron distribution (a) before the collision of the injection laser pulses (in the untrapped fluid orbit of the wake) at $\omega_p t = 36$, (b) during the collision (crossing the wake separatrix) at $\omega_p t = 39$, (c) after the collision at $\omega_p t = 50$ and (d) the resulting energetic electron bunch at $\omega_p t = 150$ ($z = 0.7$ mm). Also shown is the 1-D wake separatrix. The parameters are $a_1 = a_2 = 0.32$, $L_0 = 4L_1 = 4L_2 = \lambda_p = 40$ μm, $\phi_0 = 0.7$, $\lambda_0 = \lambda_2 = 0.8$ μm, $\lambda_1 = 0.83$ μm and $r_0 = r_1 = r_2 = 15$ μm, with the position of the forward injection pulse centered at $\psi_{\text{inj}} = -12.6$. After $z \simeq 0.7$ mm of propagation following the collision, Fig. 2.64(d), the bunch length is 1 fs with a mean energy of 38 MeV, a fractional energy spread of 0.2% and a normalized transverse emittance of 0.9 mm·mrad. The trapping fraction f_{trap} is 3%, corresponding to 2.6×10^6 bunch electrons. Here, f_{trap} is defined as the fraction of electrons trapped that were initially loaded in a region of length $\lambda_p/4$ with $r \leq 2$ μm (simulations indicate that electrons loaded outside this region are not trapped). Note that the bunch number can be increased by increasing the laser spot sizes (i.e. laser powers). For example, when the laser spot sizes are doubled to $r_i = 30$ μm in the simulation of Fig. 2.64 (with all other parameters as in Fig. 2.64), the number of trapped electrons increases to 1.5×10^7 and the normalized transverse emittance increases to 3.9 mm·mrad. Estimates indicate that space charge effects can be neglected while the bunch remains inside the plasma [Schroeder *et al.* (1999)].

Experiments on laser injection methods are being pursued at several laboratories worldwide. For example, at Lawrence Berkeley National Laboratory (LBNL), experiments are underway on the colliding pulse method. The initial set of experiments uses only two pulses: a pump pulse for wake field generation and a single backward propagating injection pulse (see Fig. 2.65). Here the pump and injection pulses have the same polarization such that injection results from the slow ponderomotive beat wave that is produced when the injection pulse collides with the tail of the pump pulse. Experimentally, the use of collinear pulses is technically challenging as the counter-propagating pulse must be reflected by a mirror through which the electron beam must propagate and from which the high power drive pulse must be reflected. Ultrathin, dielectrically coated substrates are being developed, but substrates that can handle the high fluence in these experiments are not presently commercially available. Therefore the current experimental implementation uses non-collinear injection of the drive and colliding beams. Non-collinear injection has been explored theoretically and shown to provide nearly the same beam quality as collinear injection [Fubiani *et al.* (2003)].

In these experiments two intense short laser pulses were produced by a

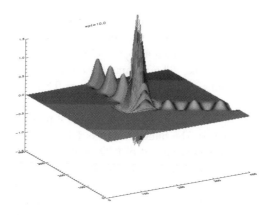

Fig. 2.65 Electric field profiles and corresponding wake fields, plotted during the collision of the drive pulse and colliding pulse versus x and $z - ct$ (where z is the drive pulse propagation direction), for the two-pulse configuration with a 30° interaction geometry.

10 Hz, Ti:Al$_2$O$_3$, CPA laser system [Leemans *et al.* (2001)]. Low energy pulses ($\lambda \simeq 0.8$ μm) from a laser oscillator were first temporally stretched, amplified to 1 mJ using a regenerative amplifier, split into two pulses and then amplified to 1 J/pulse and 0.3 J/pulse, respectively. Each pulse was then compressed using its own grating-based optical compressor (installed in a vacuum chamber) to pulse widths as short as 45 fs with an overall power transmission efficiency of about 50% onto a gas jet target.

Following compression, the main drive laser beam was focused to a 6 μm spot size with a 30 cm focal length (F/4) off-axis parabola (OAP) mirror onto the pulsed gas jet. With this single drive beam, electron bunches have been produced through self-modulation containing up to 5 nC charge with electron energy in excess of 40 MeV [Leemans *et al.* (2002)]. The colliding beam was focused to an 8 μm spot size with an identical OAP onto the pulsed gas jet with a 30° angle with respect to the drive beam. The layout of the experiment is shown in Fig. 2.66. The intersection of the beams was measured using a CCD camera looking from above onto the plasma region and with side-on imaging. The top view CCD camera image is shown in Fig. 2.67, indicating the spatial overlap of the colliding and drive beams.

The total charge per bunch and spatial profile of the electron beam were measured using an integrating current transformer (ICT) and phosphor screen imaged onto a 16-bit CCD camera, respectively. As shown in Fig. 2.68, preliminary results have been obtained that indicate electron

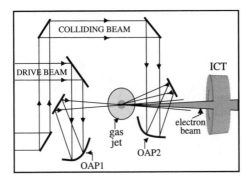

Fig. 2.66   Layout of experiment showing the drive and colliding laser beams exiting the compressor. The drive and colliding beams are focused onto the off-axis parabola mirrors OAP1 and OAP2, respectively, which focus the beams under a 30° angle onto the gas jet. The resulting electron beam charge is measured using the integrating current transformer (ICT).

yields have been affected by the second laser beam which intersected the forward-going drive laser beam at 30°. Note that the peak power in the drive beam was lowered to reduce the charge production to about 0.1 nC. The charge enhancement resulting from the second pulse could be due to several mechanisms, such as the generation of a beat wave (i.e. colliding pulse injection), heating of the background electrons or other stochastic processes. Ongoing experiments at LBNL are measuring the electron beam spectra and carrying out various parametric studies to understand in detail the underlying injection mechanisms.

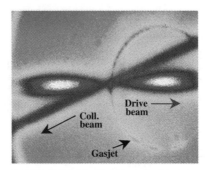

Fig. 2.67 Top view CCD camera image of the plasmas produced by two laser pulses propagating in a backfilled chamber indicating the spatial overlap of the beams, the drive beam (horizontal) and the colliding beam (30°). The measured light emission is from recombination in the helium plasma, and the shape follows the Gaussian iso-intensity contours of a focused laser beam. The drive beam was focused at the upstream edge of the gas jet with a smaller f-number lens than the colliding beam and hence has a shorter Rayleigh length. The ring passing through the intersection point is caused by laser light scattering off the gas jet edge.

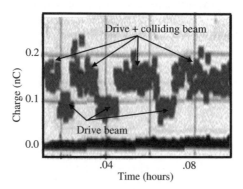

Fig. 2.68 Electron beam charge measured using the ICT versus clock time when a single drive beam is used (lower charge level) and when both drive and colliding beams are used (higher level).

2.4.1.4   *Electron injection due to Langmuir wave breaking*

S. V. BULANOV[1,2]
T. Zh. ESIRKEPOV[1,3]
[1]Advanced Photon Research Centre, JAERI-Kansai, Kisu,
KYOTO, JAPAN
[2]General Physics Institute of RAS,
MOSCOW, RUSSIA
[3]Moscow Institute of Physics and Technology,
DOLGOPRUDNYI, RUSSIA

M. UESAKA
Nuclear Engineering Research Laboratory
University of Tokyo,
22-2 Shirane-shirakata, Tokai, Naka,
IBARAKI 319-1188, JAPAN.

(a)   Langmuir wave breaking in homogeneous plasma

The basic factors of strong wake field excitation in an under-dense plasma by an intense laser pulse are the intensity and the duration $\tau_p = l_p/c$ of the laser pulse (or the scale length of the laser pulse front). The relativistically strong laser pulse in the under-dense plasma is unstable due to stimulated backward Raman scattering (SBRS). SBRS leads to erosion of the amplitude profile at the leading edge of the pulse and to the formation of a steep laser front similar to a shock in a time of the order of $\omega_{pe}^{-1}(\omega_0/\omega_{pe})^2$, where the laser pulse carrier frequency ω_0 is larger than the Langmuir frequency $\omega_{pe} = (4\pi n e^2/m_e)^{1/2}$. If the scale length of the relativistically strong laser pulse front is shorter than the Langmuir wavelength λ_{pe}, the response of the electron component is nonadiabatic and, as a result, a wake plasma wave is efficiently generated [Tajima and Dawson (1979)]. This leads to a fast depletion of the laser pulse energy on the timescale $t_{dep} \sim (\omega_0/\omega_{pe})^2\tau_p$ [Bulanov *et al.* (1992); Mori *et al.* (1994)] and to the induced focusing of the laser light behind the wave front [Bulanov and Sakharov (1991)]. The group velocity of the wake field $\partial\omega_{pe}/\partial k_p$ is zero, and its phase velocity $v_{ph} = \omega_{pe}/k_p$ is equal to the group velocity of the laser pulse $v_{ph} = v_g \approx c\left(1 - \omega_{pe}^2/2\omega_0^2\right)$.

An electron at resonance with the wake field, whose wavenumber is $k_p = c/\omega_{pe}$, can be accelerated by the wake field up to ultra-relativistic energies. The main problem of the electron injection into the wake field is the production of accelerated electron beams with low energy spread. It requires a very precise injection of extremely short electron bunches in the appropriate phase of the wake field. Since the typical length of the wake plasma wave is of the order of $2\pi c/\omega_{pe} \approx 10$–$100$ μm, the length of the injected electron bunch must be shorter than 2–20 μm.

Electron injection into the acceleration phase of the wake field can appear spontaneously due to the finite amplitude wave breaking, as has been discussed in Refs. [Bulanov *et al.* (1992); Bulanov *et al.* (1996)].

In the theory of the interaction of high intensity laser radiation with plasmas, the seminal paper by Akhiezer and Polovin [Akhiezer and Polovin (1956)] has played a key role for many years. In this paper the exact solution to the problem of the propagation of a relativistically strong electromagnetic wave in a collisionless plasma was found. If an unbounded cold collisionless plasma is described by Maxwell's equations and by the hydrodynamic equations of an electron fluid, the coupled electromagnetic and Langmuir waves are given by

$$\phi'' = \frac{V}{1 - \beta_g^2} \left(\frac{\psi_e}{R_e} - \frac{\psi_i}{R_i} \right) , \tag{2.112}$$

$$a'' + \omega^2 a = a \frac{\beta_g}{1 - \beta_g^2} \left(\frac{1}{R_e} + \frac{\mu}{R_i} \right) . \tag{2.113}$$

Here the waves are assumed to depend on the coordinates and time via the variables $\xi = x - \beta_g t$ and $\tau = t - \beta_g x$. The normalized electromagnetic and electrostatic potentials depend on ξ and on τ as $A_y + iA_z = a(\xi) \exp{(i\omega\tau)}$ and $\phi = \phi(\xi)$. The prime in Eqs. (2.112) and (2.113) denotes a derivative with respect to the variable ξ. In these equations $\beta_g = v_g/c$ is the normalized group velocity of the electromagnetic wave (equal to the Langmuir wave phase velocity), $\mu = m_e/m_i$ is the electron to ion mass ratio and we use the functions $\psi_e = \mathcal{E}_e + \phi$, $\psi_i = \mathcal{E}_i - \mu\phi$ and $R_e = [\psi_e^2 - (1 - \beta_g^2)(1 + a^2)]^{1/2}$, $R_i = [\psi_e^2 - (1 - \beta_g^2)(1 + \mu^2 a^2)]^{1/2}$. The constants $\mathcal{E}_e$ and $\mathcal{E}_i$ must be specified by the boundary conditions at infinity. The density and the energy of particles, denoted by the subscripts $\alpha = e, i$, are equal to

$$n_\alpha = \beta_g \frac{(\psi_\alpha - \beta_g R_\alpha)}{R_\alpha(1 - \beta_g^2)} , \qquad \gamma_\alpha = \frac{(\psi_\alpha - \beta_g R_\alpha)}{(1 - \beta_g^2)} . \tag{2.114}$$

The above system admits the first integral,

$$\frac{1-\beta_g^2}{2}\left(a'^2 + \omega^2 a^2\right) - \frac{1}{2}\phi'^2 + \frac{\beta_g}{1-\beta_g^2}\left(R_e + \frac{R_i}{\mu}\right) = \text{const.} \qquad (2.115)$$

For $a = 0$, Eq. (2.112) describes a longitudinal plasma wave. In this case the integral Eq. (2.115) gives the relationship between the electric field and the particle energies: $E^2 + 2(\gamma_e + \gamma_i/\mu) = \text{const.}$ For a given dependence of the electromagnetic wave on the coordinate ξ, Eq. (2.112) describes the longitudinal wave excitation by the laser pulse. The wavelength of a weak Langmuir wave is $\lambda_p = 2\pi\beta_g c/\omega_{pe}$. In the ultra-relativistic case ($\gamma_e, \gamma_g \gg 1$) the wavelength is about $4\lambda_p(2\gamma_e)^{1/2}$, where $\gamma_e \leq \gamma_g = (1-\beta_g^2)^{-1/2}$. We see that relativistic nonlinearity effects lead to an increase of the wavelength. However, the effects of ion motion decrease the Langmuir wave wavelength [Khachatryan (1998); Bulanov *et al.* (2001)].

The amplitude of the Langmuir wave cannot be arbitrarily large. It is limited by the condition $R_\alpha > 0$. At $R_\alpha = 0$ we have the wave breaking point. Close to this point the maximum of the electron density, given in Eq. (2.114), tends to infinity while the width of the density spike tends to zero:

$$n(\xi)/n_0 = 2^{1/3}\gamma_m(c\beta_m/3\omega_{pe}\xi)^{2/3} + \dots. \qquad (2.116)$$

This singularity is integrable and the total number of particles in the density spike is finite. The velocity of the electrons in the spike is maximum, it is equal to the Langmuir wave phase velocity, i.e. to the group velocity of the laser pulse $\gamma_e = \gamma_g$. These electrons are injected into the wake field.

Wave breaking imposes a constraint on the maximum value of the electric field in the wake:

$$E_m = \frac{m_e\omega_{pe}c}{e}(2(\gamma_g - 1))^{1/2}, \qquad (2.117)$$

which is called the *Akhiezer–Polovin limiting electric field*. It is crucial to know the maximum value of the longitudinal electric field, since its magnitude determines the acceleration rate of charged particles. The constraint in Eq. (2.117) implies that a stationary Langmuir wave cannot have an electric field larger than the critical value. However, in a nonstationary, breaking plasma wave the amplitude of the electric field can be substantially larger, up to $E_m = m_e c\omega_{pe}\gamma_{ph}/2e$. The effective acceleration of charged particles in such a field has been shown in numerical simulations [Bulanov *et al.* (1991)] and in experiment [Gordon *et al.* (1998)].

In the case of a laser pulse with amplitude $a > (m_i/m_e)^{1/2}$, which corresponds to the pW power range, ions can no longer be considered to remain at rest. In the case of immobile ions, the electrostatic potential is bounded by $-1 < \phi < a_m^2$, where a_m is the maximum value of the laser pulse amplitude [Bulanov *et al.* (1989)]. On the contrary, the effect of the ion motion changes the upper bound for the potential ϕ to $-1 < \phi < \min\{\mu, a_m^2\}$ (in these estimates we assume $\beta_g \to 1$). From Eqs. (2.112) and (2.113) we find that behind a short laser pulse with length $l = 2^{1/2}/a_m$ (which is the optimal length for the relativistically strong wake field excitation) the wavelength λ_w of the wake field and the maximum values of the electric field E_w and the potential ϕ_w scale as

$$\lambda_w = 2^{3/2} a_m \,, \quad E_w = a_m/2^{1/2} \,, \quad \phi_w = a_m^2 \quad \text{for } 1 < a_m < \mu^{1/2} \,, \tag{2.118}$$

$$\lambda_w = 2^{1/2} \mu/a_m \,, \quad E_w = a_m/2^{1/2} \,, \quad \phi_w = \mu \quad \text{for } a_m > \mu^{1/2} \,. \tag{2.119}$$

For $a_m > \mu^{1/2}$ the wake field wavelength decreases with increasing laser pulse amplitude, while the value of the electrostatic potential does not change. As has been shown in Ref. [Khachatryan (1998)], the ion influence on the wave-breaking limit is small and can be described by terms of the order of μ.

The equations of electron motion in an electric field of a 1-D wake plasma wave can be written in the form [Bulanov *et al.*(1997b)]

$$\frac{d}{dx}\left(\frac{\Psi}{\omega_{pe}}\right) = \frac{(m_e^2 c^2 + p^2)^{1/2} - p}{cp} - \frac{\omega_{pe}^2}{2\omega^2 c} \,, \tag{2.120}$$

$$\frac{d}{dx}(m_e^2 c^4 + p^2 c^2)^{1/2} = -eE \,, \tag{2.121}$$

where x is the particle coordinate, p is the particle momentum, E is the wake field, $\Psi = \omega_{pe}(t - t_0) = \omega_{pe}\left(t - \int^x dx'/v_{ph}\right)$ is the wave phase and t_0 is the time at which the laser pulse reaches the point x.

The electric field E depends on the coordinate x and on the phase Ψ. As follows from Eqs. (2.120) and (2.121), in a homogeneous plasma an ultra-relativistic particle in a moderately strong plasma wave acquires an energy of the order of

$$\Delta \mathcal{E} \approx eE\, l_{acc} \,, \tag{2.122}$$

where l_{acc} is the acceleration length [Tajima and Dawson (1979)]

$$l_{acc} \approx \frac{c^2}{\omega_{pe}(c - v_{ph})} \approx \frac{2c}{\omega_{pe}}\gamma_{ph}^2 = \frac{2c}{\omega_{pe}}\left(\frac{\omega}{\omega_{pe}}\right)^2 . \tag{2.123}$$

This length is $(\omega/\omega_{pe})^2$ times larger than the plasma wave length. We note that this result has been obtained in the limit of a small amplitude wake field. In the case of a relativistically strong wake field, the acceleration length is $l_{acc} \approx (2c/\omega_{pe})\gamma_{ph}^2 a$. The maximum energy of accelerated particles is limited by the constraint imposed due to plasma wave breaking: $\mathcal{E}_{max} = 4m_e c^2 \gamma_{ph}^3$ [Esarey and Pilloff (1995)].

To be effectively accelerated, a charged particle must be in phase with the wake plasma wave. In a plasma with uniform density the phase velocity of the wave does not change along the path of propagation while the particle velocity increases in the acceleration process. This leads to the break of the wave–particle resonance conditions and limits the particle energy. In inhomogeneous plasmas the group velocity and the amplitude of the electromagnetic wave packet depend on the coordinates so that the phase velocity and the amplitude of the wake plasma wave vary. With the appropriate choice of plasma density profile it is possible to increase the acceleration length significantly.

(b) Wave breaking in inhomogeneous plasma

As we have seen above, the Langmuir wave break occurs when the quiver velocity of the electrons becomes equal to the phase velocity of the wave. In a plasma with inhomogeneous density, the Langmuir wave frequency depends on the coordinates, as a result, the wavenumber of the wave depends on time through the well known relationship (see [Whitham (1974)]) $\partial \mathbf{k}/\partial t = -\nabla\omega$. The resulting growth in time of the wavenumber results in a decrease of the wave phase velocity and leads to the break of the wave at the instant when the electron velocity becomes equal to the wave phase velocity, even if the initial wave amplitude is below the wave break threshold. In this case the wave break occurs in such a way that only a relatively small part of the wave is involved. We can use this property to perform a gentle injection of electrons into the acceleration phase, as was shown in Ref. [Bulanov *et al.* (1998)] (see also Ref. [Bulanov *et al.* (1999b)]). In a similar way, Langmuir wave breaking may occur in the non-1-D configurations [Dawson (1959); Bulanov *et al.* (1997a)] due to the dependence of the relativistically strong Langmuir wave frequency on its amplitude, as analyzed in Ref. [Bulanov *et al.* (1997a)].

In an inhomogeneous plasma with a density that depends on the coordinate as $n_e(x) = n_0(L/x)^{2/3}$, $L \approx (c/3\omega_{pe})(\omega/\omega_{pe})^2$, a laser pulse with moderate amplitude $a < 1$ and length l_p excites a wake plasma wave with electric field $E(x,t) = -\omega_{pe}^2(x)(m_e l_p a^2/4e)\cos\Psi$. In this wave the acceleration length becomes formally infinite and the particle energy growth is unlimited [Bulanov *et al*(1997b)]:

$$\mathcal{E}(x) \approx m_e c^2 \left(\frac{\omega}{\omega_{pe}}\right)^2 \left(\frac{x}{L}\right)^{1/3}. \qquad (2.124)$$

As the result of the wave break, fast electrons from the wave crest, corresponding to the cusp of the electron density (Eq. (2.116)), are trapped by the wave and are pre-accelerated into the region where the phase velocity increases and the wake field has a regular and steady structure. In this way we obtain a gentle injection of electrons into the acceleration phase in the wake field far from the breaking region.

The breaking leads to a local decay of the wake plasma wave. Its energy is transported away by the fast electrons. From the energy balance we estimate the fast electron density in the breaking region to be

$$n_{inj} = n_0 \xi_m/L. \qquad (2.125)$$

(c) Numerical simulations
Electron injection due to wake field breaking is demonstrated in the 1-D particle-in-cell (PIC) simulations presented in Fig. 2.69 (see Refs. [Bulanov *et al.* (1998)] and [Bulanov *et al.* (2001)]). A circularly polarized laser pulse interacts with a weakly inhomogeneous plasma. Ions are assumed to be immobile. Asymptotically, as $x \to \infty$, the plasma is homogeneous with a density $n = (1/625)n_c r$ that corresponds to $\omega/\omega_p e = 25$. The plasma density varies smoothly from zero at $x/\lambda = 32$ to $(1/547)n_c r$ at $x/\lambda = 96$ to avoid the distortion of the plasma wave due to wave break at the vacuum–plasma interface [Bulanov *et al.* (1990)]. The plasma is homogeneous in the domain $96 < x/\lambda < 128$ and its density decreases gradually from $(1/547)n_c r$ to $(1/625)n_c r$ in the domain $128 < x/\lambda < 152$. The laser pulse length is 12λ and its amplitude is $a = 2$. We see that due to wave breaking a portion of the electrons is injected into the acceleration phase and is further accelerated up to the energy that corresponds to the expression given by Eq. (2.122). In the vacuum–plasma interface we see the formation of wave breaking towards the vacuum region discussed in Ref. [Bulanov *et al.* (1990)]. At $\omega t/2\pi = 130$ [Fig. 2.69(a)] we see the formation of the

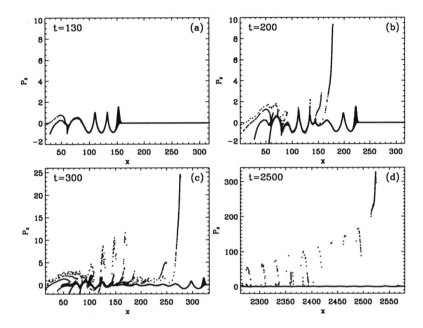

Fig. 2.69 Electron phase plane obtained in 1-D PIC simulations: (a) at $t = 130(2\pi/\omega)$, (b) at $t = 200(2\pi/\omega)$, (c) at $t = 300(2\pi/\omega)$ and (d) at $t = 2500(2\pi/\omega)$.

cusp structures that characterize the wave break described by Eq. (2.116). At $\omega t/2\pi = 200$ [Fig. 2.69(b)], during the wave break, particles are injected into the accelerating phase of the wake field. Further acceleration is seen for $\omega t/2\pi = 300$ [Fig. 2.69(c)], and $\omega t/2\pi = 2500$ [Fig. 2.69(d)]. At time $\omega t/2\pi = 2500$ the maximum energy of the fast particles is approximately $330m_ec^2$. The most energetic particles have been accelerated in the first period of the wake field behind the laser pulse.

A very important feature of this injection regime is that it provides conditions where the resonant wave–particle interaction in the region of homogeneous plasma forms electron bunches that are well localized both along the x-coordinate and in energy space, as is seen in Fig. 2.70.

A different form of wave breaking can occur in 2-D and 3-D wakes of a relativistically strong wide laser pulse, or when a pulse is propagating inside a plasma channel. The 2-D wake field in a plasma has a specific "horse-shoe" shape [Bulanov and Sakharov (1991); Bulanov *et al.* (1997a)], and the curvature of the constant phase surfaces increases with distance

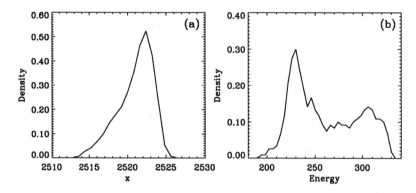

Fig. 2.70 Density inside the bunch of accelerated electrons at $t = 2500(2\pi/\omega)$: (a) vs. particle position x and (b) vs. particle energy p_x.

from the laser pulse front. The corresponding structure in the 3-D case has the form of a paraboloid. These structures improve the focusing of the accelerated particle and focus the laser radiation. The curvature radius R decreases until it become comparable to the electron displacement ξ in the nonlinear plasma wave, and self-intersection of the electron trajectories occurs.

Consider a wake field plasma wave excited by a laser pulse of finite width S. The resonance condition gives $\omega_{pe}/k = v_{ph}$. Transverse inhomogeneity of ω_{pe} is caused by the inhomogeneity of the plasma density, if the laser pulse is guided in a plasma channel, and by the relativistic dependence of $\Omega_p \approx \Omega_p(0) = \pi\omega_{pe}/2a(0)$ on the plasma wave amplitude, which is determined by the pulse transverse shape. The plasma wave frequency can be approximated in the vicinity of the axis by $\Omega_p(y) \approx \Omega_p(0) + \Delta\Omega_p(y/S)^2$. Here $\Delta\omega_{pe}$ is the difference between the plasma frequency outside and inside the channel. In the initially homogeneous plasma $\Delta\Omega_p \approx \Omega_p(0)$ if the wake field is excited by an ultrahigh intensity pulse with $a \gg 1$ and the pulse transverse profile is approximated as $a(y) = a(0)(1 - y^2/S^2)$. As a consequence, the plasma wake field wavelength $\lambda_p = 2\pi/k_p$ depends on y. From the expression of the constant phase curves $\Psi(x, y) = \omega_{pe}(y)(t - x/v_{ph}) = \text{const}$ it follows that their curvature $1/R$ increases linearly with the distance l from the laser pulse: $R = \Omega_p(0)S^2/2\omega_{pe}l$ where $l \equiv \Psi v_{ph}/\Omega_p(0)$. Thus we can write the constant phase curves as $x_0 \equiv x - v_{ph}t + \Psi v_{ph}/\Omega_p(0) = y_0^2/2R$. The real position of the constant phase curves (in a nonlinear plasma wave) is shifted from the curves given above by the oscillation amplitude ξ. Thus, when R becomes of the order of the electron displacement ξ, the wake plasma wave starts to break [Bulanov *et al.* (1997a)]. From these considerations, the distance between the laser pulse and the location, where the breaking begins, can be estimated to be $\Omega_p(0)S^2/(2\Delta\Omega_p\xi)$, and the number of regular wake field periods to be $N_p \approx \Omega_p^2 S^2/4\pi c\Delta\Omega_p\xi_m$.

The displacement ξ in the 2-D or 3-D case depends on the specific laser plasma regime under consideration. Nevertheless, a few important features of the phase surfaces in the wave-breaking regime can be understood if we assume that the displacement is perpendicular to the phase surfaces (parabolic curves in the 2-D case, $x_0 - y_0^2/2R_y = $ const, paraboloids in the 3-D case, $x_0 - y_0^2/2R_y - z_0^2/2R_z = $ const). At the wave break, singularity occurs. It corresponds to the electron trajectory self-intersection. Consider the 2-D case, where $R_z \to \infty$ and $S_z \to \infty$. If the displacement ξ is independent of the coordinate y_0, which is a plausible approximation, we obtain a singularity of the form $y \approx |x|^{3/4}$. For $R < \xi_m$, a multivalued structure appears, which is known as the *swallow tail* in catastrophe theory. Near the breaking threshold, the displacement along the axis becomes close to the curvature radius $\xi(0)/R - 1 = \varepsilon \ll 1$ and the size of the swallow tail is of the order of ε^2 along x and $\varepsilon^{3/2}$ along y. In Fig. 2.71 we show the surface of constant phase in the 3-D case for different types of wave breaking.

(a)   (b)

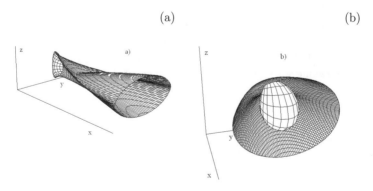

Fig. 2.71   Surface of constant phase in the 3-D case for $R \leq \xi$ with (a) $\xi = $ const and (b) ξ not constant.

The laser generated wake plasma wave in the 3-D geometry is demonstrated in the following 3-D PIC simulation. We use relativistic electromagnetic particle-mesh code (REMP), which exploits a new scheme of current assignment [Esirkepov (2001)] that reduces unphysical numerical effects of the PIC method significantly. In the simulation, the laser pulse propagates along the x-axis. The pulse is linearly polarized in the direction of the z-axis and its dimensionless amplitude is $a = eE_z/(m_e\omega c) = 2$, corresponding to the peak intensity $I = 5.5 \times 10^{18}$ W/cm^2 for the $\lambda = 1$ μm laser. The laser pulse has a Gaussian envelope with FWHM size $8\lambda \times 5\lambda \times 5\lambda$. The length of the plasma slab is 47λ. The plasma density is $n_e = 0.09n_{cr}$, i.e. for the $\lambda = 1$ μm laser, it is 10^{20} cm^{-3}. The simulation box has a size of $50\lambda \times 20\lambda \times 20\lambda$. The total number of quasi-particles is 426×10^6. The boundary conditions are periodic along the y- and z-axes and absorbing along the x-axis for both electromagnetic radiation and quasiparticles. The simulations results are shown in Fig. 2.72, where the space unit is the wavelength λ of the incident radiation. Fig. 2.72(a) shows the electron density distribution in a plasma domain $50\lambda \times 10\lambda \times 10\lambda$ near the laser pulse axis at $t = 167$ fs obtained by the ray tracing technique. We see the paraboloidal modulations of the electron density in the wake. In the subdomain $36\lambda < x < 42\lambda$, $-5\lambda < y < 5\lambda$, $-5\lambda < z < 5\lambda$ in Fig. 2.72(b) we see the 17 MeV electron bunch that appears due to 3-D wave breaking. The electron density distribution is presented in 2.72(c) and 2.72(a) in the projections on the (x, y)- and the (x, z)-planes. We see the nonlinear distortion of the wake field shape with the injection of the electron bunches along the direction of the laser pulse propagation.

In Fig. 2.72(e) we present the electron phase plane (p_x, x) at $t = 167$. It can be seen that the electrons with the maximum energy are localized in the first breaking wave and they form a μm size fast electron bunch.

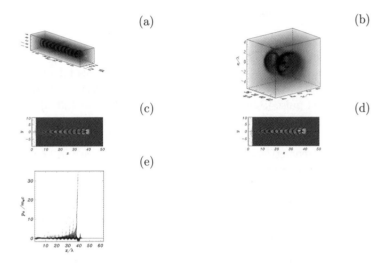

Fig. 2.72   Electron density distribution in (a) a plasma domain $50\lambda \times 10\lambda \times 10\lambda$ near the laser pulse axis at $t = 167$ fs, (b) a subdomain $36\lambda < x < 42\lambda$, $-5\lambda < y < 5\lambda$, $-5\lambda < z < 5\lambda$, (c) electron density distribution projection on the (x, y)-plane and (d) on the (x, z)-plane, and (e) electron phase plane (p_x, x) at $t = 167$.

2.4.1.5   *Plasma cathode by self-injection*

T. HOSOKAI
A. ZHIDKOV[1]
M. UESAKA
Nuclear Engineering Research Laboratory,University of Tokyo
2-22 Shirane-shirakata, Tokai, Naka,
IBARAKI 319-1188, JAPAN
[1]Division of Accelerator Physics and Engineering,
National Institute of Radiological Sciences,
4-9-1, Anagawa, Inage-ku, Chiba-chi,
CHIBA 263-8555, JAPAN

(a) Self-injection produced by laser pre-pulse
One of the simplest ways to inject energetic electrons into the wake field
for their further acceleration is wave breaking of plasma waves by a single
intense laser pulse [Bulanov *et al.* (1998); Hemker *et al.* (2002); Malka *et al.*
(2001); Malka *et al.* (2002b); Hosokai *et al.* (2003)]. Though such injection
gives a broad Maxwell-like energy distribution of accelerated electrons, the
transverse geometrical emittance of an electron bunch can be much better
than that from a conventional linac.

If the intensity of the laser pulse is not very high [Bulanov *et al.* (1998)],
wave breaking appears when the plasma wave amplitude exceeds the wave-
breaking field $E_B \sim [2(\omega/\omega_{pl} - 1)]^{1/2} mc\omega_{pl}/e$, where ω and ω_{pl} are the
laser and plasma frequencies. Wave breaking occurs in a plasma with a
rather steep density profile, $\lambda_{pl} d(\ln N)/dx \sim 1$, where $\lambda_{pl} = 2\pi\omega_{pl}/c$ is the
wavelength of the plasma wave [Hemker *et al.* (2002)]. However in a gas jet,
the density gradient is much smaller, usually $N/(dN/dx) \sim 200/500$ μm,
and injection originating from wave breaking of plasma waves rarely occurs
if there is only a single main laser pulse. In real conditions, a laser pre-pulse
of approximately a few nanosecond duration precedes the main laser pulse
[Kmetec *et al.* (1992)]. The usual contrast ratio varies from $1{:}10^5$ to $1{:}10^7$
for the fundamental laser frequency. In contrast to the plasma channel
produced by a long Rayleigh length laser pre-pulse [Malka *et al.* (2001);
Faure *et al.* (2000); Durfee *et al.* (1995); Johnson and Chu (1974)], if the
Rayleigh length L_R is short enough, the pre-pulse can form a cavity with
a shock wave in front of the laser propagation. Figure 2.73 illustrates the
concept behind the injection by wave breaking at the cavity. The pre-
pulse can be used to form the appropirate conditions for the wave-breaking

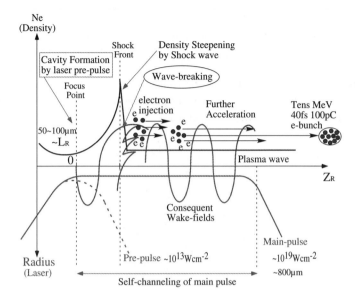

Fig. 2.73 Illustration of electron injection by wave breaking at a cavity. The cavity is produced by a laser pre-pulse.

injection of electrons into the wake field, while the main pulse can produce injection by itself due to relativistic effects. The length of the cavity is determined by the small L_R, because the energy is deposited in the plasma mostly near the focus point $x = 0$ as $W(x) \sim 1/(1+(x/L_R)^2)$, where $W(x)$ is the energy of a laser pre-pulse deposited at x, and x is the coordinate in the laser propagation direction. Since the laser pre-pulse has a low intensity, the electron temperature T_e can be estimated via the collisional absorption mechanism [Zel'dovich and Raizer (1967)]: $dT_e/dt = \Delta\varepsilon\, \nu_{ei}(1\text{eV})/T_e^{3/2}$, where $\Delta\varepsilon = 2\pi e^2 I/m\omega^2$ is the energy acquired by an electron in a collision, I is the laser intensity and ν_{ei} is the frequency of electron–ion collisions. For an intensity of $I = 10^{13}$ W/cm^2, the final ion density is $N_i=3\times10^{18}$ cm^{-3} (in the cavity), the pulse duration is $\tau = 2$ ns and the above equation gives $T_e = 150$ eV. If $x = C_s\tau > L_R$, where C_s is the ion sound speed, a shock wave can be formed in the plasma. If the shock wave relaxation depth $\Delta x \sim (M/m)^{1/2}l_i$, where M and m are the ion and electron masses and l_i is the ion free path, is less than the wavelength of plasma wave λ_{pl}, strong wave breaking of the wake field produced by the main pulse can occur and can be a good source of injection. For a temperature $T_e \sim 150$ eV in a He gas jet, the ion sound speed is $C_s \sim 5 \times 10^6$ cm/s and $x \sim 100$ μm, so that the effect appears for a laser pulse with $L_R \leq 100$ μm. The shock wave can be generated in a jet with $\omega_{pl} l_i (M/m)^{1/2}/2\pi c \leq 1$, that gives the density range of $N_i > 5 \times 10^{18}$ cm^{-3}. A typical calculated distribution of the plasma density after the laser pre-pulse is shown in Fig. 2.74. A strong shock wave is clearly seen. The density gradient at the shock wave is steep and effective wave breaking is expected.

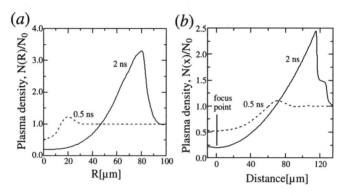

Fig. 2.74   Density distributions of He jet after laser pre-pulse, calculated by 2-D hydrodynamic simulation. The power density of the pre-pulse is 10^{13} Wcm^{-2} and the Rayleigh length is 50 μm. (a) Radial direction ($x = 0$), (b) longitudinal direction ($r = 0$).

The above conditions have been achieved in an experiment [Hosokai *et al.* (2003)]. In this experiment an intense and ultrashort laser pulse was focused on helium gas. In order to form a spatially localized gas column, i.e. to suppress the transverse expansion into vacuum due to the thermal and fluid motion of the injected gas, a supersonic pulsed gas injection was used as a target. The pulsed gas jet was produced by a device consisting of an axially symmetric *Laval* nozzle [Anderson (1989)] and a solenoid fast pulse valve. The nozzle was designed for a flow of $M_e = 4.2$ for He, where M_e is the Mach number at the exit of the nozzle, with a 2.0 mm inner diameter at the exit. The typical experimental setup is shown in Fig. 2.75. The nozzle with the pulse valve was placed inside a vacuum chamber. The pulse valve was driven for 5 ms a shot at a repetition rate of 0.2 Hz. The stagnation pressure of the valve was varied from 5.0 to 20.0 atm. With these pressures the density at the exit of the nozzle ranged from 7×10^{18} to 3×10^{19} cm^{-3}.

A 12-TW Ti:Sapphire laser system based on the CPA technique gen-

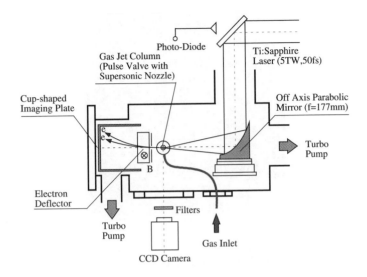

Fig. 2.75   Experimental setup with supersonic nozzle.

erates up to 600-mJ, 50-fs laser pulses at a fundamental wavelength of 790 nm with 10-Hz repetition rate. The laser power at the target in the vacuum chamber is as high as 5 TW. As shown in Fig. 2.75, a p-polarized laser pulse with a diameter of 50 mm is directed into the vacuum chamber through a vacuum laser transport line and is focused on the front edge of the helium gas jet column at a height of 1.3 mm from the nozzle exit with an $f/3.5$ off-axis parabolic mirror (OAP). Figure 2.76 gives a side-view of the interaction region, obtained by a CCD camera, and illustrates the gas jet and the focus point. Figure 2.77 shows a typical image of the laser focal spot at the target position. The spot size is 7.5 μm in full width at $1/e^2$ of maximum. The maximum laser intensity on the target was estimated to be 1.0×10^{19} W/cm^2 so that the laser strength parameter a_0 exceeds 2.0. The measured Rayleigh length was approximately 53 μm for this spot. According to the specification of the laser system, the contrast ratio of the main pulse to the pre-pulse that precedes it by 8 ns was typically greater than 10^{-6} at the fundamental wavelength. In order to investigate the laser pre-pulse effect on the ejection of electrons from the gas jet, we controlled nanosecond-order laser pre-pulses by detuning a Pockels cell of a regenerative amplifier of the laser system. The laser pulses were monitored by a photo-diode behind a dielectric coated mirror set inside the vacuum laser

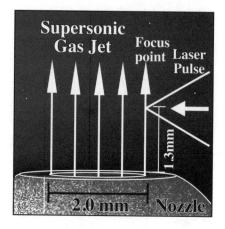

Fig. 2.76   CCD image of the interaction region in the setup in Fig. 2.75. The supersonic gas jet and the laser focus position are illustrated.

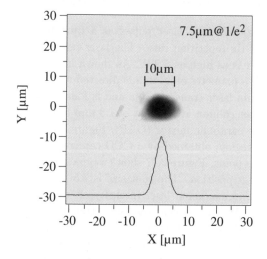

Fig. 2.77   Image of laser focal spot at the target.

transport line. The time resolution of the diode ($\sim$ 200 ps) was poor for the main pulse measurement, but was sufficient to detect changes in the nanosecond-order pre-pulse. Typical laser pulses detected by the diode are shown in Fig. 2.78(a)–(c). The pre-pulses ranged typically from -5 to

−1 ns, where 0 corresponds to the beginning of the main pulse. As shown in Fig. 2.78, the amplitude of the main pulse was maintained at a constant level for all pre-pulses by adjusting the pumping power of Nd:YAG lasers for a multi-pass amplifier of the Ti:Sapphire laser system.

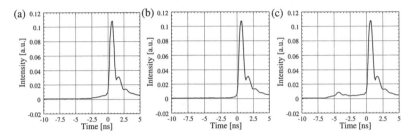

Fig. 2.78   Typical laser pulse shape detected by a photo-diode. (a) ∼ 2.5 ns pre-pulse, (b) ∼ 1 ns pre-pulse, (c) ∼ 5 ns non-monotonic pre-pulse.

The spatial distribution of the electrons ejected from the gas jet was directly measured by a cup-shaped detector consisting of imaging plates (IP) with a spatial resolution of 50 μm. The IP was a plastic film detector sensitive to high energy particles and radiation, used for electron microscopy and X-ray imaging.

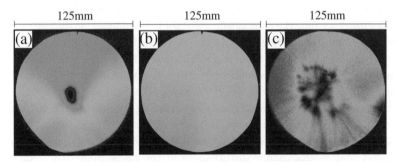

Fig. 2.79   Typical images of electrons deposited on the bottom plate of cup-shaped IPs. (a) ∼ 2.5 ns pre-pulse, (b) ∼ 1 ns pre-pulse, (c) ∼ 5 ns non-monotonic pre-pulse.

Figure 2.79(a)–(c) show spatial distributions of electrons deposited on the bottom plate of the cup-shaped IPs for different pre-pulse conditions. The spatial distribution of the ejection of electrons from the jet depends strongly on the laser pre-pulse. In the first case, shown in Fig. 2.79(a), for

Femtosecond Beam Science

a pre-pulse with contrast ratio 1:10^6, pulse duration 2–3 ns and energy of $\sim 10\%$ of the total pulse energy, corresponding to Fig. 2.78(a), a peaked spot distribution from a narrow-coned ejection of electrons is clearly seen at the center of the plate. In the second case, shown in Fig. 2.79(b), for a pre-pulse with contrast ratio 1:10^6 and pulse duration less than 1 ns with energy less than $\sim 10\%$ of the total pulse energy, corresponding to Fig. 2.78(b), no electron signal is observed on the bottom plate. In the third case, shown in Fig. 2.79(c), for a pre-pulse with contrast ratio 1:10^6, pulse duration more than 5 ns and non-monotonic pulse shape with energy more than $\sim 10\%$ of the total pulse energy, corresponding to Fig. 2.78(c), the ejected electron beam has exploded to pieces and smaller spots are detected at the bottom plate.

According to calculations, after 2 ns of irradiation by the laser pre-pulse, there is a shock wave and the density gradient become steep at the front of the shock wave. The thickness of the shock wave, $\sim 10~\mu$m, is comparable to the plasma wavelength so that strong wave breaking and electron injection is expected for this condition. A condition of pulse duration 2–3 ns matches well with the condition for shock wave formation and with consequent wave breaking of the wake field produced by the main laser pulse. For a pre-pulse intensity $I = 10^{13}$ W/cm^2, the shock wave is formed after 1 ns of irradiation. This means that if the laser pre-pulse is shortened, such as in the second case in Figs. 2.78 and 2.79, there should be no electron injection caused by wave breaking into the consequent wake field. In addition, if the laser pre-pulse gets shorter and its energy decreases, the density distribution becomes uniform. This is because there is no shock wave in the cavity and the laser intensity is not strong enough to produce the wake field with intensity higher than $E_B \sim [2(\omega/\omega_{pl} - 1)]^{1/2}mc\omega_p/e$. The spotted distribution in the case of the non-monotonic laser pre-pulse is, we believe, the result of an instability at the shock wave and laser diffraction in the cavity. Since the condition for wave breaking depends strongly on plasma dynamics, we expect that hydrodynamic disturbances in the shock front may strongly affect the process. Once it has appeared, the shock wave does not decay rapidly. However, such a hydrodynamic instability disturbs the front of the shock wave. Since the injection of the main acceleration due to wave breaking must be directed along with the density gradient, this instability leads to the formation of a spotted distribution. These results suggests that the laser pre-pulse affects the initial plasma density profile, steepened in the direction of laser propagation, due to the formation of a shock wave, which is essential for the injection into the consequent wake fields and the ejection of the narrow-coned electrons from the jet.

In order to obtain the energy distributions of the electrons, a magnetic electron deflector is set in the laser axis behind the jet, as shown in Fig. 2.75. The bottom plate of the IP cup is used as an electron detector. The magnetic field between the magnets is mapped out with a Hall probe and the maximum field strength reaches 300 mT. It has an entrance aperture of 2.0 mm with an acceptance solid angle of 1 msr. With this setup, an energy distribution of the ejected electrons up to 40 MeV can be detected. A measured distribution of narrow-coned electrons is shown in Fig. 2.80. The distribution agrees well with that obtained in 2-D particle-in-cell (PIC) calculations. This energy distribution is Maxwell-like with an effective temperature of $T_h \sim 10$ MeV, and the maximum energy observed is 40 MeV. Electrons with energy from 10 to 40 MeV constitute a bunch with duration 40 fs in the PIC simulation. The calculated charge of the electron bunch is 0.7 nC/J.

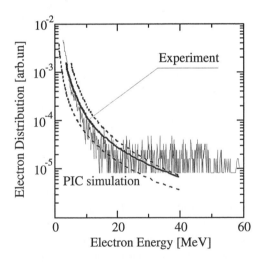

Fig. 2.80 Measured and calculated energy distributions of electron bunch for density of 1.4×10^{19} cm^{-3} and laser power 4 TW. Dotted line shows the experimental error.

(b)   Self-injection produced by main laser pulse

Wave breaking is initially a stochastic process that provokes rapid randomization in energy of the accelerated electrons. This sometimes makes such an injection inefficient and usually very sensitive to plasma parameter changes. A mechanism of electron injection derived from the relativistic character of the laser plasma interaction, can be applied to improve the energy spread of accelerated electrons. Along with the common wake field, a relativistically intense laser pulse moving in an under-dense plasma with group velocity less than light speed generates an additional electrostatic wave, which has a group velocity close to the group velocity of the laser pulse from linear theory. This wave comes from electrons temporally trapped and accelerated directly by the laser pulse forming a bunch at the front of the laser pulse. This bunching of electrons creates a potential difference, forming a potential cavity [Pukhov and Meyer-ter-Vehn (2002)], behind the laser pulse due to the evacuation of electrons. The number of electrons in this bunch is limited by repulsion; the repulsed electrons are accelerated to an energy equal to the potential difference in the cavity and thus can be efficiently injected for further acceleration if a proper matching condition occurs. However, this injection, which can be considered a self-injection, contends with injection after wave breaking. At low plasma density, the self-injection produces efficient acceleration with a low energy spread, while at high density wave breaking dominates, producing a Maxwellian distribution of energetic electrons with an effective temperature. Propagating in an under-dense plasma with group velocity v_g, an intense laser pulse can accelerate plasma electrons, which are at rest initially, up to an energy of $\varepsilon_{e\,max} = mc^2a_0^2/2$, where $a_0 = eE/mc\omega$, with E the laser electric field and ω the laser frequency. If the velocity of these electrons exceeds the group velocity of the laser pulse, they can be trapped and move with the pulse, forming an electrostatic wave. The matching condition can be written in the following form [Bulanov *et al.* (1991)]: $\gamma_{e\,max} = a_0^2/2 = \gamma_g = 1/(1 - v_g^2/c^2)^{1/2} \sim \omega\gamma_e^{1/2}/\omega_{pl}$, where ω_{pl} is the plasma frequency and $\tilde{\gamma}_e \sim \sqrt{1 + a_0^2/2}$ is the electron quiver energy. (For a laser intensity of $I = 10^{20}$ W/cm^2, $\lambda = 1$ μm, $a_0 = 8$ and $N_{em} = 5 \times 10^{18}$ cm^{-3}. Without the relativistic effect, N_{em} could be equal to 10^{18} cm^{-3}. For $a_0 \gg 1$, matching occurs for $a_0 \geq \sqrt{2}(\omega/\omega_{pl})^{2/3}$.) In the plasma, the number of trapped electrons cannot grow infinitely. The potential difference $\Delta\phi$ produced by electrons directly accelerated by the laser at the front of the pulse cannot exceed the ponderomotive potential $mc^2a_0^2/2$. As a result, we get in the 1-D case, $e|\Delta\phi| = 2\pi z e^2 N_i d^2 mc^2 a_0^2$, where z is the ion charge, N_i is the ion density and d is the length of a cavity behind the laser pulse. (We assume that the length d_e of the electron cloud generated at the front of the pulse is such that $d_e \ll d$.) From the equation one can easily estimate the maximum charge of trapped electrons, $Q = eN_e\lambda_p a_0 S/2\pi$, with λ_p the wake field wavelength and S the square of the laser focus spot. Setting $S = 5 \times 10^{-6}$ cm^2, we get $Q \sim 10$ nC/J of the laser energy for $N_e = 10^{19}$ cm^{-3}. The length of the cavity behind the laser pulse is $d = \lambda_p a_0/2\pi$. At this distance behind the laser pulse the repulsed electrons acquire an energy of $\varepsilon = |\Delta\phi|$ and, moving with the group velocity, are further accelerated in the cavity. For mono-energetic acceleration, d must exceed the pulse length $c\tau$, where τ is the pulse duration. The maximum energy finally acquired by such an electron is $E_{max} = (a_0/2\pi)E_{max}^{WF}$, where E_{max}^{WF} is the corresponding maximum energy in acceleration by a longitudinal plasma wave.

(a)   (b)

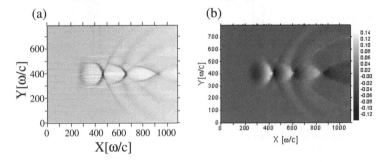

X[ω/c]

Fig. 2.81   Spatial distribution ($X = x + ct$) of (a) the normalized density of electrons with energy over 4 MeV and (b) the normalized x-component of the plasma field in the gas jet of density $N_e = 10^{19}$ cm^{-3} at $\omega t = 4000$ for $I = 10^{20}$ W/cm^2..

Results of 2-D simulation [Zhidkov *et al.* (2003)], displaying the formation of the electrostatic wave, are shown in Fig. 2.81. One can see a clear cavity structure in the electron density in Fig. 2.81(a) after the pulse propagates 0.5 mm in the plasma, though there is no wave breaking of the plasma wave according to Fig. 2.81(b). The transverse size of the first bunch in the rear of the first cavity is about 10 μm. The electron density in bunches decreases with the distance from the laser pulse; clear traces of electrons expelled at either side are produced by nonlinear ponderomotive scattering [Zhidkov *et al.* (2003)].

In lower density plasmas, when the maximum charge that can be sustained by the laser pulse is reached, a second electrostatic bunch is produced, formed by electrons repulsed from the first bunch and accelerated by the potential difference. In higher density plasmas, the wave breaking process becomes important, as presented in Fig. 2.82. One can see in Fig. 2.82(a) two bunches originating from different processes: the first bunch is due to self-injection and the second is due to the wave breaking injection. With a density increase, the energy distribution of the accelerated electrons approaches a Maxwellian distribution in the second, "wave breaking", bunch.

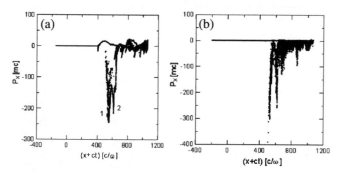

Fig. 2.82 Spatial distribution of electron longitude momentum P_x at $\omega t = 6000$ at a laser intensity of $I = 10^{20}$ W/cm^2 and plasma density (a) $N_e = 2 \times 10^{19}$ cm^{-3} and (b) $N_e = 5 \times 10^{19}$ cm^{-3}.

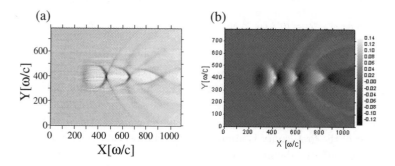

Fig. 2.83 Electron energy distribution in 2-D simulation: (a) $\omega t = 6000$ (b) $\omega t = 12000$ at a plasma density of $N_e = 10^{19}$ cm^{-3}.

The energy distribution of the electrons in the bunch after 1 mm and 2 mm in the plasma are shown in Fig. 2.83. Both distributions have a peak that corresponds to electrons accelerated after the self-injection. The energy spread (e^{-1} from the maximum density) is 5% in both cases, although the energy spread in the pedestal increases considerably. Formal calculations of the emittance in the bunch gives 0.1π mm·mrad at a total charge of $Q \sim 100$ pC.

2.4.2 *Ion*

A. ZHIDKOV,

Nuclear Engineering Research Laboratory, University of Tokyo,
2-22 Shirane-shirakata, Tokai, Naka,
IBARAKI 319-1188, JAPAN
Division of Accelerator Physics and Engineering,
National Institute of Radiological Sciences,
4-9-1, Anagawa, Inage-ku, Chiba-chi,
CHIBA 263-8555, JAPAN

2.4.2.1 *Mechanism*

The irradiation of an intense short pulse laser on matter gives rise to the efficient emission of particles such as X-rays, energetic electrons and ions [Krue and Wilks (1992)]. For a thin target, this phenomenon may be of particular interest, as the slab plasma provides perhaps the most efficient energetic ion emission in the absence of heat transfer and reduced radiative losses.

Though ions accelerated from the laser plasma to MeV energy were first observed more than 30 years ago [Krue and Wilks (1992); Ehler (1975); Gitomer *et al.* (1986); Pearlman and Morse (1978); Campbell *et al.* (1977); Wickens *et al.* (1978); Decoste and Ripin (1978); Sakabe *et al.* (1982)], interest in this process for short pulse laser irradiated plasmas has grown rapidly because of a valuable new characteristic of accelerated ions: anisotropy of their velocity distribution that provides an emittance of ion bunches comparable to that of electrostatically accelerated ion beams [Zhidkov *et al.* (1999); Hatchett *et al.* (2000); Clark *et al.* (2000); Maksimchuk *et al.* (2000); Fews *et al.* (1994); Beg *et al.* (1997); Lawson *et al.* (1997); Zhidkov *et al.* (2000a)].

The generation of fast ions of MeV/amu from a plasma corona during the interaction of intense, long (> 100 ps) pulse lasers with solids was investigated as a part of laser fusion studies. Ion energy up to 10 MeV was observed and it was found qualitatively from hydrodynamic ablation [Ehler (1975); Pearlman and Morse (1978); Campbell *et al.* (1977); Wickens *et al.* (1978); Decoste and Ripin (1978); Sakabe *et al.* (1982)] that the maximum ion energy is determined by the energy of hot electrons as ZT_h, where Z is ion charge and T_h is the hot electron temperature, which follows a scaling of $I\lambda^2$ [Gitomer *et al.* (1986)]. However, energetic ions are emitted in a 2π solid angle because the elastic collisions make even the hot electron velocity distribution isotropic during the long laser plasma interaction.

For short-pulse laser produced plasma, the first clear observation of velocity anisotropy of emitted energetic (> 1 MeV) He- and H-like ions of fluorine from a solid target irradiated by an intense, $I > 10^{18}$ W/cm^2, p-polarized laser (60 fs), inferred from blue-shifted spectra, has been reported [Zhidkov *et al.* (1999)]. The most impressive experiments have been done by the Lawrence–Livermore [Hatchett *et al.* (2000)] and Rutherford Appleton [Clark *et al.* (2000)] groups. Protons with energy up to 57 MeV have been emitted forward in a small cone of 3° from a target irradiated by a PW laser $I = 3 \times 10^{20}$ W/cm^2). Ion acceleration in the shock wave has been observed [Maksimchuk *et al.* (2000)].

So far, three mechanisms of ion acceleration from solid slabs irradiated by a short laser pulse have been recognized (see Fig. 2.84). The first is backward acceleration (BIA) driven by hot electrons from the front side of the target toward the laser pulse [Zhidkov *et al.* (1999); Clark *et al.* (2000); Fews *et al.* (1994); Beg *et al.* (1997)]. The second is forward acceleration (FIAR) from the rear side of a slab target dominated by hot electrons penetrating through the slab [Hatchett *et al.* (2000); Lawson *et al.* (1997); Zhidkov *et al.* (2000a); Esirkepov *et al.* (1999); Murakami *et al.* (2001); Zhidkov and Sasaki (2000); Wilks *et al.* (2001); Pukhov (2001); Nakajima *et al.* (2001); Bulanov *et al.* (2000); Andreev *et al.* (2002); Mackinnon *et al.* (2002)]. The third, never observed in long pulse experiments, is forward acceleration in a shock wave, which is produced by the ponderomotive and $(E \times H)$ forces from the front side of the target [Maksimchuk *et al.* (2000)] (FIAF). The efficiency, maximum ion energy, and emittance in these acceleration mechanisms vary differently with laser intensity, pulse duration and plasma density. Though having a low emittance, the ions in BIA and FIAR are exponentially distributed over energy with the exponent increasing with laser intensity. Ions in FIAF with relatively low energy spread constitute a beam with a poor emittance [Maksimchuk *et al.* (2000); Pukhov (2001)]; their maximum energy is smaller than that in BIA and

FIAR. A relatively new concept of ion acceleration that combines all the advantages of FIAR and FIAF is solitary wave acceleration [Zhidkov *et al.* (2002)]. In this scheme, the ions initially formed in the shock wave (FIAF) reach the rear side of the exploding plasma layer and are further accelerated by the plasma electric field produced by the hot electrons (FIAR).

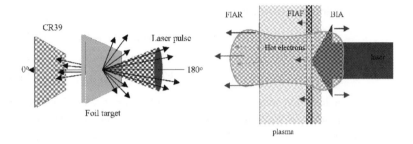

Fig. 2.84 Typical scheme of experiment for ion acceleration and various mechanisms of ion acceleration.

2.4.2.2 *Low intensity laser case*

The theory of ion acceleration by laser pulses with low intensity, $I\lambda^2 < 10^{16}$ Wμm^2/cm^2, was developed at the end of the '70's [Ehler (1975); Gitomer *et al.* (1986); Pearlman and Morse (1978); Campbell *et al.* (1977); Wickens *et al.* (1978); Decoste and Ripin (1978); Sakabe *et al.* (1982)]. The simplest equation for the ion velocity distribution can be derived for an isothermal planar expansion of a semi-infinite plasma using the so-called self-similar solution,

$$N_i(\chi) = \exp(-\eta)\,, \qquad \eta = x/Ct\,, \qquad (2.126)$$
$$u = \eta + 1\,, \qquad C = (zkT_e/M)^{1/2}\,, \qquad (2.127)$$

where N_i/N_{i0} and u are the normalized density and ion velocity, C is the ion sound velocity, T_e is the electron temperature, $T_i=0$ is the ion temperature and z and M are the ion charge and mass. From Eq. (2.126) one can get the ion velocity distribution

$$dN_i(u)/du = \exp(-u)\,. \qquad (2.128)$$

This acceleration is produced by the plasma electric field that appears in any plasma with a density gradient.

In plasmas irradiated by laser pulses, hot electrons appear with a Maxwell-like distribution with effective temperature T_h. If the ratio T_h/T_e is not too large, the ion velocity distribution can be found analytically [Pearlman and Morse (1978)]. The hydrodynamic equations with the self-similar approximation have the following form:

$$(u - \chi)\mathrm{d}N/\mathrm{d}\chi + N\mathrm{d}u/\mathrm{d}\chi = 0, \qquad (2.129)$$

$$(u - \chi)\mathrm{d}u/\mathrm{d}\chi + (S^2/N)\mathrm{d}N/\mathrm{d}\chi = 0, \quad u - \chi = S, \qquad (2.130)$$

$$S = \{z(N_h + N_e)/[M(N_h/kT_h + N_e/kT_e)]\}, \qquad (2.131)$$

where S is the local ion sound speed and N_h and N_e are the local density of the hot and plasma background electrons. Assuming that the ratio N_h/N_e is constant, one can get the equation for the ion velocity,

$$u = \left(\frac{T_h T_e}{T_h - T_e}\right)\left[\frac{C}{T_e}\ln\left(\frac{S - C}{S + C}\frac{S_0 + C}{S_0 - C}\right) + \frac{C_h}{T_h}\ln\left(\frac{C_h + S}{C_h - S}\frac{C_h - S_0}{C_h + S_0}\right)\right], \qquad (2.132)$$

where $C_h = (zkT_h/M)^{1/2}$, $C = (zkT_e/M)^{1/2}$ and S_0 is the local sound speed in the non-perturbed plasma. This equation becomes singular if $T_h/T_c > 5 + (24)^{1/2}$, which results from assuming N_h/N_e is constant. Nevertheless at lower ratios, this equation gives results that agree fairly well with experimental results [Gitomer et al. (1986)]. A suitable approximation can be used for the ion distribution:

$$N(x, t) = A\exp(-u/C_c) + B\exp(-u/C_h), \qquad (2.133)$$

with the coefficients A and B found from initial conditions. More comprehensive hydrodynamics research on ion acceleration has been done, as reported in Ref. [Sakabe et al. (1982)].

2.4.2.3   Moderate intensity laser case

For a pulse laser with non-relativistic intensity $I\lambda^2 < 10^{18}$ Wμm^2/cm^2, efficient emission of energetic ions can be obtained for oblique incidence pulses because superthermal electrons are produced via resonance absorption. For example in the case of a foil target, ion acceleration is an effective mechanism of electron cooling, and the absorption efficiency, in principle, could be as high as that of solids, up to 70%, if the foil thickness is larger than the skin layer.

The process of ion acceleration by such short laser pulses can be subdivided into two steps. Initially, the ion acceleration from the steep density surface is dominated by a beam-like population of fast electrons produced by vacuum heating. The maximum ion energy is determined by the electron quiver energy and ion charge [Zhidkov *et al.* (2000a)]. With the appearance of a critical density plasma, a bi-Maxwellian distribution is formed due to resonance absorption (see Eq. (2.133)). The ion acceleration is then sustained by the potential difference produced by escaped superthermal electrons and by rarefaction due to the rest of the superthermal electrons locked electrostatically in the plasma with a density gradient. The latter process has already been considered in the context of the self-similar expansion of a plasma with bi-Maxwellian electrons given above.

The equation for ion velocity derived in Ref. [Pearlman and Morse (1978)] becomes singular if the ratio T_h/T_c exceeds 10, indicating a breakdown of plasma quasi-neutrality. There is no simple expression for such a case. The non-local distribution of high energy electrons in foils due to the potential difference can also affect the ion acceleration [Sakabe *et al.* (1982); Zhidkov *et al.* (2000a)] so that to study ion acceleration processes at $T_h/T_c > 10$, a self-consistent particle-in-cell simulation coupled with atomic kinetics has to be applied [Zhidkov *et al.* (2000a)].

Usually, because the foil thickness is greater than the skin depth, the evolution of the absorption efficiency is similar to that in the case of a solid target [Zhidkov *et al.* (2000a)]. In the case of an obliquely incident *p*-polarized pulse, the absorption is initially determined by the anomalous skin effect, vacuum heating and collisions. Since fast electrons do not lose their energy outside the foil, the absorption efficiency is less than that of a solid target [Zhidkov *et al.* (2000a)]. The temporal evolution of plasma with and without elastic collisions are qualitatively similar. In both cases the time of optimum absorption is about 1000ω [Zhidkov *et al.* (2000a)]. This optimal time is determined by the formation of an optimal density gradient in the critical point with a scale length $L \sim 0.1\lambda$. This is at least twice as long as the simple time of an ion traveling a length L. The plasma ionization delay also has an influence on density gradient formation. After approximately half a picosecond, the absorption efficiency is saturated both in collisional and collisionless plasmas. More than 50% of the absorbed laser energy is used to accelerate ions after 0.5 ps. The mass of ions accelerated over $E > 1$ MeV exceeds 2% of the foil mass and increases with the laser intensity. A typical ion velocity distribution is shown in Fig. 2.85 for intensity $I = 4 \times 10^{16}$ W/cm^2.

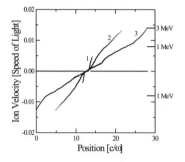

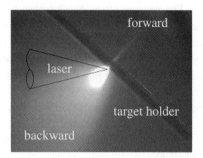

Fig. 2.85 x-component of Al ion velocity at different irradiation times by an obliquely incident pulse (see Eq. (2.126)): (1) $t = 0.16$ ps, (2) $t = 0.5$ ps, (3) $t = 0.8$ ps.

Fig. 2.86   Snapshot of a short pulse laser irradiated foil [Wada et al. (2002)].

The maximum ion energy for backward and forward acceleration increases nonlinearly with laser intensity $I^{1/2}$ and exhibits a linear dependence of fast ion energy on the hot electron temperature. A typical view of a laser-irradiated foil is shown in Fig. 2.86 [Wada et al. (2002)]. The forward and backward accelerated ions are clearly seen.

The most important characteristic of short laser pulse interaction with a solid target is the anisotropy of the velocity distribution of hot electrons [Zhidkov et al. (2000a)]. This makes such an interaction very different from that of a long laser pulse. Due to this anisotropy, emission of energetic particles including ions is small and comparable to that of common particle accelerators. The most visible consequence of this anisotropy is a change in the hot electron temperature. Assuming a Maxwell-like distribution, one can easily find that for the same total energy E the temperature $T_h = 2E/3$ for the isotropic distribution and $T_h = 2E$ for a fully anisotropic distribution. This fact explains the observed change in the hot electron temperature scaling from $T_h \sim 30$ keV $(I\lambda^2)^a$ to $T_h \sim 100$ keV $(I\lambda^2)^a$ ($a \sim 0.43$) in Refs. [Fews et al. (1994); Beg et al. (1997)] with the increase in laser intensity. The anisotropy of the hot electrons originates from the fact that electrons acquire energy from the laser pulse only normal to the target surface direction and there is no elastic collision during this process [Zhidkov et al. (2000a)].

2.4.2.4   *Ultra-intense laser case*

Absorption of very intense normal incident pulse lasers by thin foils of carbon and aluminum have been calculated in Refs. [Lawson *et al.* (1997); Zhidkov *et al.* (2000a); Esirkepov *et al.* (1999); Murakami *et al.* (2001); Zhidkov and Sasaki (2000); Wilks *et al.* (2001); Pukhov (2001); Nakajima *et al.* (2001); Bulanov *et al.* (2000)] by the particle-in-cell method, assuming full plasma ionization. An absorption efficiency of about 5–10% has been found and an increase in efficiency of ion acceleration for thinner foils has been calculated [Zhidkov and Sasaki (2000); Mackinnon *et al.* (2002)]. Energetic ion emission in the forward and backward directions is observed and attributed to acceleration by the ponderomotive force and surface electrostatic field [Zhidkov *et al.* (2000a)]. It has been shown [Esirkepov *et al.* (1999); Murakami *et al.* (2001); Zhidkov and Sasaki (2000); Wilks *et al.* (2001); Pukhov (2001)] that more energetic ions are produced with forward acceleration than with backward ion acceleration from fully ionized plasmas of carbon foils. All ion acceleration mechanisms drastically depend on ion charges. The effects of transient plasma ionization, such as plasma (PFI) and optical (OFI) field ionization, have been included in PIC simulations in Ref. [Zhidkov and Sasaki (2000)]. The effect of OFI and PFI on the interaction of an intense laser pulse with a thin Al foil are shown in Fig. 2.86, where the spatial ion charge with and without the field are presented. In both cases the charge distribution is almost uniform in the plasma bulk and asymmetric on the plasma surface.

A strong electrostatic field is produced on a plasma surface by hot electrons because of the density gradient. These hot electrons are generated via beam-like acceleration by the ponderomotive force made turbulent due to the small thickness of the plasma. The charges of the ions accelerated forward due to the PFI and backward due to the OFI increase significantly. Clearly seen in Fig. 2.87 are four regions: the skin layer, plasma bulk and the rear and front of the foil. In the skin layer both OFI and collisional ionization are important, while in the plasma bulk only collisional ionization occurs.

The maximum energy of forward accelerated Al ions is greater than 30 MeV, while the maximum energy of backward accelerated ions is less than 10 MeV. In the front of the foil the ion acceleration is a Coulomb explosion, while on the rear side of the foil the ion acceleration is dominated by hot electrons [Gitomer *et al.* (1986)].

Ion acceleration is the only mechanism of electron cooling during laser

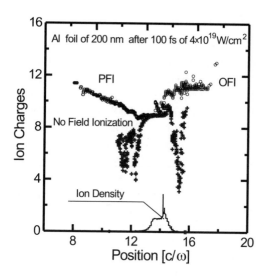

Fig. 2.87   Calculated spatial distribution of ion charges 100 fs after irradiation of an Al foil of 200 nm thickness by $I = 4 \times 10^{19}$ W/cm^2, $\lambda = 800$ nm pulse laser with 80 fs duration. The circles show the calculation with PFI and OFI, and the crosses show the calculation without field ionization.

irradiation. The process of field ionization drastically affects the ion acceleration and hence results in absorption efficiency. The dependence of absorption efficiency on the foil thickness and laser intensity has been calculated in Ref. [Zhidkov and Sasaki (2000)] for Al.

Field ionization increases the absorption efficiency by a factor of 1.5 to 2.5 for thinner foils. It should be noted that the skin layer $d = c/\omega_{pl}$ is smaller than the foil thickness even for $d = 0.025$ mm, due to plasma ionization. The total absorption efficiency decreases with foil thickness, while the absorption of electrons increases. At laser intensities of 10^{18} W/cm^2, the absorption efficiency is still dominated by collisions [Zhidkov *et al.* (2000a)], while at higher intensity the relativistic resonance process becomes effective, significantly increasing the absorption rate. In Fig. 2.88, the experimental dependence of the maximum ion energy with the foil thickness from Ref. [Mackinnon *et al.* (2002)] shows that the most efficient ion acceleration occurs in thin foils.

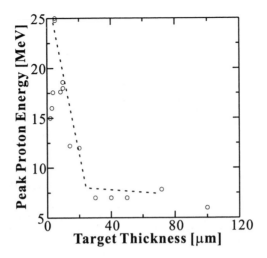

Fig. 2.88 Measured peak proton energy vs. target thickness. The intensity of the 100 fs laser pulse exceeded $I = 10^{20}$ W/cm^2.

2.4.2.5 *Ion acceleration in a solitary wave*

For ion acceleration in a solitary wave[Zhidkov *et al.* (2002)], ions initially formed in the shock wave reach the rear side of the exploding plasma layer and are further accelerated by the plasma electric field produced by the hot electrons [Zhidkov *et al.* (2002)] (see Fig. 2.89). Decaying, the shock wave transforms to an ion acoustic wave moving with velocity $\mathbf{u}(\mathbf{r}) = anc_s(\mathbf{r}) + \mathbf{v}(\mathbf{r})$, where c_s is the local ion sound speed, a is a coefficient and $\mathbf{u}(\mathbf{r})$ is the local velocity of the expanding plasma. Therefore, the maximum velocity of ions approaching the rear side of the target is $u_{max} = ac_s + v_{max}$ with v_{max} the maximum velocity of ions in FIAR. This kind of acceleration can appear only in relatively thin over-dense plasmas, $L < (zm/M)^{1/2}a_0^2(\omega\tau)c/\omega_p$, where L is the thickness of the plasma slab, M and z are the ion mass and charge, ω and ω_p are the laser and plasma frequencies, τ is the pulse duration and a_0 is the normalized laser pulse amplitude.

A typical energy distribution of ions accelerated from the rear side of a slab target and at the front of a shock wave are shown in Fig. 2.89. As the laser intensity I increases, the distribution $dN_i\varepsilon/d\varepsilon$ of ions in the rear decreases, while that for ions in the shock wave does not, though the maximum energy of ions in the shock wave increases.

The propagation of ion sound waves generated by the shock wave at

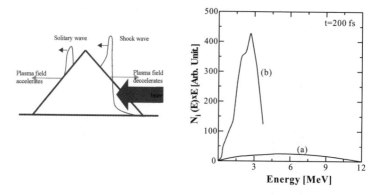

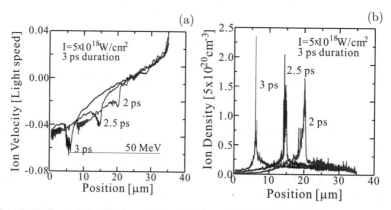

Fig. 2.89 Scheme of ion acceleration in a solitary wave and a typical ion energy distribution for a laser pulse with $I = 10^{19}$ W/cm^2 and $\tau = 100$ fs (a) at the target rear and (b) in the shock wave.

a laser intensity $I = 5 \times 10^{18}$ W/cm^2 is illustrated in Figs. 2.90(a) and (b). One can see that the ion acoustic wave propagates in the plasma with acceleration. The maximum energy in the ion bunch approaches 50 MeV at 3 ps at this laser intensity.

Fig. 2.90 Spatial distribution of (a) ion velocity and (b) ion density in an Al slab irradiated by 3 ps laser pulse with intensity $I = 5x10^{18}$ W/cm^2 for $t = 2$ ps, 2.5 ps and 3 ps.

As the laser intensity increases, the physical picture is not changed qualitatively. The maximum ion energy in the solitary wave for a laser intensity $I = 10^{19}$ W/cm^2 and pulse duration 2.5 ps rises to 250 MeV, exceeding the maximum energy of ions accelerated from the rear side of the target, as shown in Figs. 2.90 and 2.91.

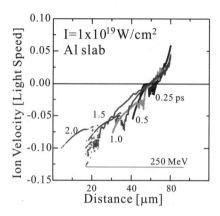

Fig. 2.91 Spatial distribution of ion velocity in an Al slab irradiated by 2 ps laser pulse with intensity $I = 10^{19}$ W/cm^2 for $t = 0.25$, 0.5, 1, 1.5 and 2 ps.

The acceleration length for this particular case is $l = 60$ μm, less than the Rayleigh length for focused laser beams. Since the ion acceleration in a solitary wave is maintained by hot electrons as well, the emittance of ions in the wave may be as good as that of ions accelerated from the rear side of the target $(1–10)\pi$ mm·mrad for a laser spot size comparable to the acceleration length.

This new mechanism of ion acceleration is based on forced solitary wave motion over an expanding plasma produced by a picosecond laser pulse. With a laser intensity above 5×10^{18} W/cm^2, an ion bunch with relatively low energy spread and ion energy $\varepsilon > 3$ MeV/nucleon, which is necessary for ion injection, can be produced in a laser pulse of duration $\tau = 2$–3 ps. The number of ions depends on their maximum energy and can be $N \sim 10^{10}$ particles/J of laser energy ($E \sim 30/50$ MeV) and $N \sim 10^9$ particles/J ($E \sim 200$ MeV) with an energy spread better than 10%. This mechanism of ion acceleration can be seen also for much shorter laser pulses. With higher laser intensity, one may get higher energy ions in the bunch and a shorter acceleration length with better emittance if the effect of the laser prepulse is suppressed.

2.4.3   X-ray

T. OHKUBO

M. UESAKA

Nuclear Engineering Research Laboratory, University of Tokyo
2-22 Shirane-shirakata, Tokai, Naka,
IBARAKI 319-1188, JAPAN

2.4.3.1   *Mechanism of laser plasma short X-ray generation*

Laser plasma X-rays (LPX) are emitted from a solid target irradiated by a
focused laser pulse. The laser pulse generates an abundance of hot electrons,
which propagate into the solid and excite the inner-shells of the target
atoms. Along with auto-ionization, exited atoms and ions radiate X-rays.
The laser pulse usually consists of an amplified spontaneous emission (ASE)
pre-pulse and a main pulse, as shown in Fig. 2.92. First, the pre-pulse heats
the target surface and forms a plasma density distribution. In the vicinity of
the critical point, the following main pulse deposits its energy to the plasma
on account of resonant absorption and produces hot electrons. Then, these
electrons emit continuous X-rays due to bremsstrahlung radiation and bring
about inner-shell excitation of the target atoms; characteristic X-rays are
emitted by the transition from the excited state to the ground state. The
$K\alpha$ X-rays are in an ultrashort pulse with a duration less than or about
10 ps and are useful for time-resolved measurements [Kinoshita *et al.* (2003);
Rischel *et al.* (1997); Rose-Petruck *et al.* (1999); Hironaka *et al.* (1999)].
Although the intensity of this radiation is smaller than of other, non-plasma,
sources, as shown in Table 2.5, the narrow line width and simplicity of
production of such X-rays make this source very practical.

2.4.3.2   *Measurement*

Figure 2.93 shows a schematic view of a typical experimental setup for
$K\alpha$ measurements. A laser pulse is guided into a vacuum chamber and
is focused on a copper target with an off-axis parabolic mirror, in order
to generate X-ray pulses. Characteristic X-rays, $CuK_{\alpha 1}(8.04778$ keV) and
$CuK_{\alpha 2}(8.02779$ keV), are obtained by Bragg diffraction at a LiF(200) crystal and detected with a PIN photodiode (IRD, AXUV) outside the chamber
through a Be window to filter out less than 3-keV X-rays. The LiF crystal
is surrounded by lead plates for shielding from background X-rays scattered
from the circumference.

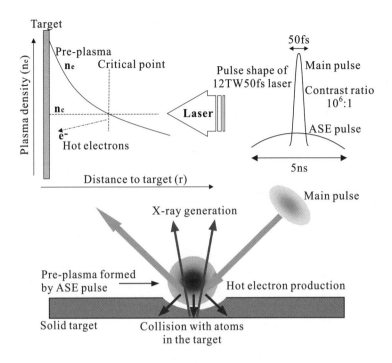

Fig. 2.92 Illustrations of the electron density distribution of a plasma produced by a ASE pre-pulse with a contrast ratio of $1:10^6$ and the mechanism of X-ray generation. At the critical point, the main pulse energy is absorbed into the plasma due to resonance of the laser frequency with the plasma frequency.

Figure 2.94 shows a laser spot size measured by a cooled CCD camera (Hamamatsu, C4880). We obtained a minimum spot size of 7.8 μm$\times$8.0 μm ($1/e^2$). The difference from the diffraction limit of 3.6 μm may be due to the quality of the laser, the optics or miss-alignment. The power density of the laser is around 10^{18} W/cm^2 in this experiment. The laser power density dependence on generated X-ray intensity is shown in Fig. 2.95. The vertical axis represents the signal height of the photodiode. The maximum number of generated X-rays photons is about 4.0×10^9 photons/shot/4πsr, estimated considering the quantum efficiency of the photodiode, the attenuation of the X-rays, decrease in air, signal efficiency and gain and the distance between the detector and the target. The energy conversion efficiency from the laser pulse to the X-ray pulse is 3.4×10^{-5}. We can adjust the power density by moving the target stage toward the focal depth axis. The

Table 2.5   Parameters of various X-ray sources for X-ray imaging.

	Intensity [photon/s/cm^2]	Time resolution	Spatial resolution	Application
Laser plasma X-rays	$< 10^9$ at 1 m from source	10 ps	200 μm	Time-resolved analysis
Synchrotron radiation	$\sim 10^{14}$	$\sim$ ms	$\sim \mu$m	Subjects requiring high intensity
Micro-focus X-rays	$\sim 10^9$	—	$\sim \mu$m	High spatial resolved imaging
Inverse Compton scattering X-rays	$\sim 10^8$	20 ps	100 μm	Time-resolved analysis
Nonlinear Thomson scattering of LFWA electrons	$\sim 10^{11}$	< 50 fs	$\sim \mu$m	Time-resolved measurements of ultrafast processes in liquid matter

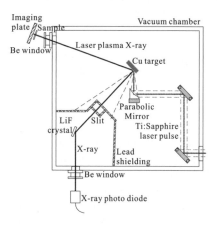

Fig. 2.93   Schematic view of the experimental setup for radiography by imaging plate and X-ray intensity measurement with X-ray PIN photodiode.

horizontal axis represents the target position. The Cu target was mounted on an automatic stage in order to set an appropriate focus position.

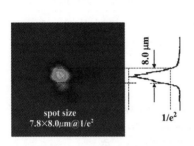

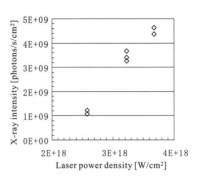

Fig. 2.94 Laser spot size in vacuum measured by a cooled CCD camera, and its transverse profile.

Fig. 2.95 Laser power density dependence on X-ray intensity. The number maximum of generated X-ray photons is 4.0×10^9 photons/shot/4πsr and the energy conversion efficiency from the laser pulse to the X-ray pulse is 3.4×10^{-5}.

2.4.3.3 *Numerical analysis to enhance intensity*

Recent calculated results of the interaction of a laser pulse with a solid target show that the X-ray intensity is very sensitive to the distribution of the plasma density formed by a long laser pulse (~ 5 ns) of ASE, which is accompanied by a main pulse [Zhidkov *et al.* (2000b)]. We have carried out simulations of the influence of the X-ray CuK$_\alpha$ intensity on the plasma density distribution formed by a 12-TW, 50-fs laser, using three codes (the procedure is shown in Fig. 2.96) [Ohkubo *et al.* (2002)]. The laser pulse contained a pre-pulse of ASE with a focused power density of $\sim 10^{13}$ W/cm^2 and pulse duration of 5 ns. The power density and pulse duration of the main pulse were $\sim 10^{19}$ W/cm^2 and 50 fs, respectively. We calculated the plasma density distribution using the hydrodynamics simulation code HYADES [Larsen and Lane (1994)]. This code solves the following equations in the Lagrangian coordinate system: the equation of continuity, the equation of motion, the equation of state and the energy equation. Figure 2.97 shows the result of calculations of the plasma density distribution formed by an ASE pre-pulse with power densities of 10^{13} W/cm^2 (solid bold line), 10^{12} (black dotted line) and 10^{11} (gray dotted line) at 2.5 ns from the beginning of the pre-pulse irradiation. At this point, these distributions are to be applied in PIC calculations. In Fig. 2.97 hot electron distributions (> 8 keV for CuKα emission) obtained using a code based on the collisional particle-in-cell (PIC) method [Zhidkov and Sasaki (1999)],

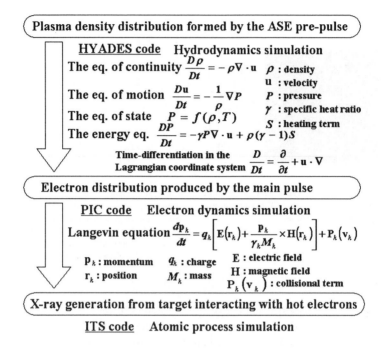

Fig. 2.96 Simulation procedure of LPX generation, consisting of three steps.

are given different initial plasma density profiles. In the PIC simulation, the Langevin equation of 1-D in space and 3-D in velocity is solved under non-local thermodynamic equilibrium (non-LTE), taking into account the interaction of the plasma with the main pulse. The outputs of the electron distributions with a main pulse power density of 10^{19} W/cm^2, just after the end of the main pulse irradiation, are shown in Fig. 2.98. As the initial plasma distribution spreads, the electron energy decreases. Finally, we have carried out calculations of CuK$_{\alpha 1}$ and CuK$_{\alpha 2}$ emissions from electron distributions by the Monte Carlo code ITS-3.0 [Seltzer (1991)], for atomic processes. As shown in Fig. 2.99, the number of photons in (c) is larger than in (b); the optimal plasma density distribution (c) generates more than ten times as many X-rays. These results show that the X-ray intensity decreases as the initial plasma distribution spreads. By decreasing the ASE intensity, we can control the plasma distribution and increase the X-ray intensity.

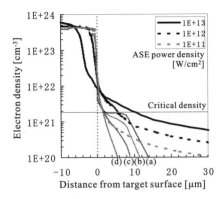

Fig. 2.97 Numerical result of the plasma density distribution with pre-pulse power densities of 10^{13} W/cm^2 (bold), 10^{12} (black dotted) and 10^{11} (gray dotted) at 2.5 ns from the beginning of the pre-pulse irradiation. The other four distributions (fine lines) are the initial conditions for the PIC calculations, labeled (a), (b), (c) and (d) in order of plasma length.

2.4.3.4 *Nonlinear Thomson scattering*

If free electrons in a plasma interact with a low powered laser, they oscillate linearly and emit radiation of the same frequency as the laser. This is linear Thomson scattering. As the laser power increases, the electrons radiate at harmonics of the Doppler shift laser frequency [Sarachik and Schappert (1970)]. However, this emission has a large angular spread so that the X-ray flux is quite low. At high intensities of $a_0 \sim 1$ ($a_0 = eE/m\omega c$, where E and ω are the electric field and frequency of the laser, e is the electron charge, m is the electron mass and c is the speed of light), the electron motion is influenced by the magnetic field of the laser and the electron trajectory becomes a "figure-8". This is nonlinear or relativistic Thomson scattering. Recently, Umstadter *et al.* have demonstrated this scattering by measuring the second and third harmonics and the typical angular distribution that comes from nonlinear electron motion [Chen *et al.* (1998); Chen *et al.* (2000)]. However, the radiation is still visible. Further, they also measured well-collimated radiation in the forward direction in the VUV region. Imaging of the harmonic beam showed that it was emitted in a narrow cone with a divergence of 2° to 3° [Banerjee *et al.* (2002)].

L'Huillier *et al.* used ultrafast laser pulses from a Nd:glass laser centered at 1053 nm, with 1 ps pulse duration, and generated coherent soft X-ray harmonics. The maximum orders were up to the 29th in Xe, the 57th in

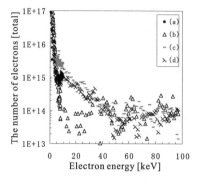

Fig. 2.98 Numerical results of the electron distributions by PIC simulation with a main pulse power density of 10^{19} W/cm^2, just after the end of the main pulse irradiation, corresponding to the four distributions (a) to (d) in Fig. 2.97. The distributions (c) and (d) show more hot electrons than those of (a) and (b).

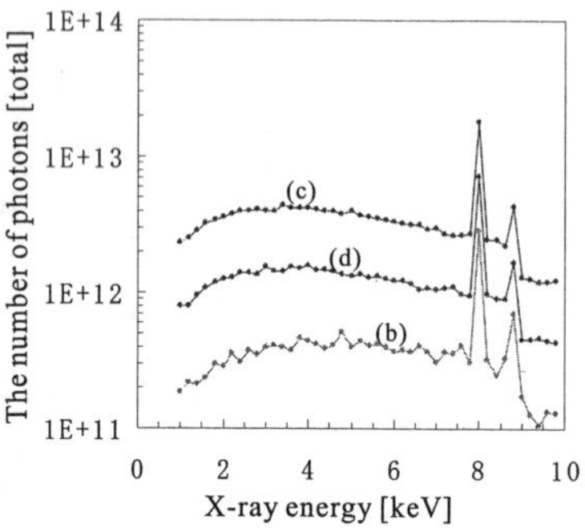

Fig. 2.99 Calculated spectra of the generated X-rays using the electron distributions (b), (c) and (d) shown in Fig. 2.98 as initial conditions of a Monte Carlo simulation. The number of photons in (c) is larger than in (b) or (d).

Ar and the 135th in Ne and He gases (7.8 nm, 160 eV). This harmonic emission was directional and of short pulse duration of a few hundred fs (shorter than the pump pulse). The instantaneous power generated at 20 eV reached about 30 kW, with a conversion efficiency of 10^{-6} [L'Huillier and Balcou (1993)]. Macklin *et al.* observed the 109th harmonic in Ne (7.3 nm, 170 eV) using a 125-fs, 800-nm Ti:Sapphire laser. Depletion of neutral atoms by ionization reduced phase matching due to free electrons and made it impossible to observe higher order harmonics than the 109th [Macklin *et al.* (1993)]. Preston *et al.* reported high harmonic generation from a 380-fs, 248.6-nm KrF laser pulse giving harmonic orders up to the 37th in a He gas jet (6.7 nm, 185 eV) and the 35th in Ne (7.1 nm, 175 eV). The source of the harmonics was shown to be the ion species He$^+$, Ne$^+$ and Ne^{2+}. They also demonstrated that He could produce very high conversion efficiencies for the lower harmonics of 248.6 nm laser radiation, giving energies up to 31 nJ and powers up to 80 kW for the 7th harmonic (35.5 nm, 35 eV) [Preston *et al.* (1996)]. Chang *et al.* used a 26-fs, 800-nm Ti:Sapphire laser to generate harmonics at wavelengths down to 2.7 nm (460 eV) in He and 5.2 nm (239 eV) in Ne [Chang *et al.* (1997)]. In the future, this very compact femtosecond X-ray source, driven by kHz repetition rate lasers, may be very important for applications such as imaging through aqueous solutions or time-resolved photoelectron spectroscopy of organic molecules and solids.

Thomson scattering using electron beams produced via laser wake field acceleration (LWFA) can be an efficient source of femtosecond hard X-rays [Uesaka *et al.* (2003)]. In contrast to electron beams generated by common accelerators, bunches produced via LWFA are short, < 100 fs, and very dense [Hosokai *et al.* (2003)]. A typical momentum distribution of such a bunch generated via a wave-breaking mechanism is shown in Fig. 2.100. The duration of the bunch is 40 fs and its diameter is <10 mm.

The efficiency of Thomson scattering strongly depends on the electron

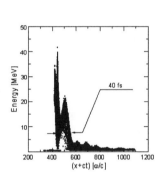

Fig. 2.100   Typical distribution of electron momentum for LWFA.

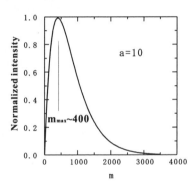

Fig. 2.101   Spectrum of X-rays produced via nonlinear Thomson scattering, $a = eE/m_e\omega c$, where E and ω are the electric field and frequency of the laser and m is the number of X-ray harmonics.

beam size. At an equal total charge, more dense beams radiate more photons since the laser pulse can be tightly focused. Moreover, the spectrum of X-rays strongly depends on the laser intensity, as shown in Fig. 2.101 [He *et al.* (2003)]. At higher intensities, shorter wavelength X-rays are emitted. About 10^{12} photons per second in a 10–20-keV range can be produced by a 1-J laser pulse with a duration of 50 fs and 10-Hz repetition rate. The experimental setup and results of the first measurements performed by Leemans' (USA) and Malka's (France) groups have recently been presented at the Laser–Particle Accelerator Workshop in Oxford, 2003.

2.4.4   *Terahertz (THz) radiation*

H. OHTAKE,
Assistant Manager,
Fiber Laser Bussiness Group,
AISIN SEIKI Co., Ltd.,
17-1 Kojiritsuki, Hitotsugi, Kariya,
AICHI 448-0003, JAPAN

H. MURAKAMI,
Department of Quantum Engineering and Systems Science,
School of Engineering, University of Tokyo,
7-3-1 Hongo, Bunkyo-ku,
TOKYO 113-8656, JAPAN

N. SARUKURA,
Laser Center, Institute for Molecular Science,
38 Nishigonaka, Myodaiji, Okazaki,
AICHI 444-8585, JAPAN

Y. KONDO,
Department of Applied Physics, Tohoku University,
Aoba, Aoba-ku, Sendai,
MIYAGI 980-8579, JAPAN

A useful application of femtosecond lasers is the generation of terahertz (THz) radiation. After the publication of Ref. [Auston *et al.* (1984)], interest in applications of THz radiation for sensing, imaging and spectroscopy has been growing for the last few decades. However, a severe sensitivity to alignment and insufficient power has prevented wide popularity of these devices. To overcome these limitations, simple and intense THz radiation sources have been studied eagerly. In this section, practical methods for generating intense THz radiation are briefly introduced. One method is the magnetic field enhancement scheme and another is the terawatt laser (TW laser) excitation scheme.

2.4.4.1   *Magnetic field enhancement scheme*

A mode-locked Ti:Sapphire laser is used as an excitation source, as described in Refs. [Sarukura *et al.* (1998); Ohtake *et al.* (2003)]. The sample used is undoped bulk InAs with a (100) surface. The average power for excitation is about 700 mW with a 3-mm diameter spot size on the sample. A split coil superconducting magnet with cross room temperature bores can provide a magnetic field up to 5 T. A liquid helium cooled silicon bolometer is provided for detecting the power of the total radiation and a wire-grid polarizer is placed in front of the bolometer, as shown in the insets of Fig. 2.102. With this specially designed magnet, five different optical geometries, as described in [Ohtake *et al.* (2000)], can be compared by changing the magnetic field, magnetic field direction and excitation laser incident angle. Among these geometrical layouts, an interesting magnetic field dependence on THz radiation power is observed in two cases, as shown in Fig. 2.102(a) and (b). In the case of Fig. 2.102(a), with the magnetic field parallel to the surface and a laser incident angle of 45° to the surface normal (G-1), the THz radiation polarizes almost horizontally. From low magnetic fields up to near 2 T, the THz radiation power shows a quadratic magnetic field dependence [Sarukura *et al.* (1998)]. A saturation at around 3 T and a reduction of the radiation power above 3 T are observed.

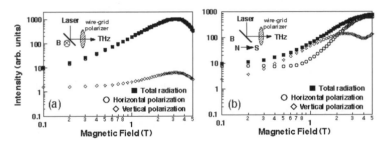

Fig. 2.102   Magnetic field dependence of THz radiation power (a) for the G-1 geometrical layout and (b) for the G-5 geometrical layout. Clear saturation is observed at a magnetic field around 3 T in the case of the G-1 geometrical layout. The insets show the experimental setup for the G-1 and the G-5 geometrical layouts.

On the contrary, clear saturation is not observed in the case of G-5, where the magnetic field is parallel to the THz radiation propagation direction and a laser incident angle of 45° to the surface normal, as shown in Fig. 2.102(b). In this layout, the polarization of the THz radiation changes dramatically as the magnetic field increases. Photoexcited electrons are accelerated in the plane perpendicular to the propagation direction of the THz radiation. Since the polarization of the THz radiation reflects the projection of the carrier accelerating direction to each direction, both components of polarization should be observed. Figure 2.103 illustrates the time domain measurement of the THz radiation electric field from InAs in the G-1 geometrical layout with a 1-T permanent magnet.

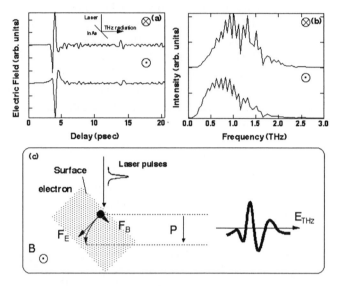

Fig. 2.103   Time domain measurement for THz radiation. (a) Delay dependence of electric field from InAs in 1.0-T magnetic field. The direction of the magnetic field is shown in the insets above the plots. The phase of each plot is completely different from each other. (b) THz radiation spectra. The interference patterns are due to dipole antenna, the substrate of which is transparent in the terahertz region. (c) Generation mechanism. An electron (solid circle) is accelerated by a depletion field F_E and a magnetic field F_B. This acceleration induces a transient dipole P that radiates THz radiation (E_{THz}).

A dipole antenna is used as a receiver for the THz radiation. There is a very clear difference between the two cases. The pulse width of the THz radiation is estimated to be less than 1 ps, as shown in Fig. 2.103. In Fig. 2.103(a), the phase of the two plots of field oscillation is completely opposite, because the photo-excited electrons are accelerated in opposite directions by the magnetic field. Thus, the spectral shapes in Fig. 2.103(b) show a clear difference. Additionally, in the G-1 geometrical layout, we have observed a remarkable magnetic field direction dependence of the THz radiation power. This result also originates from the difference of the electron acceleration direction as shown in Fig. 2.103(c). This process reduces the number of electrons in the Γ-valley contributing to the generation of THz radiation.

To explore a much higher magnetic field dependence, a cryogen-free superconducting magnet with sufficient bore size should be prepared. For this purpose, a cryogen-free superconducting magnet with a 52-mm diameter room temperature bore was designed [Watanabe *et al.* (1998)]. The magnet generates a magnetic field up to 15 T. The geometrical layout is the same as the G-1 geometrical layout. Anomalous magnetic field direction dependent saturation, decrease and recovery of THz radiation power from femtosecond laser irradiated InAs in a strong magnetic field was observed, as shown in Fig. 2.104.

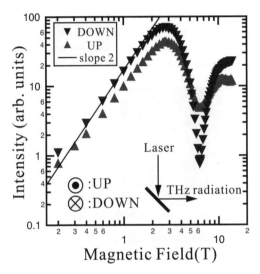

Fig. 2.104 Anomalous magnetic field direction dependent saturation decrease, and recovery of THz radiation power from InAs irradiated by a femtosecond laser in a strong magnetic field. The radiation intensity has an asymmetric dependence on the magnetic field inversion, as observed in the low field case. At around 6 T, the radiation intensity reaches a minimum value and recovers slowly.

The radiation intensity demonstrates an asymmetrical dependence on the magnetic field inversion, as observed by [Zhang *et al.* (1993)] for low magnetic field. The maximum intensity was obtained at approximately 3 T. This can be achieved even with a permanent magnet incorporating a special design. The total radiation power is estimated to be approximately 80 μW on average, after the InAs emitter. From the viewpoint of applications, this is a significant finding for practical light source design. The theoretical approach to understanding the quadratic magnetic field dependence is given in Ref. [Weiss *et al.* (2001)]. The power enhancement factor η is written as

$$\eta \propto (\frac{e}{m^*})^2 B^2 , \qquad (2.134)$$

where e, m^* and B are the elementary electric charge, effective mass and magnetic field, respectively. The magnetic field dependence of the radiation power shows good agreement with Eq. (2.134) for magnetic fields smaller than 3T. Above 3T, the linear approximation leading to Eq. (2.134) is no longer valid because the magnetic field strength becomes very high. As Weiss expected in the case of InSb, a deviation from a quadratic dependence is clearly seen in case of InAs, as shown in Fig. 2.104. Concerning the saturation, decrease and recovery, there are lots of theoretical works, however, a clear explanation has not been proposed.

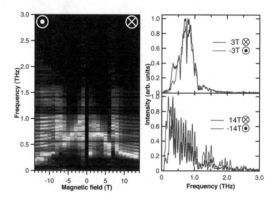

Fig. 2.105   Radiation spectrum exhibiting an interesting magnetic dependence and periodic spectral structure. This may be attributed to interference of the radiation from different holes, since the radiation originating from different carriers should have the same phase difference.

The THz radiation spectrum was obtained by a polarizing Michelson interferometer. The radiation spectrum exhibited an interesting magnetic dependence and periodic spectral structure, as shown in Fig. 2.105. The cyclotron frequency was 3.36 THz for a 3-T magnetic field, and hence the spectral structure could not be explained by this process. At around 6 T, the radiation intensity reached a minimum value and recovered slowly. This drastic change of magnetic field dependence may be attributed to the change of radiation mechanism at higher fields. An explanation for this experimental result is that there is a change of dielectric tensor originating from the existence of strong magnetic fields. According to Ref. [Palik and Furdyna (1970)], diagonal and off-diagonal components of the dielectric tensor depend on the magnetic field. At low magnetic fields, the off-diagonal component should be neglected because the magnetic field is almost zero. On the contrary, the off-diagonal component is not negligible at higher magnetic fields and the optical properties of the semiconductor change dramatically.

2.4.4.2 *Terawatt laser excitation scheme*

Another approach for generating intense THz radiation is TW laser excitation. Ref. [Sarukura *et al.* (1998)] reports a quadratic excitation power dependence of THz radiation power and suggests that THz radiation power might increase with increasing excitation power. From a practical point of view, a TW laser system is an important system for generating intense THz radiation. A 10-Hz repetition rate Ti:Sapphire amplifier system delivers nearly transform limited 60-fs pulses at 800 nm. The geometrical layout is the same as the G-1 geometrical layout and maximal pulse energy is achieved at 10 TW. The spot size is approximately 10 mm in diameter on the sample, which avoids laser-induced breakdown of the sample surface. The sample used is InAs with (100) surface in a 1-T magnetic field, as shown in the inset of Fig. 2.106.

At lower excitation powers, the THz radiation power for a magnetic field in the direction surface-to-back ($\otimes$) is larger than that for back-to-surface ($\odot$). It is consistent with our results discussed above. On the other hand, at higher excitation powers, the radiation power for $\odot$ becomes larger than that for $\otimes$. Moreover, the THz radiation power does not show a quadratic excitation power dependence and seems to saturate above 0.1 TW, as shown in Fig. 2.106. Above 0.1-TW excitation, laser-induced breakdown occurs in both cases on the sample surface. We are conducting research to clarify the mechanism of the different excitation power dependences between $\otimes$ and $\odot$ [Murakami *et al.* (2003)]. The properties of THz radiation generated by the magnetic field enhancement and the terawatt laser excitation schemes are summarized in Table 2.6.

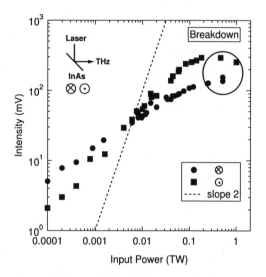

Fig. 2.106 Excitation power dependence of THz radiation power. Inset indicates the experimental setup. Dashed line shows slope of gradient 2. The solid circles and squares show the magnetic field direction, surface-to-back and back-to-surface, respectively. Laser-induced breakdown is observed inside the open circle.

Table 2.6   Parameters ([†] estimated value)

Parameters	Magnetic enhancement	TW laser excitation
Magnetic field	~ 3 T	~ 1 T
Excitation energy	~ 0.1 MW	~ 1 TW
Repetition rate	80 MHz	10 Hz
Pulse width (THz radiation)	≥ 1 ps	≥ 1 ps[†]
Pulse energy (THz radiation)	~ 1 pJ	~ 1 nJ[†]
Spectral width	0.1–3 THz	0.1–3 THz[†]

2.4.5   *Neutron*

Y. KISHIMOTO
Department of Fusion Plasma Research,
Naka Fusion Research Establishment,
Japan Atomic Energy Research Institute (JAERI)
801-1 Mukouyama Naka-machi, Naka,
IBARAKI 311-0193, JAPAN

2.4.5.1   *Cluster science*

The cluster medium has been recognized as a *small particle system* in material science, which is intermediate between the *condensed state*, like solids and liquids, and the *isolated state*, like atoms and molecules [Kawabata and Kubo (1966); Castleman and Keesee (1988)], and exhibits many prominent dynamics (cf. enhanced fluctuation and transverse polarization through the surface, which does not take place in ideal gas and plasma states [Tajima *et al.* (1999)]). Further, based on advanced technologies for nano-size processing and making fine particles, the preparation of a wide range of cluster sizes, materials, shapes, spatial distributions and so on, is now available. As a result, it has become possible to choose parameters that are much wider than for interactions of ideal gasses and/or the solid state. Research using such cluster mediums will open up many innovative applications. In addition, a number of salient phenomena have been observed in laser–cluster interactions. These include Coulomb explosion of clusters [Gotts *et al.* (1992)], enhanced emission of X-rays [Ditmire *et al.* (1995)] and the generation of energetic electrons and energetic ions [Ditmire *et al.* (1996)].

Perhaps, the most spectacular demonstration of enhanced laser–cluster interactions among recent experiments, is the report of copious fusion neutron generation upon laser irradiation of deuterium clusters [Ditmire *et al.* (1999)]. It was shown that a deuterium cluster explosion under ultrashort intense laser irradiation can lead to ion energies sufficient to drive nuclear fusion reactions. The resultant neutron emission may also reveal short-pulse characteristics with a temporal duration of a few hundred picoseconds. Such a bright neutron pulse can play an important and unique role in studying ultrafast phenomena in many fields for different applications, and is expected to be widely applied to material science, soft X-ray lithography, radiography and keV X-ray spectroscopy.

2.4.5.2   *Characteristics of laser–cluster interaction*

A cluster medium and its interaction with a laser field are characterized by:

(1) its material state (charge state Z and density $n_e = Zn_{cl}$, where n_{cl} is the cluster ion density)

(2) spatial size (radius a)

(3) packing fraction (ratio between the volume the clusters occupy to the system volume) $f \equiv 4\pi a^3 N_{cl}/3V$, where N_{cl} is the number of clusters in volume V

(4) cluster size to laser wavelength a/λ_{laser}

(5) electron skin depth to cluster size δ_e/a ($\delta_e \equiv \omega_p/c$)

(6) electron excursion length to cluster size ξ_e/a

(7) inter-cluster distance to electron excursion length ξ_e/R ($R \equiv (3V/4\pi N_{cl})^{1/3}$)

(8) laser cut-off density to local cluster density Zn_{cl}/n_c.

There exist two possible regimes in characterizing the laser–cluster interaction depending on the above relations specifically (5) and (6), i.e. the so-called *Coulomb explosion* initiated in the regime $\xi_e \gg \delta_e \geq a$ and ambi-polar expansion initiated in the regime $\xi_e \sim \delta_e < a$. In a Coulomb explosion, laser fields penetrate into the interior of a cluster and a large fraction of the cluster electrons leaves the host cluster within its optical cycle. As a result, the cluster becomes highly non-neutral so that ions simultaneously explode due to the Coulomb repulsive force. In ambi-polar expansion, the laser field does not fully penetrate into the interior of the cluster and a coronal plasma with a hot electron temperature is produced around the cluster surface. Then, the ion front is accelerated by a strong ambi-polar field and expands into vacuum. These phenomena are schematically illustrated in Fig. 2.107. Typical ion energy distributions in the Coulomb explosion regime reveal an inverted structure with an abrupt cutoff, indicating that energy is concentrated near the expansion front, whereas the distributions in the ambi-polar expansion show a monotonic exponential decay profile with a self-similar characteristic [Kishimoto *et al.* (2002)]. Note that the ion energy distribution is an important key factor in whether efficient cluster fusion is achieved, as we discuss in the following.

2.4.5.3   *Neutron generation*

Here, we estimate fusion neutrons in the Coulomb explosion regime realized using a high power laser [Kishimoto *et al.* (2002); Kishimoto *et al.* (a)]. We

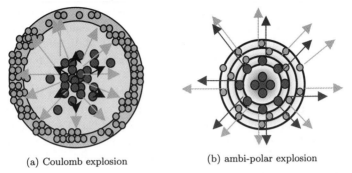

(a) Coulomb explosion　　　　　(b) ambi-polar explosion

Fig. 2.107 Schematic illustration of laser–cluster interactions for different parameter regimes: (a) Coulomb explosion in the high laser power or small cluster regime and (b) ambi-polar expansion in the low laser power or large cluster.

restrict our attention to deuterium–deuterium (DD) fusion employing deuterium clusters. However, it can be straightforwardly applied to deuterium–tritium (DT) fusion where the cross-section is much higher than for DD fusion. It is known that the neutron yield is maximized when cluster ions are coherently accelerated up to around 1 MeV, although a wide range of deuterium temperatures around $0.1\,\mathrm{MeV} \leq T_i \leq 10\,\mathrm{MeV}$ contribute to producing neutrons. Since the obtainable ion energy is related to the Coulomb energy stored in the cluster, we evaluate the average potential energy assuming that all electrons are expelled from the host cluster and that the cluster only consists of pure ions. However, it is natural to consider that the expelled electrons from clusters are redistributed in the system and play a role in partially canceling the electrostatic Coulomb field. As a simple case, we assume uniformly distributed electrons in the system, such that the expelled electron temperature is extremely high, going as $n_0 \exp(e\phi/T_e) \cong n_0$. The electrostatic field is then given by $E_r(r) = 4\pi Ze(n_{cl} - n_{av})r/3$ for $0 \leq r < a$ and $E_r(r) = 4\pi Ze(n_{cl}a^3/r^3 - n_{av})r/3$ for $a \leq r < R$. Then, the potential energy per ion, which corresponds to the maximum average ion energy, is given by

$$w_{ion} = \frac{3}{4\pi a^3 n_{cl}} \frac{1}{8\pi} \int_0^R E_r^2(r) 4\pi r^2 \mathrm{d}r = \left(\frac{2\pi Z^2 e^2}{5}\right) n_{cl} a^2 (2 - 3f^{1/3} + 3f).$$
$$(2.135)$$

Here, n_{av} is the average electron density and the inter-cluster distance, R, is regarded as a boundary beyond which the electrostatic potential is set to zero. The result is illustrated in Fig. 2.108(a), where w_{ion} is plotted as a function of the cluster radius for different values of the packing fraction. Note that w_{ion} is proportional to a^2. For example, in order to maximize the fusion cross section around $E_{ion} \sim 1$ MeV, electrons have to be simultaneously expelled from the larger cluster of size approximately $a \cong 100$ nm.

The laser intensity necessary to fully expel electrons from the cluster may be evaluated by equating the electron kinetic energy with the average Coulomb energy of the cluster as $(\gamma - 1)mc^2 = (Ne^2/2a)(1 - f^{1/3})$, where N represents the number of electrons inside the cluster. This leads to the relation between the cluster radius a and the normalized laser intensity a_0 $(= eA/mc^2)$,

$$a = \left[\frac{3}{2\pi} \frac{mc^2}{n_{cl}e^2} \frac{\sqrt{1 + a_0^2/2} - 1}{(1 - f^{1/3})} \right]^{1/2}. \tag{2.136}$$

This is illustrated in Fig. 2.108(b) in the case of a solid deuterium cluster. It is found that the laser amplitude for expelling electrons scales as $a_0 \propto a$ in the low power regime ($a_0 \ll 1$) and $a_0 \propto a^2$ in the high power regime ($a_0 \gg 1$).

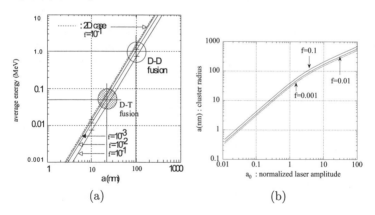

Fig. 2.108   (a) Maximum average deuterium kinetic energy obtained by Coulomb explosion as a function of cluster radius. (b) Cluster radius as a function of laser normalized amplitude needed to fully expel the electrons from the cluster.

For a given DD cross section, the total neutral yield can be roughly estimated to be

$$N^{(neutron)} \sim \int dt \int \langle \sigma_{DD} v \rangle n_{av}^2 dx \sim \frac{\tau_d}{2} \langle \sigma_{DD} v \rangle f^2 n_{cl} V, \qquad (2.137)$$

where $\langle \sigma_{DD} v \rangle$ is an average over the obtained deuterium energy distribution function, and V and τ_d represent the effective interaction volume and disassembly time of the cluster medium respectively. The packing fraction f is defined by $f = a^3/R^3$. The disassembly time τ_d is approximately estimated by $\tau_d \sim V^{1/3}/C_s$, where $C_s = \sqrt{T_e/M}$ is the deuterium ion sound speed with an electron temperature T_e. Note that since the deuterium energy distribution from a complicated laser–cluster interaction is not generally described by a Maxwellian distribution, direct velocity space integrals have to be performed to evaluate the fusion cross section. As found from Eq. (2.137), in order to increase the neutron yield (besides $\langle \sigma_{DD} v \rangle$) the packing fraction f, the effective interaction volume V and also the disassembly time τ_d have to be increased. Note that the interaction volume might not be restricted to the region where the laser directly interacts with the clusters, but can extend to the surrounding region secondarily heated by accelerated high energy electrons and ions, as discussed in Sec. 2.4.5.5.

In the experiment in Ref. [Ditmire *et al.* (1999)], deuterium clusters with a typical diameter of 5 nm ($a = 2.5$ nm) were irradiated with a laser intensity of 1.5×10^{16} W/cm^2 per shot, yielding around 10^4 neutrons. Employing the packing fraction $f \cong 0.32 \times 10^3$ and the cluster radius $a \cong 2.5$ nm, the laser amplitude necessary for expelling electrons from the cluster is estimated from Eq. (2.136) to be $a_0 = 0.08$–0.1, which is roughly the same as that found in the experiment. From Eq. (2.135), the average ion energy for a cluster radius of $a \cong 2.5$ nm is around $E_{ion} \cong 1$ keV, which is of the same order as that in Ref. [Ditmire *et al.* (1999)], i.e. $E_{ion} \cong 2.5$ keV. The discrepancy may come from an uncertainty in the measurement of the cluster radius. Using an estimated disassembling time of $\tau_d \cong 200$ ps, an interaction volume of $V \cong 6 \times 10^{-11}$ cm^{-3} (200-mm diameter and 2-mm length) and $\langle \sigma_{DD} v \rangle \cong 10^{-20}$ cm^3/s, a neutron yield $N^{(neutron)} = 1.5 \times 10^4$ is obtained.

2.4.5.4 *Numerical simulation*

Figure 2.109 shows a typical 2-D particle-in-cell (PIC) simulation of a laser–cluster interaction in the case of a normalized laser amplitude of $a_0 = 0.25$ ($I_L \cong 1.2 \times 10^{17}$ W/cm^2) [Kishimoto *et al.* (a); Kishimoto *et al.* (b)]. Solid deuterium clusters of radius $a \cong 32$ nm are employed with a packing fraction

$f \cong 4.8 \times 10^{-3}$.

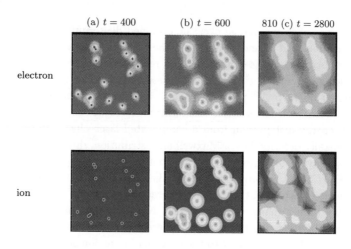

Fig. 2.109   Electron and ion number density distributions in configuration space at three different times. Clusters are initially distributed in a disordered manner with a packing function of $f = 4.78 \times 10^{-3}$.

The relative relation between the cluster radius, electron excursion length and skin depth is given by $\delta_e < a \sim \zeta_e$, so that the interaction regime may be close to that of a Coulomb explosion. Heated electrons from one particular cluster immediately interact with adjacent clusters that are probabilistically distributed. As a result, localized colony-like structures in which several clusters are contained are established. Cluster ions are subsequently exploded on a faster time scale and then start to overlap with adjacent clusters. Then, a fusion reaction is initiated through the DD collision process.

The dependence of the cluster radius on average electron and ion energies, and fusion cross section averaged over the obtained energy distribution, i.e. $\langle \sigma_{DD} v \rangle$, are illustrated in Fig. 2.110 in the case of $a_0 = 0.25$ [Kishimoto *et al.* (2002)]. The crosses $(+)$ and the line that passes through them represent the average ion energy obtained from Eq. (2.135), which is proportional to a^2. In the regime of a small cluster radius $(a < 10$ nm$)$, the ion energy increases according to Eq. (2.135), but tends to saturate as the cluster radius further increases $(a > 20$ nm$)$, suggesting that the expansion characteristics are gradually changing from the Coulomb explosion to the ambi-polar expansion. The details of the interaction are described in Ref. [Kishimoto *et al.* (2002)]. It is found that the experiment in Ref. [Ditmire *et al.* (1999)] falls in the circled region in Fig. 2.110, which provides a low level of neutron yield. A large size cluster irradiated with a higher power laser may be preferable to obtain a higher fusion cross section.

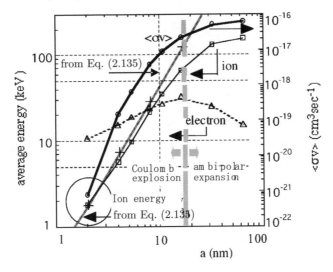

Fig. 2.110   Electron and ion kinetic energies and DD fusion cross section as a function of cluster radius in the case of $a_0 = 0.25$ and $f = 1.2 \times 10^{-3}$. The hatched vertical line corresponds to the critical cluster radius that demarcates ambi-polar expansion and Coulomb explosion.

2.4.5.5   *High efficiency neutron source*

In order to obtain a high efficiency neutron source, careful set up and optimization of the laser–cluster interaction are necessary. Some ideas have been proposed. Generally speaking, the number of deuterium ions that contribute to the nuclear fusion is limited to a small number compared with heated deuterium. Therefore, in order to efficiently utilize heated high energy electrons and ions, it is desirable to introduce a solid deuterium (and/or doped deuterium) collar surrounding the cluster medium. This is schematically illustrated in Fig. 2.111. In this configuration, the feasibility to realize 10^{12}–10^{13}/cm^3 neutrons per laser shot has been discussed

[Kishimoto *et al.* (a); Kishimoto *et al.* (b)].

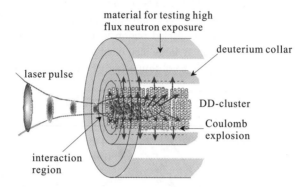

Fig. 2.111   Schematic picture of the setup of laser–cluster interaction for producing fusion neutrons. The cluster medium is surrounded by a deuterium collar in order to enhance the obtained neutron yield. Material for testing high flux neutron exposure is placed outside the collar.

Not only the configuration in Fig. 2.111, but the cluster design is also an important issue. There is a proposal to employ molecular clusters, in particular CD_4 instead of cryogenated D_2. CD_4 has a higher polarizability, allowing strong Van der Waals forces to bind the cluster and give good clustering in a CD_4 gas jet, even at room temperature [Balcou *et al.*].

2.4.6   *Positron*

K. NEMOTO
Central Research Institute of Electric Power Industry (CRIEPI)
2-11-1, Iwado Kita, Komae-shi,
TOKYO 201-8511, JAPAN

2.4.6.1   *Processes of positron production using lasers*

There are several ways to produce electron–positron (e^+e^-) pairs using lasers, as shown in Fig. 2.112. For example, a super high intensity laser with intensity higher than 5×10^{28} W/cm^2 can produce positrons directly from the vacuum. This process is thus called *vacuum breakdown*. However, the required laser intensity is much higher than the state-of-the-art of laser technology. On the other hand, when photons collide with a high energy photon of a few tens of GeV, the required intensity of the photons can be greatly reduced. 'Bruke *et al.* produced electron–positron pairs by the two-step Briet–Wheeler process [Burke *et al.* (1997)]. The first step is nonlinear Compton scattering (e + $n\omega_0 \rightarrow$ e* + ω), where e and e* are a high energy electron and a scattered electron, and ω_0 and ω are a laser photon and a scattered high energy photon, respectively. The second step is the multi-photon Breit–Wheeler reaction ($\omega + n\omega_0 \rightarrow$ e$^+$ + e$^-$). In the case of the experiment by Bruke *et al.*, high energy photons with an energy of 29.2 GeV were generated by the scattering of 46.6 GeV electrons accelerated by a linear accelerator with photons generated from a visible TW laser (527 nm, $\sim 1.3 \times 10^{18}$ W/cm^2). These high energy photons collide with the visible laser photons, and 106 ± 14 positrons (~ 0.2 positrons/laser shot) were observed. However, this experiment is only possible in a huge facility.

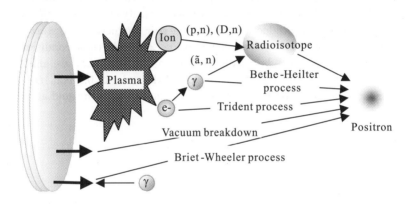

Fig. 2.112   Processes of positron production using lasers.

Indirect processes, on the other hand, can be performed in smaller facilities. Interactions between intense laser fields and plasma generate high energy electrons that have sufficient energy for positron production; high energy photons are also produced by bremsstrahlung. When the electron energy distribution is Boltzmann-like, the temperature of the bremsstrahlung photons T_γ is close to that of the electrons, T_e [Oishi *et al.* (2001)], for example, $T_\gamma = 0.25$ MeV for $T_e = 0.35$ MeV. There are two dominant indirect processes; one is the trident pair production process (e$^-$ + Z $\rightarrow$ e$^+$ + 2e$^-$ + Z) and the other is the photo pair production Bethe–Heilter process (γ + Z $\rightarrow$ e$^+$ + e$^-$ + Z). In the case of the Bethe–Heilter process, the threshold energy of γ is $E_{min} = 2m_0c^2(1 + m_0/M) \sim 1.022$ MeV, where

m_0 is the electron rest mass, c is the speed of light and M is the mass of the target atom. The cross sections of pair production of these processes and bremsstrahlung depend strongly on the atomic number of the target. Therefore, high Z targets such as Pb and W are desirable for positron production.

2.4.6.2 *Laser–solid interaction*

Liang *et al.* calculated positron production yields from relativistic electrons generated by irradiation of ultra-intense laser pulses onto solid targets [Liang *et al.* (1998)]. They proposed double side irradiation to confine energetic electrons and to increase pair production. They mainly considered the trident process, which was appropriate for their experiments, with $T_{hot} \sim 5$ MeV and a thin gold target. A laser intensity around 10^{20} W/cm^2 was adequate for pair production because the pair production rate starts to saturate at around this value, as shown in Fig. 2.113. At these conditions, positrons are expected to be produced more rapidly than they are annihilated. The estimated pair growth rate was of the order of $0.1(N_{Au})$/ns at a laser intensity of the order of 10^{19} W/cm^2, where N_{Au} is the gold atomic density.

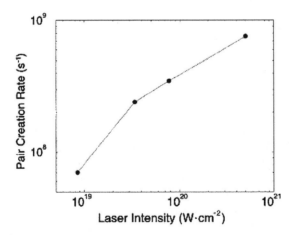

Fig. 2.113 Dependency of pair creation rate on laser intensity, calculated by Liang *et al.*

Gryaznykh *et al.* estimated positron yields to be 10^9–10^{11} from high Z

targets such as Au thin plates with thickness 0.1–1 mm when irradiated with a laser with a power of 10^2–10^3 TW [Gryaznykh *et al.* (1998)]. In their estimation, the Bethe–Heilter process is dominant because a thicker target was used. Nakashima *et al.* found that the dominant process for pair production depends on the thickness of the target and the atomic number of the target material [Nakashima and Takabe (2002)]. In the case of a thin target, photons escape from the target; therefore, a thicker, higher Z target is more suitable for the Bethe–Heilter process.

In any case, whether the dominant process is the trident process or the Bethe–Heilter process, the key issues for positron production experiments using laser plasma interaction are the generation of energetic electrons and the detection method. In these processes, many positrons are produced simultaneously, therefore conventional positron detection methods, such as pulse height analysis or coincidence measurement, which is very effective for positron-emitting isotope production experiments, do not work. If a high repetition laser is used, measurement with a detection rate of less than 1 per shot should be effective, as was done for fusion neutron detection by Ditmire *et al.* [Ditmire *et al.* (1999)].

Cowan *et al.* produced positrons by irradiating the Lawrence Livermore National Laboratory (LLNL) petawatt laser onto a thin gold target with a laser intensity of 10^{20} W/cm^2 and a pulse duration of 450 fs [Cowan *et al.* (1999)]. They observed an electron energy distribution with two components, a low energy component of $T \sim 3$ MeV accelerated by the ponderomotive force and a high energy component caused by parametric instability or relativistic self-focusing [Cowan *et al.* (1992)] (see Fig. 2.114). The maximum electron energy was higher than 100 MeV. Their measured conversion efficiency of electrons to positrons was 10^{-4}, which was comparable to predictions. Cowan *et al.* used a magnet spectrometer with nuclear emulsion track detectors for positron and electron measurement. The detected positron energies were between 3 and 9 MeV. as shown in Fig. 2.114.

2.4.6.3 *Laser–gas-jet interaction*

A supersonic gas jet target was also used for positron–electron production. In this case, energetic electrons are accelerated by laser wake field acceleration, as described in a previous Section and Ref. [Umstadter (2001)]. Gahn *et al.* produced high energy electrons with an effective temperature of 2.7 MeV by irradiating an ultrashort laser pulse (790 nm, 220 mJ, 130 fs,

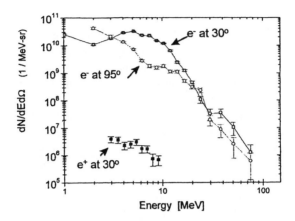

Fig. 2.114   Energy spectrum of positrons and electrons produced by the LLNL petawatt
laser.

10 Hz) onto a supersonic helium jet [Gahn *et al.* (2002)]. The obtained
effective temperature was comparable with the value expected from the
ponderomotive force obtained by the petawatt laser. This temperature was
several times higher than the 0.35 MeV obtained by irradiating a laser with
a similar peak power (790 nm, 90 mJ, 50 fs, 10 Hz) onto a copper thin tape
[Oishi *et al.* (2001)]. Therefore, a supersonic gas jet target is superior to a
solid target for positron production when using a small sized laser. Parametric instability or self-focusing, as observed by Malka *et al.* [Malka *et al.*
(2002a)], would be necessary to produce positrons effectively if a solid target
is used. Gahn *et al.* detected ~ 30 positrons per shot, although $\sim 2 \times 10^7$
positrons were estimated to be emitted quasi-isotropically. Therefore, high
S/N measurement is necessary.

2.4.6.4   *Radioactive isotopes*

Radioactive isotopes that emit positrons, such as ^{11}C, ^{13}N, ^{15}O, ^{18}F and
^{62}Cu, were also produced by nuclear reactions, such as (i,n) and (γ,n),
using lasers [Nemoto *et al.* (2001); Santala *et al.* (2001); Leemans *et al.*
(2001)], where the projectile particles i and γ are an energetic ion and an
energetic photon, respectively. As described in a previous Section, when
a thin target is irradiated by an ultrashort and intense laser, high energy
ions, most of them protons (H^+), are accelerated in the forward direction.
Other ions are also accelerated. Nemoto *et al.* produced ^{11}C, which decays

to ^{11}B emitting a positron with 20.39 min half-life, by generating energetic ions using a 400-fs, 4-J laser pulse and irradiating ions onto a boron pellet behind a laser target made of 6-μm mylar film [Nemoto *et al.* (2001)]. A ^{11}C yield of 2 nCi/laser shot was observed. If the isotope production is accumulated for one hour with 10 Hz or more, this yield should be sufficient for real applications. The positron emitter is easily detected by measuring coincidence signals resulting from the annihilation of electrons and positrons and the production of two gamma quanta each with an energy of 0.511 MeV. This process might be useful for the production of positron emitters for positron emission tomography (PET).

An ultrashort intense laser can also generate bremsstrahlung photons that have enough energy to cause (γ,n) reactions. Leemans *et al.* generated electrons with an energy in excess of 25 MeV by irradiating a laser pulse (50 fs, 8–10 TW, 10 Hz) onto a highly dense ($> 10^{19}$ cm^{-3}) gas jet target, and the high energy electrons generated high energy bremsstrahlung photons [Leemans *et al.* (2001)]. In this process, radioactive materials including a positron emitter (^{62}Cu) and other radioactive materials (for example ^{203}Pb) were produced. Neutrons produced by the reaction (γ,n) were also detected. The threshold γ energy for the (γ,n) reaction in ^{63}Cu is 10.8 MeV and the maximum cross section is more than 70 mbarn at 17 MeV. Usually, cross sections for (γ,n) are smaller than (i,n) reactions. However, high energy electron generation seems easier than high energy ion acceleration, so the comparative advantages of those processes must be further investigated. Recently, Malka *et al.* generated high energy electrons up to 60 MeV with a very narrow emission angle within 2.5° by irradiating a 1-J, 30-fs, 2–4 $\times$ 10^{19}-W/cm^2, 10-Hz laser onto a thin 6-μm polyethylene target and produced positron emitters ^{11}C and ^{62}Cu by the (γ,n) reaction [Malka *et al.* (2002a)]. The yields for one laser shot were 0.7 $\times$ 10^3 and 9 $\times$ 10^4, respectively. The electron temperature was 9.3 MeV, which was several times higher than the value expected by the ponderomotive force. Although the energy conversion efficiency from a laser pulse to energetic electrons was lower than that for the gas jet target experiment, a solid target is also attractive for energetic electron generation and production of positron emitting isotopes.

2.5   Inverse Compton Scattering X-ray Generation

2.5.1   *Laser synchrotron source and its applications*

T. HIROSE

Advanced Research Institute for Science and Engineering, Waseda
University
389-5 Shimooyamada-machi, Machida,
TOKYO 194-0202, JAPAN

2.5.1.1   *Laser synchrotron source*

Recently a new light source called the *laser synchrotron source* (LSS) has
attracted particular attention in various research fields. The LSS is based
on Compton backscattering of lasers from relativistic electrons from a linear accelerator (linac). The LSS can create X-rays or γ-rays, and has a
number of attractive features over conventional synchrotron light sources
(SLS), such as the ability to produce output with considerably shorter wavelengths and shorter pulses, i.e. femtosecond pulses can be produced by the
LSS vs. approximately tens of ps for a typical SLS. There are also several
good characteristics of the LSS, namely high brightness, quasi monochro-
maticity, high polarization and others. The LSS does not have as good a
repetition rate and average intensity as the SLS, which operates at mega-
hertz frequencies. However, the LSS has the potential to generate much
higher photon numbers per pulse than the SLS. Such LSS photons would of-
fer a wide range of applications, including X-ray diagnostics, time-resolved
observation of chemical and physical processes, high contrast imaging of
living cells in the water window spectral region, hadron physics based on
quantum chromo-dynamics, realization of a $\bar{\gamma}$–γ collider and a polarized
positron source for linear colliders.

Figure 2.115 illustrates Thomson scattering with parameters used in
this article. A relativistic electron interacting with a laser of wavelength λ_0
emits a photon of wavelength λ_1 given by

$$\lambda_1 = [(1 + a^2/2)/2\gamma^2 n(1 + \cos\theta)]\lambda_0 , \qquad (2.138)$$

where γ is the relativistic Lorentz factor, $n = 1, 2, 3, \ldots$ is the harmonic
number, θ is the incident angle of the laser photon (see Fig. 2.115) and
the parameter a is the normalized amplitude of the vector potential A of
the incident laser field and is defined as $a = eA/m_e c^2$, where m_e denotes

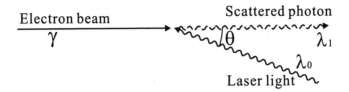

Fig. 2.115   Photon generation through Thomson scattering of laser on electron.

the electron rest mass, e the electric charge and c the light velocity [Esarey *et al.* (1993)]. Using the laser intensity I_0 and power P_0, a is given by

$$a = 0.85 \times 10^{-9} \lambda_0 \, [\mu m] I_0^{1/2} \, [\text{W/cm}^2] \,. \qquad (2.139)$$

For $a \ll 1$, Thomson scattering occurs in the linear regime, whereas for $a > 1$, the Thomson process enters the nonlinear regime in which multiple laser photons interact simultaneously with an electron and hence photons are generated at harmonics in addition to the fundamental. If the energy of the generated photons for the n-th harmonic is denoted by E_n, the relation $E_n = nE_1$ is derived. For usual solid state lasers of $\lambda_0 \sim 1$ μm, $a > 1$ requires $I_0 > 10^{18}$ W/cm^2.

The LSS can produce short wavelength photons through the relativistic Doppler factor that arises from backscattering laser photons off the counter-propagating relativistic electron beams. Putting $n = 1$ in Eq. (2.138), we obtain the wavelength of the fundamental,

$$\lambda_1 = [(1 + a^2/2)/4\gamma^2]\lambda_0 \,. \qquad (2.140)$$

Thereby, for $a^2 \ll 1$, $\lambda_1 = \lambda_0/4\gamma^2$, giving rise to the generation of extremely short wavelength photons. Practically the photon energy is given by

$$E_1 \, [\text{keV}] = 0.019 \, E_b^2 \, [\text{MeV}]/\lambda_0 \, [\mu m] \,, \qquad (2.141)$$

where E_b is the electron beam energy. For a conventional SLS using an undulator magnet with a period of λ_u, we obtain $\lambda_1 = \lambda_u/2\gamma^2$ or $E_1 \, [\text{keV}] = 0.019 \, E_b^2 \, [\text{GeV}]/\lambda_u \, [\text{cm}]$. The similarity of these expressions for the LSS and SLS indicates that the laser acts on relativistic electrons as an electromagnetic undulator with a period of 10^3–10^4 times shorter than magnetic undulators. Hence, compared to an SLS, the LSS can create photons with considerably higher energies at a given electron energy, thus leading to a 100-fold reduction in size of electron accelerators. As an example, for the generation of 30-keV photons, an LSS using a $\lambda_0 \sim 1$ μm laser needs an electron energy of 40 MeV, typical of the energies of a compact accelerator, whereas an SLS using $\lambda_u \sim 3$ cm requires an electron energy of 9.3 GeV.

LSS tunability can be achieved by adjusting the electron energy or the collision angle between the electron and laser beams. Head-on collisions are widely exploited and the bunch width of the electron beam is far shorter than the laser beam so as to accomplish effective production of photons. In this case, the pulse width of the photon beam is roughly given by the electron beam bunch width, which is of picosecond order for typical electron linacs with a laser driven photo-cathode called an RF gun. For generating an extremely short pulse length, there is a special configuration in which a femtosecond laser pulse crosses a tightly focused electron bunch at right angles [Kim *et al.* (1994)]. Actually, 280-fs X-rays are generated using 100-fs laser pulses and 3-ps electron bunches [Yorozu *et al.* (2002)]. Note that a short bunch electron beam of 200 fs was demonstrated by using a specially designed RF gun [Uesaka *et al.* (1999)], so that even in a head-on collision, short pulse photon beams comparable to those obtained for right angle scattering can be generated [Pogorelsky(1998)].

2.5.1.2 *Fundamental aspects of laser synchrotron source*

(a) Ultrahigh intensity photon beams
One of the prime objectives of the LSS is to realize an ultrahigh intensity tunable photon beam. For the head-on configuration, the electron beam should be focused to the same or a smaller spot size as compared with the laser beam spot size, in order to achieve a good overlap between the two beams. Because of the finite magnitude of the emittance of both the electron and laser beams, we maintain two beams tightly focused only within a limited distance called the *waist distance*. For a Gaussian shape laser beam, this distance limited by diffraction is related to the Rayleigh length l_R as

$$L_L \sim \pi l_R = \pi^2 r_L^2 / \lambda \,, \qquad (2.142)$$

where r_L is the laser beam radius at the focus. The physical meaning of l_R is the distance between the focus and the point where the laser beam is expanded twice in a cross-sectional area. Since the electron beam is usually focused more tightly than the laser beam, the laser beam eventually gives a limitation to the interaction length. For the efficient use of the laser pulse, L_L is extended over the overlap distance between the electron and laser beams, defined as $c(\tau_L + \tau_b)/2$, where τ_L and τ represent the time duration of the laser and electron beams, respectively. If this was not so, a proportional drop of the photon production rate would take place because of loss of laser energy. The most important factors for a high photon peak power are a high total laser energy per pulse, a high density of electrons within the bunch and a short time duration of the electron bunch. Note that the number of generated photons is proportional to the laser wavelength and thus to laser energy; hence a CO_2 laser of $\lambda_0 \sim 10$ μm can provide an order larger number of photons than a solid state laser of $\lambda_0 \sim 1$ μm.

An effort to produce ultrahigh intensity X-rays has been pursued at Brookhaven National Laboratory through Japan/US joint work. Figure 2.116 shows the principle diagram of the CO_2 LSS experiment [Pogorelsky et al. (2000)]. The electron beam is produced by a photocathode RF gun and accelerated to 60 MeV ($\gamma = 120$) through two linac sections. Typical electron beam parameters at the interaction point are a beam charge of 0.5–1.0 nC, an energy spread of 0.15%, a normalized emittance of 2–4 mm·mrad, a bunch duration of 3.5 ps FWHM and an rms radius of 32 μm. A 0.6-A mode-locked Nd:YAG laser is used for both electron bunch production at the photocathode RF gun and optical switching of the CO_2 laser. This scheme allows us to adjust precisely the timing between the CO_2 laser and the electron beams with a smaller timing jitter than the electron bunch width of 3.5 ps.

The 0.6-GW, 180-ps pulses, generated at the BNL-Accelerator Test Facility CO_2 laser (ATF-CO_2 laser), travel to the interaction point. In order to achieve a tight focus of the CO_2 laser beam, short focal length optical element are installed, i.e. copper parabolic mirrors of $F = 15$ cm with a hole of 5 mm diameter at the center, which serves to transmit downstream the electron beams and the back scattered X-rays (see Fig. 2.116). The maximum energy of the X-rays is 6.5 keV and the spread of the emission angle is $1/\gamma \sim 8.5$ mrad. An X-ray yield of 2.8×10^7 photons/pulse, or 8×10^{18} photons/s, has been demonstrated. A theoretical study of the Thomson process has been made using the Monte Carlo simulation code CAIN [Yokoya], in which various parameters of the electron and laser beams clarified beforehand are input and the intensity distributions of the electron and laser beams are assumed to be Gaussian in both the transverse and longitudinal directions. The CAIN code gives a total of 2.9×10^7 photons/pulse, which closely matches the experimental results. After the initial stage of proof-of-principle experiments, an upgrade of the CO_2 laser was pursued and a photon yield of 1.8×10^8 was accomplished using a

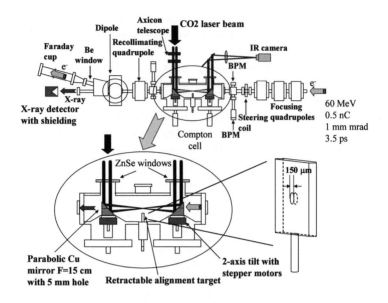

Fig. 2.116 Diagram of CO_2 LSS experiment.

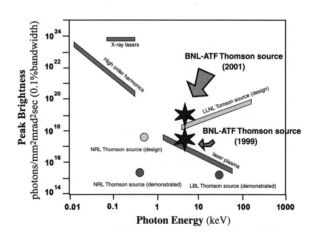

Fig. 2.117 Peak brightness of existing and designed laser driven X-ray sources.

15-GW CO_2 laser. Furthermore, a 1-TW, 3-ps CO_2 laser will be soon in operation and an X-ray flux up to 4×10^{21} photons/s can be expected. Figure 2.117 shows the brightness of the ATF LSS together with previous results obtained at the Lawrence Berkeley National Laboratory and the Naval Research Laboratory. Since the photon yield is proportional to the laser wavelength [Pogorelsky*et al.* (2000)], a CO_2 laser whose wavelength is one order of magnitude longer than a solid state laser, is favorable for generating a higher photon flux.

(b)   Nonlinear Thomson scattering

If the CO_2laser power increases to 1 TW and the laser is focused to an rms radius of 30 μm, we can achieve an intensity of 10^{16} W/cm^2 corresponding to $a \sim 1$, where a significant magnitude of the nonlinear effect is expected. At such high laser intensity, counter-propagating electrons reach a relativistic transverse quiver motion, so that the effective mass of the electron is shifted to $m(1 + a^2/2)^{1/2}$. Accordingly, the maximum energy of the backscatter photons decreases, as given by $E_1 = 4\gamma^2 E_0/(1 + a^2/2)$, where E_0 and E_1 are the energies of the laser and generated X-rays. Furthermore, a high intensity laser provides the opportunity for an electron to absorb multiple laser photons before radiating a single photon.

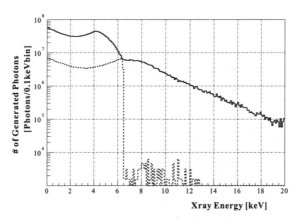

Fig. 2.118   Simulated photon density (photons/keV).

The CAIN simulation was carried out for laser intensities of $a = 0.06$ and 0.75 using the same conditions for the electron and CO_2 laser beams as those given on p. 181. Figure 2.118 shows a clear decrease of the high

energy peak (6.5 keV) of the fundamental and a gradual increase of high
energy photons arising from harmonics beyond the high energy peak. The
nonlinear effect for $a = 0.75$ is represented more distinctively, as shown in
Fig. 2.119, in the 2-D plot of the production angle vs. X-ray energy. Here
one can see clearly several bands corresponding to each harmonic $n = 1$–5.
It should be remarked that for circularly polarized laser beams, there are
no events in the forward region for $n \geq 2$.

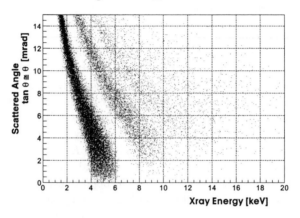

Fig. 2.119   Simulated 2-D plot of X-rays in terms of production angle and energy.

To observe the higher order harmonics, we have designed an X-ray spectrometer using a multilayer crystal and a 2-D position-sensitive detector.
Figure 2.120 shows a schematic view of the spectrometer. The multilayer
crystal consists of 45 silicon/molybdenum layer pairs curved cylindrically
with a radius of 4.8 m. The curvature causes the incident angle to vary
from 25 mrad to 10 mrad along the crystal circumference. For the multilayer crystal described above, this corresponds to Bragg-reflected energies
of 6 keV to 15 keV. The angular divergence in the vertical plane results
in a negligible change of the Bragg angle, and thus allows us to record the
angular distribution of X-rays at each energy. Consequently, energy and
angular distributions of X-rays can be observed using the 2-D detector for
each collision of the laser and electron pulses. The 2-D detector used is a
CCD based X-ray detector (MarCCD X-ray detector, Mar USA Inc.).

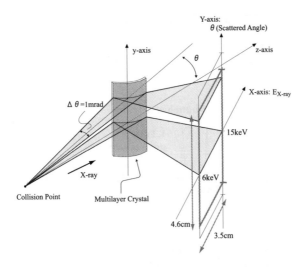

Fig. 2.120 Schematic view of the 2-D X-ray spectrometer with cylindrically curved multilayer crystal.

2.5.1.3 *Application of LSS: Polarized photon and positron production*

(a) Polarized positron source for a future linear collider

The advanced technology of the LSS may be fully utilized for producing short bunch, intense polarized positron beams of future linear colliders (LC), such as the Global Linear Collider (GLC), for which the requirements on the LSS photons are exceptionally severe in terms of pulse duration, intensity, polarization and so on. Furthermore, the nonlinear effect is closely related to the magnitude of the photon polarization and therefore a detailed characterization, especially of spin dependent phenomena, of the nonlinear processes is required.

A new method for polarized positron production was proposed based on two fundamental processes; Compton scattering of a circularly polarized CO_2 laser and the successively occurring electron–positron pair creation of circularly polarized γ-rays produced in Compton backscattering [Okugi et al. (1996)]. Based on this proposed method, basic studies have been pursued to accumulate information on advanced LSS technologies and to make the conceptual design of the polarized positron source [Hiroseet al. (2000); Omori et al. (2003)]. Figure 2.121 shows a positron or electron

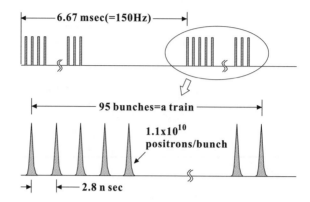

Fig. 2.121   Bunch structure of positron beam required for the GLC.

bunch structure of the GLC. Picosecond micro-bunches containing 1.1×10^{10} positrons are generated every 2.8 ns; 95 micro-bunches form one train whose repetition is 150 Hz. As illustrated in Fig. 2.122, circularly polarized CO_2 laser photons scattered on a 5.80-GeV electron beam produce circularly polarized γ-rays with a maximum energy of 60 MeV. The choice of electron energy is based on the fact that the total cross section of the Compton scattering is large, i.e. 658 mb, and the maximum energy of γ-ray results in a large cross section of the electron–positron pair creation, i.e. $\sim 10\,000$ mb. In addition, the maximum energy of the positrons, i.e. 60 MeV, thus created is also suitable for capturing positrons into subsegment accelerators.

A great challenge in designing a polarized positron source is to generate a huge number of polarized positrons, which requires ultra-intense polarized γ-rays. This requires a special configuration of electron–laser collisions, as described in Ref. [Omori *et al.* (2003)], as well as precise diagnostics of both of the electron and laser beams in order to enhance the luminosity.

(b)   Short pulse polarized γ-rays

For verifying the proposed scheme, the first step was taken in producing short pulse polarized γ-rays. In this case, 1.28-GeV electron beams and second-harmonic laser photons of $\lambda_0 = 532$ nm from a Nd:YAG laser were used. This combination of electron and laser beams gives rise to a maximum γ-ray energy of 56 MeV, close to the energy in the GLC conceptual design [Sakai *et al.* (2002)].

A series of experiments have been conducted at the High Energy Accelerator Research Organization, Accelerator Test Facility (KEK-ATF), which

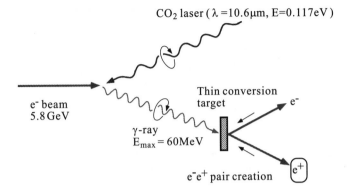

Fig. 2.122 Diagram of the polarized positron source consisting of two fundamental processes: Compton scattering of a circularly polarized CO_2 laser and electron–positron pair creation from circularly polarized γ-rays.

incorporates an S-band linac, a damping ring and an extraction line [Kubo *et al.* (2002)]. The parameters of the electron and laser beams are shown in Table 2.7, where $\sigma_{x,y}$ are the transverse beam sizes at the collision point, $\varepsilon_{x,y}$ are the emittances, $\beta_{x,y}$ are the beta functions and σ_p/p is the momentum spread [Fukuda *et al.* (2003b)]. The laser light is converted into circularly polarized light while passing through the quarter-wave plate. Generated γ-ray beams, which are longitudinally polarized, have a time duration of 31 ps, the same as that of the electron beams.

Figure 2.123 shows an apparatus called a Compton chamber, in which laser–electron collisions take place, together with the laser beam optics and the elements of the γ-ray polarimeter, i.e. a photodiode, an aerogel Cherenkov counter, a magnetized iron and an air Cherenkov counter. In order to achieve precise diagnostics of both the electron and laser beams, the Compton chamber, consisting of three cells, contains screen monitors, a wire scanner and a knife edge scanner. The screen monitors are mounted in the central cell located at the collision point and in the side cells placed at a distance of 265 mm from the collision point, so as to adjust transverse positions and angles of both beams. Furthermore, the wire and knife edge scanners are set in the central cell; the former allows the measurement of the electron beam size at the collision point with an accuracy of 3 μm, the latter determines the laser profile to $\sim$ 15 μm. The laser beam is transported to the collision point over a distance of about 10 m using six mirrors coated with a multilayer dielectric. The downstream three mirrors

Femtosecond Beam Science

can be remotely controlled to adjust the laser beam at the collision point.

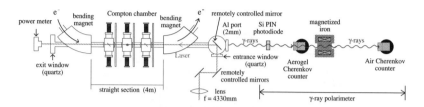

Fig. 2.123 LSS set up at KEK including Compton chamber with three cells, laser optics with three remotely controllable mirrors and γ-ray polarimeter consisting of Cherenkov counter and magnetized iron.

To maximize the luminosity of the laser and electron beam collisions, the electron and laser optics are optimized as follows. The original laser beam is expanded to an rms spot size of 4.7 mm and transported to the last lens whose focal length is relatively long, i.e. 4330 mm, resulting in a collision distance of about 10 cm (see the large values of the laser β_x in Table 2.7). In order to realize head-on collisions, the last mirror, of 3 mm thickness, is placed on the axis of the γ-rays, as shown in Fig. 2.123. In order to maximize the spatial overlap between the electron and laser beams, the position and angle of the laser beam are precisely adjusted by the remotely controlled mirrors. The collision timing was set so that the γ-ray signals are maximized by changing the timing of the laser Q-switch. The average number of γ-rays thus obtained was $(1.1 \pm 0.2) \times 10^6$ γ-rays/bunch. This is in good agreement with the estimated values of 1.0×10^6 obtained based on the Monte Calro simulation specially developed for this experiment [Fukuda *et al.* (2003b)].

(c) γ-ray polarimetry

An important step in the course of the polarized positron project is to establish the polarimetry of short bunch γ-rays and positrons whose time duration is a few tens of ps. Figure 2.124 shows the energy dependence of the cross section of inverse Compton scattering in which a right-handed polarized laser beam of $\lambda = 532$ nm collides with a 1.28-GeV electron beam. Note that γ-rays with helicity -1 are highly polarized in the high energy region.

The γ-ray polarization can be determined using spin-dependent Compton processes of γ-rays on 3-D electrons in an iron magnet, as shown in Fig. 2.125. The cross sections of the Compton process are given separately

Table 2.7 Parameters of the electron and laser beams. $\sigma_{x,y}$: transverse beam sizes at collision point, $\varepsilon_{x,y}$: emittance, $\beta_{x,y}$: beta function, σ_p/p: momentum spread.

Parameters	Electron beam	Laser beam
energy	1.28 GeV	2.33 eV (532nm)
intensity	0.65×10^{10} e$^-$/bunch	400 mJ/pulse
bunch length (rms)	31 ps	3.6 ns
σ_x	87 μm	154 μm
σ_y	72 μm	151 μm
ϵ_x	1.94×10^{-9} rad·m	11.5×10^{-8} rad·m
ϵ_y	2.36×10^{-11} rad·m	12.9×10^{-8} rad·m
β_x	0.513 m	0.104 m
β_y	52.859 m	0.058 m
rep. rate	3.12 Hz	1.56 Hz
σ_p/p	8.2×10^{-4}	—

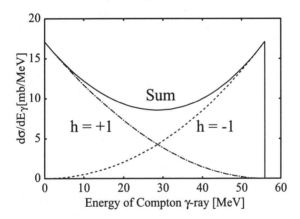

Fig. 2.124 Differential cross section of inverse Compton scattering in which a right-handed polarized laser photon of $\lambda_0 = 532$ nm collides with a 1.28-GeV electron beam. The dashed and the dash-dotted curves correspond to the γ-ray helicities of -1 and $+1$.

for the cases where the γ-ray spins are parallel ($h = +1$) and anti-parallel ($h = -1$) to the electron spin. The γ-ray bunch width of 31 ps is so short that each Compton event cannot be measured separately. Furthermore, as seen in Fig. 2.125, the electron and positron pair creation causes a considerably large background, which is a serious obstacle to the polarization measurement. Consequently a *transmission method* is adopted, wherein at a position downstream of the iron magnet, only the intensity of the transmitting γ-rays N_+ and N_- are measured for the parallel and anti-parallel cases, respectively, in order to determine the asymmetry, defined as $A = (N_+ - N_-)/(N_+ + N_-)$. In this experiment, the length of the magnetized iron is determined to be 15 cm, based on considerations of the efficiency of the asymmetry measurement as well as the background at the experimental area. The transmission of the γ-rays is measured for different magnitudes of laser polarization. One example is given in Fig. 2.126 for laser polarization of (a) -80% and (b) 80%. In measuring the asymmetry by varying the laser polarizations, the asymmetry for 100% laser polarization was obtained to be 1.29 ± 1.2, where the uncertainty includes statistical and systematic uncertainties. It has been demonstrated that the measured asymmetries are in good agreement with the simulated result of 1.3% [Fukuda *et al.* (2003b)].

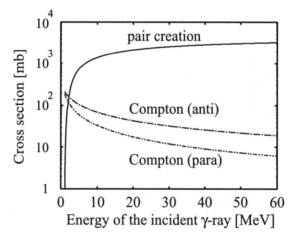

Fig. 2.125 Differential cross section of pair creation and Compton scattering for the cases in which the γ-ray spin is parallel or anti-parallel to that of the electron in an iron magnetized iron.

The transmission method can be applied to polarization measurements of γ-rays with any time structure and is influenced little by the background because the transmitting γ-rays are collimated to a narrow forward-cone (3 mrad in this case) while Compton scattering and pair creation emit, respectively, γ-rays and electron–positron pairs, which have a wide angular-spread. The successful performance of the transmission method pushes the polarized positron project of the linear collider one step forward.

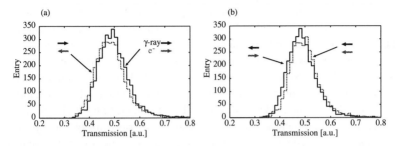

Fig. 2.126 Transmission of polarized γ-rays passing through an iron magnet for the cases where the γ-ray spin is parallel or anti-parallel to the magnetization. The laser polarization is (a) −80% and (b) 80%.

(d) Positron polarimetry

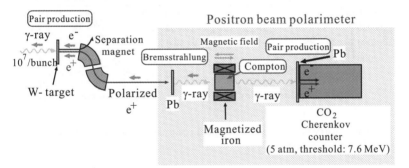

Fig. 2.127 Schematic design of positron polarimetry measurements. A γ-ray polarimeter is used for measuring the bremsstrahlung γ-rays generated from polarized positrons injected on a lead plate.

On the basis of the transmission method, we have designed a positron polarimeter, the schematic diagram of which is depicted in Fig. 2.127. Polarized γ-rays of 10^7/bunch are incident on a W plate of 1-mm thickness, and as a result, 3×10^4 e^+/bunch are created. As shown in Fig. 2.127, a pair of dipole magnets, called the *separation magnet*, is installed after a Pb plate to efficiently extract high energy positrons, which should have a high degree of polarization. Figure 2.128(a) shows simulated results of polarized positron energy distributions before the separation magnet, for left-handed and right-handed polarized positrons. When these positrons have passed through the separation magnet, their energy distributions are drastically modified, as given in Fig. 2.128(b). We can obtain high energy positrons of around 37 MeV with passing rate 4.1%. It is noted that we can achieve a high degree of polarization of 77%. Thus, separated positrons are injected into a Pb converter and circularly polarized γ-rays are generated via the bremsstrahlung process. On the assumption that left-handed positrons injected into the Pb converter have an energy of 40 MeV, we obtain, as shown in Fig. 2.129, simulated results of energy distributions (a) for left-handed (L) and right-handed (R) γ-rays and (b) the magnitude of the positron polarization. The polarization of the generated γ-rays can be measured in a similar manner to the γ-ray polarimetry, except using a CO_2 Cherenkov counter with a pressure of 5 atm so as to accept γ-rays whose energy is higher than 10 MeV (see Fig. 2.129). The technical details of the positron polarimeter are given in [Fukuda *et al.* (2003a)].

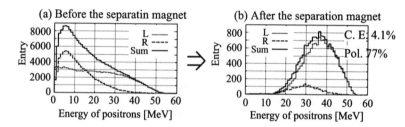

Fig. 2.128   Energy distributions of polarized positrons (a) before and (b) after the separation magnet.

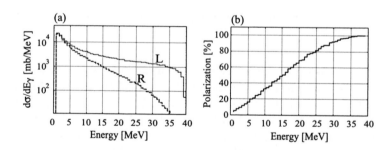

Fig. 2.129   Calculated (a) differential cross section (L: left-handed polarized, R: right-handed polarized) and (b) polarization of polarized γ-rays generated through bremsstrahlung of polarized positrons in lead.

Acknowledgments
The author wishes to thank I. Ben-Zvi, M. Fukuda, Y. Kamiya, T. Kumita,
Y. Kurihara, T. Okugi, T Omori, I.V. Pogorelsky, J. Urakawa and M.
Washio for providing various data before publication.

2.5.2   *Intra-cavity Thomson scattering*

J. R. BOYCE

Thomas Jefferson National Accelerator Facility (Jefferson Lab)

12000 Jefferson Avenue, Newport News,

VA 23606, USA

2.5.2.1   *Thomson scattering in the Jefferson Lab infrared FEL*

As is evident elsewhere in this book, several groups have demonstrated the
use of Thomson scattering of short pulses of laser photons off electrons to
produce sub-picosecond X-ray bunches. Though these experiments generate short enough X-ray pulses, more work is needed to improve the resulting X-ray fluxes for routine use in femtosecond science. Prior to the
construction of Jefferson Lab's high average power infrared free electron
laser (FEL), Krafft predicted [Krafft (1997)] high flux X-rays from intra-
cavity Thomson scattering of infrared light off the electron beam used to
generate the infrared light. After the FEL was commissioned, a simple experiment demonstrated the existence of the X-rays at significantly higher
fluxes and brightness than previous Thomson experiments. A related experiment showed the X-ray bunches to be sub-picosecond in length.

In this section, we will present the concepts of intra-cavity Thomson
scattering and initial experiments, and we will give an overview of the near
future program based on the Upgrade FEL (completion in 2003) that is ten
times more powerful than the initial machine.

Jefferson Lab's infrared free electron laser (IRFEL) uses superconducting RF technology to accelerate electron bunches to multi-MeV energies,
guides them through a wiggler where the free electron lasing action produces
infrared light pulses of less than 1 ps, amplified to saturation by bouncing
between two mirrors forming an 8-meter optical cavity. The design of the
machine is discussed in more detail elsewhere [Bohn (1997)], and the layout
of the IRFEL is shown in Fig. 2.130. The IRFEL produces high average
power coherent infrared (IR) light by combining continuous wave (CW)
operation (with a bunch repetition rate of 34.7 MHz or 75 MHz) of su-

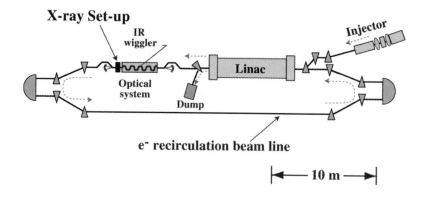

Fig. 2.130 Layout of the Jefferson Lab infrared free electron laser (IRFEL). A photocathode-based injector generates sub-picosecond duration, 10-MeV energy, electron bunches for the superconducting RF (srf) linear accelerator (linac) that increases electron energies up to a tunable final energy between 25 MeV and 48 MeV. Passing through the optical (cavity) system, the electrons produce infrared (IR) radiation as they undulate in the wiggler region's alternating magnetic fields. The IR is captured and stored between the high reflectivity mirrors, shown on either side of the wiggler, of the optical (cavity) system. The electrons are transported back around to the linac for re-insertion into the linac at 180° out of phase with the accelerating RF, de-accelerated back to 10 MeV, and finally deflected into a beam dump. This efficient design recovers RF energy and residual radiation produced in the dump is minimized. The X-ray detection setup is in the optical system downstream from the wiggler.

perconducting radio frequency (SRF) accelerator cavities with a technique that recovers the "waste" energy of the electron beam after it has been used for lasing [Neil *et al.* (2000)]. The IRFEL has lased at *average* powers up to 2.1 kW extracted from the optical cavity at 3.1 μm wavelength, a full two orders of magnitude higher extracted power than the previous average power record for FELs [Brau (1992)] set in 1990. An important feature relevant to Thomson X-rays is that the IR intra-cavity average power (the non-extracted power inside the cavity), is ten times greater than the extracted power.

The intra-cavity Thomson scattering process is illustrated in Fig. 2.131. The FEL optical cavity is symmetrically placed around the center of the wiggler. The optical cavity length was chosen to be the spacing between two electron bunches at 34.7 MHz. Thus, two optical pulses circulate in the optical cavity at all times. Just before a collision between an electron bunch and an IR bunch, overlapping IR and electron pulses move downstream (right to left in the figure) while a single IR pulse moves upstream (left to right). Just after the collision, three overlapping bunches, IR, electron and X-ray, move downstream, while the IR light that passed through the electron bunch without scattering continues upstream. Forward scattering at $\theta = 0°$, yielding the maximum X-ray energy, is defined by the direction of electron motion at the point of collision. For the IRFEL this direction is 18 mrad up from the beam centerline due to the undulating trajectory of the electrons in the wiggler and the exact location in the wiggler where the IR collides with the electrons. Thus the X-ray distribution in the laboratory frame is angled about 1° up from the centerline of the vacuum beam pipe. This allows us to insert downstream above the electron beam a diffracting crystal to separate out the X-rays from the IR and electron bunches.

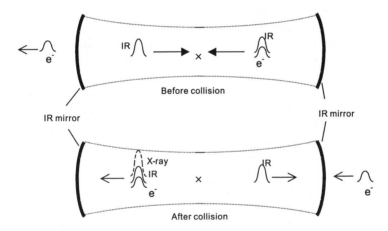

Fig. 2.131 Schematic of the basic idea of the Thomson scatter source. The scale of the pulse lengths in the cavity is enhanced considerably for clear presentation. In reality, the pulses are $\sim 200~\mu\mathrm{m}$ long.

It is important to recognize two aspects of the production of X-rays by head-on collisions of photon bunches (or pulses) with electron bunches: 1) In the rest frame of the electron, scattered photons have a dipolar distribution, but Lorentz transformation to the laboratory frame means that the scattered photons are X-rays azimuthally invariant about the electron beam direction and peaked in the electron beam direction, and 2) electrons with tens of MeV energy travel essentially at the speed of light. Thus all X-rays produced in the head-on collision of an IR pulse with a relativistic electron bunch will travel in a group peaked in the direction of the electrons and furthermore have the same longitudinal dimension (i.e. in the direction of the electron beam) as the electron bunch. The photon pulse can be shorter or longer than the electron bunch, but regardless, the X-ray group will be the same length as the electron bunch.

2.5.2.2 *Measurements of intra-cavity Thomson X-ray*

In the optical cavity of the IRFEL, the IR pulse passes through the electron bunch at the exact center of the wiggler. At this spot, the electrons are at their maximum wiggling angle of 18 mrad ($\sim 1°$) with respect to the e-beam centerline. This means that about a meter downstream from the interaction point, the X-rays are separated from the electron beam direction by ~ 2 cm, still within the 2-inch diameter vacuum beam pipe. By locating a port at that location and inserting a crystal with suitable inter-planar dimensions we were able to diffract the X-rays into a detector for the first proof-of-principle measurements.

A LiF crystal with (100) orientation, mounted on a remotely controlled rotatable rod, was placed 1/2-inch above the electron beam centerline (see Fig. 2.132). The IRFEL was set to lase at a 5-μm wavelength and the electron beam energy was set to 36.7 MeV. A cooled Si-PIN diode detector was positioned in the vacuum a few inches from the crystal to detect X-rays diffracted by the LiF crystal.

With the FEL lasing at a modest 250-W extracted IR power (10% of the intra-cavity power), the LiF crystal was rotated in small increments until a signal appeared in the multi-channel analyzer at the predicted X-ray energy. Once such X-rays were found, confirmation that they were only from Thomson scattering of IR light was obtained by collecting two energy pulse height spectra, the first with the FEL lasing and the second with the FEL not lasing. In Fig. 2.133 we show a typical pair of 60-s live time run spectra that clearly demonstrate the copious production of Thomson X-rays. The X-ray detector was calibrated with [55]Fe and [241]Am sources before and after the run. The signal to noise ratio between the Thomson X-ray flux and the accelerator room background flux was up to 90:1 in the energy region of interest. Direct comparison of the Thomson X-ray flux between the predicted and actual measurements was not possible due to the equipment used and the experimental setup.

Since the X-ray bunches have the same longitudinal time distribution

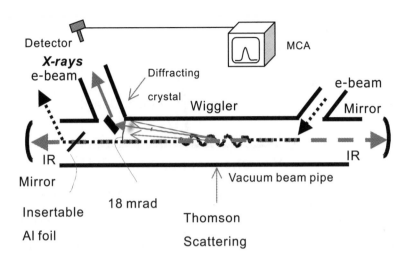

Fig. 2.132 Schematic representation of the production and detection geometry for intra-cavity Thomson X-rays. An IR pulses, traveling from left to right, collides with the electron bunch, traveling from right to left, in the center of the wiggler. X-rays resulting from the collision travel at an angle of $\sim 1°$ above the electron beam centerline.

as the electron bunch at the collision point, we used the electron bunch distribution as a measure of the X-ray distribution. This was determined by an autocorrelation measurement of coherent transition radiation (CTR) [Brau (1992)] emitted as an electron bunch crossed a very thin (1.5 μm) aluminum foil remotely inserted into the beam path (see Fig. 2.132). The CTR is transported to a Michelson-type interferometer (not shown in the figure). A typical measured autocorrelation distribution is shown in Fig. 2.134. This was obtained under accelerator operating conditions identical to the operating conditions for the X-ray measurements.

Because the longitudinal length of the electron bunch does not change inside the optical cavity, and since a Thomson X-ray bunch can be no longer than the electron bunch, we conclude that the Thomson X-rays are generated in a train of bunches, each 300-fs long, at the rate of 37.4 MHz.

Over the two-year operating period of the first Jefferson Lab FEL, other experiments demonstrated tunability over the X-ray energy range 3.5 keV to 18 keV by various combinations of electron beam energy, FEL wavelength or wiggler parameters. We also demonstrated a unique capability, namely of producing simultaneous pulses of sub-picosecond photons at three wavelengths, IR, X-rays and THz, offering the possibility of femtosecond pump-

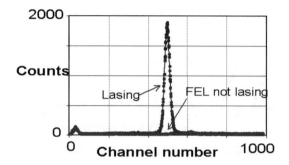

Fig. 2.133 Two spectra demonstrating the production of Thomson X-rays. Signals from a cooled Si-PIN diode sent through a multichannel analyzer were accumulated for 60 s and plotted in a pulse height distribution with channel number proportional to energy. The first spectrum was taken with the FEL lasing and the second with a shutter preventing IR lasing. X-ray energy calibrations with ^{55}Fe and ^{241}Am place this peak at 5.12 keV with a 320 eV FWHM.

probe experiments spanning 10^7 in photon wavelengths. This is shown in Fig. 2.135.

2.5.2.3 *FEL upgrade Thomson X-ray possibilities*

The year 2002 marked an exciting new phase in the Jefferson Lab FEL program: disassembly of the first FEL and construction of an upgrade FEL designed to produce 10-kW average extracted IR light and 1-kW average UV light. This new accelerator (completion in 2003) has three linac modules on one straight section producing tunable electron energies from 80 MeV to 210 MeV. Two return straight sections, one for IR and one for UV, contain new wigglers and IR and UV optical cavities, and enable both THz and X-ray capabilities. Energy recovery of the electron beam is also built into this accelerator.

Using the design specifications for the upgrade [Douglas *et al.* (2002); JLab web site (2003)] we have calculated the brightness of the X-rays from Thomson scattering in both the IRFEL and the UVFEL [Boyce *et al.* (2002)]. These are shown in Fig. 2.136.

2.5.2.4 *Conclusions and future program*

We have presented the concept, initial measurements and possibilities of intra-cavity Thomson X-ray sources at Jefferson Lab's FEL facility. This

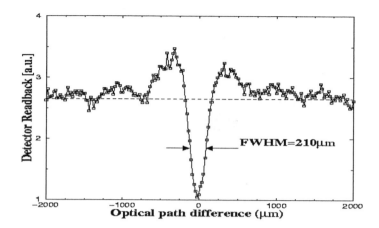

Fig. 2.134 A typical measurement of the electron bunch length using coherent transition radiation (CTR) from thin Al foil in an optical cavity (see Fig. 2.132). (The foil is retracted for lasing.) The FWHM implies a bunch length rms value of 89.4 μm or 298 fs for the electron bunches, and since the X-rays are traveling with ($\sim$ 40 MeV) electrons at c, their bunch rms distribution is also $\sim$ 300 fs.

type of photon source produces IR and X-ray radiation with unique characteristics. The IR and X-rays are highly time-correlated and both of sub-picosecond time duration. In the first IRFEL, we measured the electron bunch length, and thus the X-ray bunch length, to be $\sim$ 300 fs in duration. We have verified that the overall X-ray production rate is consistent with predictions based on beam current, on the IR and electron beam spot sizes, and on the re-circulating power in the FEL. The X-ray pulses are produced in 37.5 MHz or 75 MHz streams, and the signal to noise ratio is high, even in our first experiments. We believe that this type of source complements synchrotron-based fs X-ray sources (due to simultaneity with IR) and to fourth-generation light sources (due to the high repetition rate). In 2003 the FEL upgrade is scheduled for completion. (For current status see Ref. [JLab web site (2003)].) This higher power FEL will produce correspondingly higher brightness sub-picosecond X-rays. Furthermore, the possibilities of pump-probe studies using such a source may well have important applications in the burgeoning field of "femtosecond science".

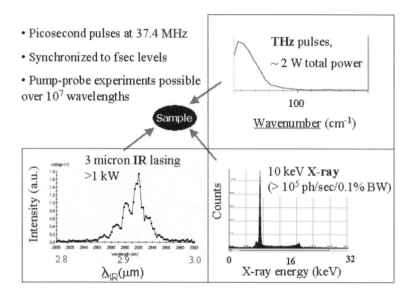

Fig. 2.135   Simultaneous high flux photon spectra demonstrating pump–probe possibilities.

Acknowledgments

The author would like to thank K-J. Kim, G. Krafft and U. Happek for valuable planning discussions and the entire Jefferson Lab FEL Team for assistance in the experimental aspects of this program.

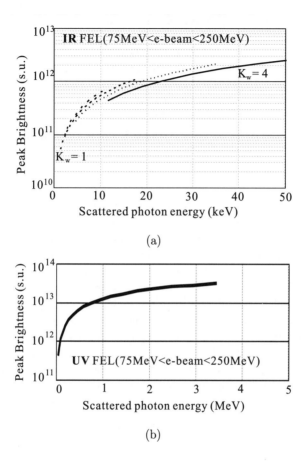

(a)

(b)

Fig. 2.136 Calculated photon brightness in standard units (s.u.) of photons/s/mm^2/mrad^2/(0.1%bandwidth) as a function of scattered photon energy (a) for the IR upgrade FEL (the upgrade wiggler is electromagnetic with variable wiggler parameter K_w from 1 to 4, hence the four curves) and (b) for the UVFEL. Note the change from keV to MeV.

2.6   Beam Slicing by Femtosecond Laser

M. ZOLOTOREV
A. A. ZHOLENTS
Accelerator and Fusion Research Division
R. W. SCHOENLEIN
Materials Sciences Division
Lawrence Berkeley National Laboratory (LBNL)
1 Cyclotron Road, Berkeley,
CA 94720, USA

The brightest sources of X-rays currently available are modern synchrotrons [Mini (1998)]. However, the pulse length of such sources, > 30 ps, is determined by the duration of the stored electron bunches. Attempts to shorten the duration of an electron bunch in a storage ring to the sub-picosecond timescale have encountered problems due to bunch instabilities, which result from bunch wake fields and coherent synchrotron radiation [Limborg (1998)]. An alternative to shortening the entire bunch is to take radiation from a femtosecond slice of the bunch. To get X-ray pulses of femtosecond duration, we need to separate a short (30 μm length or equivalently 100 fs duration) slice of electrons from the rest of the bunch.

One approach is to provide a short transverse kick to a temporally thin slice of electrons. The interaction must be in vacuum and sufficiently removed from any electrodes to avoid wake fields and beam degradation. Only interaction with a freely propagating electromagnetic wave can satisfy these requirements. In counter-propagating beams the forces from the laser E (electric) and B (magnetic) fields add, but there is a very limited interaction time. Furthermore, because all the electrons in the bunch see the same kick, no ultrashort slice will be created. In the case of a co-propagating wave the interaction time is equal to the pulse length divided by the relative difference in electron and light velocities. For ultra-relativistic particles ($\gamma \gg 1$) this difference is very small, equal to $1/2\gamma^2$. Accordingly, compared to the counter-propagating case, the interaction time is increased dramatically by a factor $2\gamma^2$. However, in the co-propagating case the force from B nearly cancels the force from E; the net transverse force is reduced by $1/4\gamma^2$ relative to the counter-propagating case. This cancellation makes it impractical to impose an appreciable transverse kick for ultra-relativistic particles.

Is there any way we can prevent the cancellation of the electromagnetic forces and still maintain a long interaction time? If an electron's trajectory makes a non-zero angle with a light wave vector k, then E is not orthogonal to the electron velocity v. In this case, the force from the electric field is not canceled by the magnetic force. There is a net force in the direction of the velocity, which is proportional to the angle between k and v. Even if we choose a small angle, $\sim 1/\gamma$, the net force is still larger by a factor of γ compared to the transverse kick case. The interaction with the light produces a longitudinal rather than a transverse force.

Since light moves faster than the electron, the light will eventually pass ahead by half a wavelength and at this point the force on the electron will change sign. One may stop the interaction after one such period of phase slippage. Alternatively, the light–electron interaction length can be increased by changing the electron trajectory so that the electron sees a constant optical phase. In this case the electron must move along a sinusoidal trajectory and this motion can be imposed by a device called a *wiggler* (see Fig.2.137(a)). The condition above known as *resonance* is given by $\lambda_L = \lambda_W/2\gamma_{\text{eff}}^2$, where λ_L is the laser wavelength and λ_W is the wiggler period. The Lorentz factor for longitudinal motion γ_{eff} is given by $\gamma_{\text{eff}}^2 = \gamma^2/(1 + K^2/2)$, where the deflection parameter (normalized vector potential of the wiggler) is $K = eB_0\lambda_W/2\pi mc^2$, where B_0 is the peak magnetic field of the wiggler, e is the electron charge, m is the electron mass and c is the speed of light. The longitudinal force changes the electron energy by an amount proportional to the electric field seen by the electron $E\cos\phi$, where ϕ is the phase of the electromagnetic wave that the electron sees.

As an alternative way to look at the interaction, acceleration or deceleration linear with the electromagnetic field is always, respectively, the result of destructive or constructive interference between spontaneous radiation and the applied electromagnetic field. For efficient interference, the electrons' spontaneous radiation must match that of the short laser pulse temporally and spatially. An effective method to match the spontaneous radiation of the wiggler to the laser pulse is to make a wiggler that has the same number of wiggles as the laser pulse cycles and is tuned to the same frequency. It is also important to have spatial overlap of the radiation from the wiggler and the laser pulse in the far field. One can show that optimal spatial overlap occurs when a Gaussian laser beam is focused in middle of the wiggler with a Raleigh range Z_R equal to one quarter of the wiggler length $L_W Z_R = L_W/4$. The electron and electromagnetic wave co-

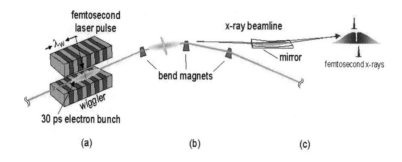

Fig. 2.137   Schematic of the laser-slicing method for generating femtosecond synchrotron
pulses.

propagate through the wiggler and depending on the optical phase (ϕ seen
by the electron), it is either accelerated or decelerated. Accordingly, some
electrons in the beam are accelerated while others are decelerated, causing
an energy modulation. The change in the electron energy ΔU is given by
$\Delta U = \Delta U_0 \cos \phi$ [Zholents and Zolotorev (1996)]. A simple way to calculate
ΔU_0 is as follows. One can look at the radiation from an electron passing
through a wiggler with no laser field present. We referred to this previously
as spontaneous radiation to make an analogy with spontaneous emission in
atomic physics. When one adds the laser field, the total field seen in the far
field is $E_{\text{tot}} = E_s + E_l$. The total energy is equal to the Poynting vector,
proportional to E_{tot}^2, integrated over solid angle and time. The total energy
therefore consists of three terms: spontaneous energy $U_s \sim E_s^2$, laser en-
ergy $U_l \sim E_l^2$ and a cross term $2(E_s \times E_l) \cos \phi \sim \sqrt{(U_s \times U_l)} \cos \phi \eta$. This
interference term describes the change in the electron energy. In comparing
this with the above we see that $\Delta U_0 = 2 \times \sqrt{(U_s \times U_l)} \times \eta$. Since energy is
conserved, this cross term represents an energy exchange between the laser
field and the electron. It is the same phenomenon known in atomic physics
as stimulated emission or absorption. The factor η comes from the overlap
integral of E_s and E_l. Perfect matching of laser and wiggler radiation gives
$\eta = 1$. It is surprising that a Gaussian laser beam spatial profile matches
wiggler radiation to $\eta = 0.98$.

The energy radiated spontaneously by a single electron for a plane wiggler (in the limit of $K \gg 1$) is $U_s \sim 2\alpha h\nu$. A mode-locked Ti:Sapphire laser
with $\lambda = 800$ nm and pulse energy $U_l \sim 200$ μJ, will give $\Delta U_0 \sim 10$ MeV,

which is independent of the initial electron energy, number of wiggler periods and K parameter when $K \gg 1$. This number at first seems very small when compared to synchrotrons with energies of a few GeV. But the relevant comparison is not to the overall beam energy but to the energy spread σ_{gf} of the beam. The square of the relative energy spread of the beam is roughly the X-ray critical energy divided by the electron energy. For synchrotron sources, the X-ray critical energy is a few keV and the electron energy is a few GeV, which yields an energy spread of 10^{-3} relative to the total electron energy. This means that the energy spread σ is a few MeV while ΔU_0 is more than a few times this energy spread. We would like to clarify that there is no spatial separation immediately after the wiggler, there is only energy modulation. The spatial separation (slicing) occurs when the electron beam comes to a bend magnet, see Fig.2.137(b), which causes the X-ray radiation. We can see that the transverse spatial separation of the energy-modulated electrons will be a few times the horizontal size of the beam as they traverse the bend magnet. It is then possible to select X-rays originating from the femtosecond slice, as in Fig.2.137(c).

Although we see radiation from only a tiny fraction of the entire bunch (10^{-3}), the peak brightness of the X-rays that come from the femtosecond slice of electrons is only a few times smaller than the peak brightness of the entire bunch (smaller by a factor of $\sim U_0/\sigma$). The sliced electrons return to the equilibrium energy after a few damping times. Thus, the repetition rate is limited by the damping time and the number of bunches in the accelerator. For typical synchrotrons this limit is a few hundred kHz. If this is exceeded the beam will warm up (the energy spread will increase) and the brightness will degrade as a result.

Proof-of-principal experiments were conducted at the Advanced Light Source (ALS) at Lawrence Berkeley National Laboratory. The ALS storage ring operates at $E = 1.5$ GeV and beam energy spread $\sigma = 1.2$ MeV. A wiggler with $M_W = 19$ periods and $\lambda_W = 16$ cm, with the gap tuned ($K \cong 13$) to match the resonance wavelength 800 nm of the femtosecond laser pulses ($\tau_L = 100$ fs, $U_l = 400$ μJ and $\lambda = 800$ nm, $f = 1$ kHz,). A Ti:Sapphire laser oscillator is synchronized to the storage ring RF using phase-locking techniques. Amplified pulses with a repetition rate of 1 kHz are directed into the vacuum chamber and co-propagate with the electron beam through the wiggler. Following the interaction region, a mirror directs the fundamental spontaneous wiggler emission and the laser beam out of the storage ring to enable direct measurements of the temporal and spectral overlap, and the spatial mode matching between the laser pulses and the wiggler radiation.

When the femtosecond slice of electrons separates from the main bunch, in the vicinity of the bend magnet, it leaves behind a femtosecond hole. The longitude distribution of electrons in Fig.2.138(a) shows an absence of electrons in the central core of the beam that corresponds to a few hundred femtoseconds. This was measured by using a frequency up-conversion process in a BBO crystal to correlate in time the synchrotron radiation intensity with the femtosecond laser pulse used to create the slice. In Fig.2.138(b) we see that there are electrons in the tail of the horizontal distribution.

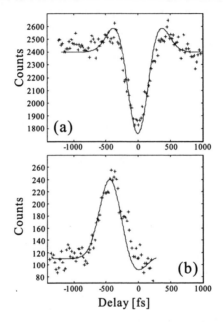

Fig. 2.138 Cross-correlation measurements between a delayed laser pulse and synchrotron radiation originating from an energy-modulated electron bunch. In (a), synchrotron radiation from the central core ($\pm 3\sigma_x$) of the electron bunch is selected. In (b), synchrotron radiation from the horizontal wings ($+3\sigma_x$ to $+8\sigma_x$) of the electron bunch is selected. Solid lines are from a model calculation of the spatial and temporal distribution of the energy-modulated electron bunch following propagation through 1.5 arc-sectors at the ALS.

These electrons are shifted in time because they pick up or loose energy during the modulation and take a longer or shorter path around the bend magnet, respectively. Because the femtosecond laser pulse used to measure the distribution is the same pulse used to create the beam slice, there is no jitter between them.

Slicing is a relatively inexpensive and simple way to provide existing synchrotrons with new functionality. Since this approach creates femtosecond time structure on the electron beam, standard radiating devices, such as bend magnets, wigglers or undulators, can be designed to emit femtosecond X-ray pulses with the desired properties, bandwidth, tunability, brightness, etc. The X-ray flux and brightness levels available from such a femtosecond synchrotron source will enable a wide range of X-ray techniques, including diffraction and EXAFS, to be applied on a 100-fs timescale to investigate atomic motion associated with phase transitions in solids, making and breaking of bonds during chemical reactions and possibly ultrafast biological processes.

2.7 Free Electron Lasers

R. HAJIMA

Advanced Photon Research Center,

Japan Atomic Energy Research Institute (JAERI)

Tokai, Naka,

IBARAKI 319-1195, JAPAN

2.7.1 *Femtosecond infrared free electron laser*

A free electron laser (FEL) is a device that produces coherent electromagnetic light (laser) from a relativistic electron beam. Figure 2.139 shows a schematic view of an FEL oscillator, which consists of an electron accelerator, an undulator and an optical resonator. An electron beam injected into the undulator experiences a periodically alternating magnetic field and acquires a wiggling motion and emits forward radiation, *undulator radiation*. This undulator radiation is stored in the optical cavity and interacts with electron bunches repeatedly, developing a large amplitude. Longitudinal and transverse coherence of the radiation is established through the interaction with the electrons.

The principle of light amplification in an FEL oscillator is explained by cooperative radiation emission from an electron bunch under a resonant condition. The undulating motion of the electrons enables them to exchange their energy with the radiation wave, having a transverse electric field. The FEL resonant condition is given by $\lambda = \lambda_w(1 + a_w^2)/(2\gamma^2)$, where

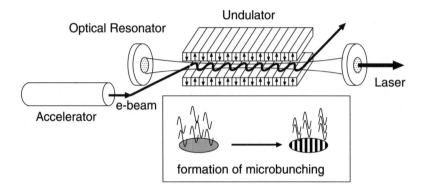

Fig. 2.139 Schematic view of an FEL oscillator. Formation of electron beam microbunching through the interaction with the radiation field results in cooperative emission from the electron bunch.

λ is the radiation wavelength, λ_w is the undulator period, $\gamma \equiv E/mc^2$ is the electron energy and $a_w \equiv eB_w\lambda_w/(2\pi mc)$ is the undulator parameter determined by the on-axis rms magnetic field amplitude B_w and the undulator period. The undulator parameter is $a_w \sim 1$ for an ordinary configuration with permanent magnets. For the resonant radiation wave, electrons slip back one radiation wavelength exactly with respect to the radiation wave after traveling one undulator period. If an electron feels the radiation wave as a deceleration field every undulator period, the electron loses energy and its longitudinal velocity decreases. Electrons with the opposite phase, at which they feel an accelerating field from the radiation wave, gain energy from the radiation and increase their longitudinal velocity. This modulation of longitudinal velocity introduced in the electron bunch is converted into a microscopic density modulation, the interval of which is equal to the radiation wavelength. The undulator radiation from this modulated electron bunch has a large amplitude with temporal coherence due to cooperative emission of the electrons, as shown in Fig. 2.139. A further detailed description of FEL physics including gain and saturation mechanisms and coherent properties are found elsewhere [Brau (1990)].

Since the wavelength of the amplified radiation is determined by the FEL resonant condition, we can choose an arbitrary radiation wavelength by changing the electron energy or the undulator configuration. The development of FEL devices has, however, been focused on the infrared (IR) and X-ray regions, where conventional lasers are not available. We describe femtosecond IRFELs in this section and X-ray FELs in the next section.

An infrared FEL oscillator operated with bunched electrons from a radio frequency accelerator is capable of producing optical pulses with picosecond or subpicosecond duration. Table 2.8 gives a list of femtosecond IRFELs currently in operation.

Table 2.8 Femtosecond infrared free electron lasers. (Lengths of electron bunch and optical pulse are FWHM.)

	energy	electron bunch charge	length	optical pulse wavelength	length
FELIX[1]	15–45 MeV	200 pC	3 ps	10–25 μm	220–500 fs
CLIO[2]	40 MeV	600 pC	10 ps	8.5 μm	700 fs
Stanford[3]	32 MeV	17 pC	1 ps	5 μm	600 fs
IR-demo[4]	48 MeV	60 pC	0.4 ps	1-7 μm	700 fs
JAERI[5]	17 MeV	500 pC	5 ps	23 μm	320 fs

[1][Knippels *et al.* (1995)]
[2][Glotin *et al.* (1993)]
[3][Crosson *et al.* (1995)]
[4][Neil *et al.* (2000)]
[5][Hajima and Nagai (2003)]

The IR-demo FEL in the Thomas Jefferson National Accelerator Facility (TJNAF) was developed to demonstrate a high average power FEL based on a superconducting linear accelerator (linac) and energy recovery technology [Neil *et al.* (2000)]. Electron microbunches of 60 pC and 20 ps rms are extracted from a GaAs photocathode installed in a 320-kV DC gun at 75-MHz repetition and accelerated up to 48 MeV by a superconducting linac operated at 1.5 GHz. The electron bunch is compressed to 400 fs by a chicane bunch compressor and transported to the FEL undulator. The IR-demo delivers subpicosecond ($\sim$ 700 fs) IR pulses at 3–7 μm with 2-kW average power. This unique property, high average power with ultrashort pulses, makes FELs useful for various industrial applications, such as pulsed laser ablation and deposition, laser nitriding, synthesizing carbon nanotubes and micromachining.

Femtosecond FEL pulses can be generated even from picosecond electron bunches if an FEL oscillator is operated in the super radiant regime with high gain and large slippage parameters. During FEL interaction, electrons have a longitudinal velocity slightly less than the vacuum speed of light due to the undulating motion, and slip backwards in the rest frame of the optical pulse. The distance the electrons slip back throughout the undulator is called the *slippage distance* and is given by $L_s = N_w \lambda$, where N_w is the number of undulator periods. When an electron bunch is shorter than the slippage distance and the FEL gain is large enough, a super radiant FEL spike formed in the slippage region grows continuously and becomes narrower than the electron bunch after the onset of saturation (see Fig. 2.140).

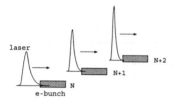

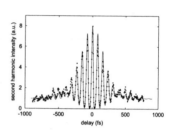

Fig. 2.140 Super radiant FEL interaction between a laser pulse and electron bunches for Nth–(N+2)th round trips. The laser pulse slips forward in the electron bunches and is amplified via the FEL interaction through many round trips.

Fig. 2.141 Second-order autocorrelation signals of an ultrashort FEL pulse generated in the JAERI-FEL. The pulse duration is found to be 340 fs FWHM.

This type of lasing has been clearly demonstrated at the Japan Atomic Energy Research Institute (JAERI). In the JAERI-FEL, electron bunches of 0.5 nC are extracted from a thermionic gun of 230 kV at 10-MHz repetition and accelerated up to 17 MeV by a superconducting linac. The electron bunch is compressed to 5 ps FWHM by velocity bunching. It produces 16–23-μm FEL over 2-kW average power with femtoseconds pulse duration [Hajima and Nagai (2003)]. Figure 2.141 shows the fringe resolved second-order autocorrelation signal for FEL pulse duration measurement at the JAERI-FEL. Detailed analyses have revealed that the FEL pulse has a frequency down-chirp and a pulse duration of 320 fs FWHM, which corresponds to 4.1 optical cycles. Femtosecond FEL pulses were also obtained at the FOM Institute for Plasma Physics 'Rijnhuizen' (FELIX), Netherlands, the LURE laboratory (CLIO), France, and the Stanford FEL, Stanford Univ., USA.

2.7.2 Femtosecond X-ray free electron laser

Exploring coherent radiation shorter than 100-nm wavelength is another trend in FEL source development. Since a mirror of good reflectivity in this wavelength region does not exist, FELs must be operated in a high-gain single-pass configuration, i.e. self-amplified spontaneous emission (SASE). In a SASE-FEL, small shot noise originating from population fluctuation of electrons in the microscopic scale of laser wavelengths is amplified exponentially via the FEL interaction. Figure 2.142 shows the evolution of a SASE-FEL, in which the radiation signal from shot noise grows exponentially and reaches saturation.

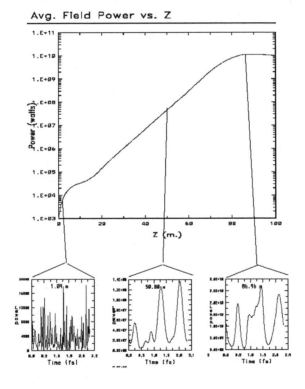

Fig. 2.142 SASE-FEL output power and optical pulse structure along an undulator obtained by numerical simulation. The FEL signal starts from shot noise and grows exponentially to reach saturation. Longitudinal coherence is also established through the evolution.

Lasing behavior of SASE-FELs, such as gain length, conversion efficiency and saturation power, are determined by a single parameter ρ called the *high-gain FEL parameter* and is defined to be $\rho \simeq [(a_w^2 \lambda_w^2 r_e n_e)/(16\pi\gamma^3)]^{1/3}$, where r_e is the classical electron radius and n_e is the electron density of the beam. The exponential power gain length is given by $L_g = \lambda_w/(4\sqrt{3}\pi\rho)$. We need an undulator as long as 10–15 gain lengths to obtain full saturation. The FEL parameter is also equal to the gain bandwidth, which is the conversion efficiency from electron beam power P_{beam} to laser power P_{sat}. The saturation FEL power is, thus, $P_{sat} \sim \rho P_{beam}$.

After saturation, an FEL pulse contains lots of spikes, as shown in Fig. 2.142 [Arthur *et al.* (1998)]. The temporal duration of each spike is determined by the electron slippage distance until the onset of saturation and is given by $2\pi L_c$, where $L_c = \lambda/(2\pi\sqrt{3}\rho)$ is the cooperation length of the SASE-FEL. With the parameters of the 1.5 Å X-ray FEL at the Linac Coherent Light Source (LCLS), U.S.A., the spike structure length is of the order of 1 fs.

We can see from the definition of ρ that shorter FEL wavelengths require higher electron beam currents in order to obtain laser saturation with an undulator of a reasonable length. Moreover, critical criteria of electron-bunch emittance and energy spread exist for a SASE-FEL. The spatial divergence of the electron beam must be smaller than the intrinsic divergence of the optical wave to make the FEL interaction efficient. This condition requires geometrical transverse emittance smaller than the diffraction of the optical beam: $\varepsilon_n/\gamma < \lambda/4\pi$. The energy spread must be smaller than the FEL gain bandwidth: $\Delta E/E < \rho$. The electron bunch parameters required for a typical FEL in the hard X-ray region $\lambda \sim 1$ Å are: normalized emittance $\varepsilon_n \leq 1$ mm·mrad, peak current $I > 1$ kA and energy spread $\Delta E/E < 0.05\%$.

Realizing such extremely high brightness electron bunches has been a great challenge for accelerator physicists since the first proposal of angstrom FELs in 1992. Extensive research efforts have, however, resolved most of the problems. For example, emittance compensation by solenoid magnetic fields and pulse shaping of drive lasers have improved the performance of photocathode RF guns drastically, and electron bunches of normalized rms emittance 1.2 mm·mrad at 1 nC charge are now available [Yang *et al.* (2002)]. Another critical task, the suppression of the emittance growth due to coherent synchrotron radiation (CSR) in a magnetic bunch compressor, is also in progress (see Sec. 2.2.2).

Table 2.9   X-ray FELs under development.

	energy	electron bunch charge	length (rms)	X-ray wavelength
LCLS[1]	14 GeV	1 nC	77 fs	1.5 Å
TTF-FEL (phase-I)	240 MeV	2.8 nC	100 fs	80 nm
TTF-FEL (phase-II)[2]	1 GeV	1 nC	160 fs	6.4 nm
TESLA-FEL[3]	25 GeV	1 nC	83 fs	1 Å
SCSS[4]	1 GeV	1 nC	250 fs	3.6 nm

[1][LCLS (2002)]
[2][Rossbach (1996)]
[3][Wiik (1997)]
[4][Matsumoto *et al.* (2002)]

Table 2.9 shows X-ray FEL projects under development around the world. In the United States, the Linac Coherent Light Source (LCLS) is an X-ray FEL using the Stanford Linear Accelerator Center (SLAC) linac and is ready for construction [LCLS (2002)]. In Deutches Elektronen-Synchrotron (DESY), Germany, an X-ray FEL driven by a superconducting linac is planned as a parasite of a linear collider TESLA [Wiik (1997)]. A 240-MeV, 80-nm FEL at the Tesla Test Facility (TTF) has already been constructed as an R&D program for TESLA, and will soon be upgraded to a 1-GeV, 6-nm FEL [Rossbach (1996)]. A Japanese program, SPring-8 Compact SASE Source (SCSS), is a soft X-ray SASE-FEL facility based on a high gradient C-band (5.712 GHz) linac and a short-period in-vacuum undulator [Matsumoto *et al.* (2002)].

Figure 2.143 shows the accelerator system proposed for LCLS [Arthur *et al.* (1998)]. An electron bunch from a photocathode RF gun, which has an energy of 7 MeV and longitudinal duration of 2.7 ps, is accelerated to 14 GeV by a 1-km long S-band (2856 MHz) linac, a part of the SLAC two-mile accelerator. Two bunch compressors, BC-1 and BC-2, are installed to shorten the bunch to 77 fs and increase the peak current to 3.4 kA. The bunch compressors are optimized for small emittance in both transverse and longitudinal phase space. An X-band (11.424 GHz) accelerator is added before the first compressor to compensate longitudinal phase distortion arising from RF curvature of the S-band frequency.

X-ray FELs can produce coherent X-ray pulses with 10–100-fs temporal duration, the peak brilliance of which exceeds the existing synchrotron radiation facilities by 10 orders of magnitude. These high peak power X-rays will open the door to completely new scientific applications, such as

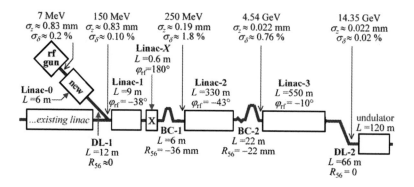

Fig. 2.143 Accelerator system proposed for LCLS. A 1-km long S-band linac is part of the SLAC injector. Two bunch compressors, BC-1 and BC-2, are installed to obtain electron bunches of 77 fs rms at the undulator.

structural studies on single bio-molecules, nanoscale dynamics in condensed matter, femtochemistry and others.

2.8 Energy Recovery Linac

R. HAJIMA
Advanced Photon Research Center,
Japan Atomic Energy Research Institute (JAERI)
Tokai, Naka,
IBARAKI 319-1195, JAPAN

An energy recovery linear accelerator (linac) (ERL), a novel device for producing femtosecond high brightness electron beams, is now expected to be the driver of future light sources in the wide wavelength region from terahertz radiation to hard X-rays.

The principle of the ERL is presented in Fig. 2.144. Electron bunches from an injector are accelerated by a superconducting linac and transported to an undulator. The electron bunch after the emission of the undulator radiation is then re-injected into the linac 180° out of phase, so that the linac acts as a decelerator. We can retrieve, in the decelerator, the energy of the electrons back into the RF power, which can be recycled for accelerating succeeding electron bunches. This energy recycling procedure combined with superconducting accelerator cavities, where heat loss is negligible, enables us to produce relativistic electron beams of high average current with small RF power from klystrons. The ERL also eliminates radioactive and thermal problems at the beam dump, because the electron bunch is decelerated to an energy almost the same as the injection energy, 2–10 MeV typically.

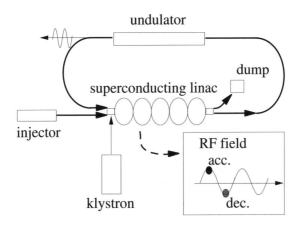

Fig. 2.144   Schematic view of an energy recovery linac (ERL).

ERLs can be adopted in high power free electron lasers (FEL). In an FEL oscillator only a small fraction of the electron energy is converted into laser power. If we recycle the rest of the electron energy using an ERL, the capacity and running cost of the RF klystrons can be much reduced and the total efficiency of the FEL is improved. Two ERLs have been developed for high power FELs: the IR-demo at the Thomas Jefferson National Accelerator Facility (TJNAF), USA, and the Japan Atomic Energy Research Institute ERL (JAERI-ERL) in Japan. The IR-demo is a 48-MeV ERL with 5-mA average current and generates a 2-kW FEL at 1–7-μm wavelength, and will soon be upgraded to a 210-MeV/10-mA ERL for a 10-kW FEL [Neil *et al.* (2000); Benson *et al.* (2001)]. The JAERI-ERL is a 17-MeV/5-mA machine and an upgrade program for increasing the beam current to 40 mA is in progress [Hajima *et al.* (2003)].

The research development of ERLs has been greatly improved since the first successful demonstration of energy recovery operation in the IR-demo in 1999. Worldwide, large scale accelerator projects utilizing superconducting linacs, such as the Continuous Electron Beam Accelerator Facility

(CEBAF), USA, the Tera Electron-Volt Energy Superconducting Linear Accelerator (TESLA), Germany, and the Spallation Neutron Source (SNS), USA, have also contributed to the rapid progress of superconducting linac technologies. The ERL is now considered to be a promising device not only for high power FELs but also for next generation light sources in the VUV and X-ray wavelength regions.

For the radiation wavelengths between VUV and hard X-rays (100 nm–1 Å), third-generation synchrotron light sources based on high energy electron storage rings are now the frontier of radiation brilliance, which provide X-ray pulses having peak brilliance $\sim 10^{23}$ photons/mm^2/mrad^2/s/0.1%bandwidth at 1 Å and a time duration of 10–100 ps. In the electron storage rings, the spatial and temporal distribution of electron bunches are determined from the radiation equilibrium state of the electron dynamics through a number of circulations. The geometrical (not normalized) emittance in the transverse phase plane and the rms bunch duration at the European Synchrotron Radiation Facility (ESRF), a third-generation 6-GeV storage ring, have values $\varepsilon_x = 4$ nm, $\varepsilon_y = 0.025$ nm and $\sigma_t = 20$ ps. The asymmetric property of emittance in the horizontal and vertical planes is due to an emittance dilution mechanism in the storage ring, i.e. quantum excitation of an electron by synchrotron radiation in horizontally bending paths. The electron bunch length is limited by the potential well of synchrotron oscillation in the storage ring, and an electron bunch converges to an equilibrium distribution with picosecond duration. Recently, a bunch slicing method using an external femtosecond laser was developed at the Advanced Light Source (ALS) at LBL, USA, to generate a 100-fs X-ray pulse from a picosecond electron bunch in a storage ring (see Sec. 2.6). The brilliance of sliced X-rays is, however, limited by the electron charge ratio between the sliced part and the whole bunch.

An electron bunch in an ERL is, on the contrary, free from radiation equilibrium and able to have a more flexible profile in both the transverse and longitudinal phase space of motion. Dilution of the electron distribution in phase space, i.e. the growth of entropy, due to quantum excitation and the space charge repulsion force is quickly drawn off from the system, because every electron bunch goes to the beam dump after a single recirculation for energy recovery. As a result, the emittance and time duration of the electron bunch in ERLs can be designed flexibly by the injector configuration. When photo-cathode guns are adopted to the injector, a normalized emittance of $\varepsilon_n < 1$ mm·mrad is reasonably attainable for a 50–100-pC bunch charge. The geometrical emittance decreases by linear acceleration according to the scaling $1/\gamma$ and reaches $\varepsilon_n/\gamma < 0.1$ nm at 6 GeV. Moreover, the bunch duration can be shortened to 100 fs by magnetic bunch compression before or after the main linac. In Fig. 2.145, we compare a beam-damping mechanism in a storage ring and the ERL described in the preceding paragraphs.

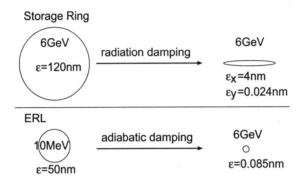

Fig. 2.145 Damping of electron beam emittance in a storage ring and an ERL.

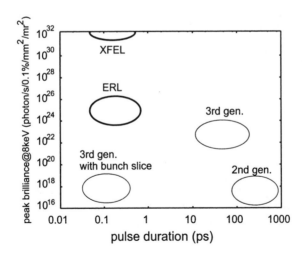

Fig. 2.146 Time duration and peak brilliance of existing and future light sources.

Figure 2.146 shows the time duration and peak brilliance of existing and future light sources. ERLs and X-ray free electron lasers (XFEL) are called *next generation light sources*, which provide brighter and shorter X-ray pulses than existing light sources (see Sec. 2.7 for XFELs). The properties of ERLs, such as small emittance in both horizontal and vertical planes and femtosecond bunch duration, will be exploited for novel applications of synchrotron radiation research, such as spectroscopy with spatial resolution of nanometer scale, X-ray tomography utilizing better transverse coherence and time-resolved pump-and-probe studies in the femtosecond regime, all of which are quite difficult with existing light sources based on storage rings. Several proposals of X-ray ERL light sources are now being discussed around the world (see Table 2.10).

Table 2.10   X-ray ERL light sources proposed around the world. (Some of the parameters are tentative.)

	energy (GeV)	bunch charge (pC)	av. current (mA)	bunch length (ps)
Cornell ERL[1] (Cornell Univ., USA)	5	7.7–77	10–100	0.1–2
PERL[2] (BNL, USA)	3–7	75–450	100–200	0.1–0.4
KEK-ERL[3] (KEK, Japan)	2.5–5	77	100	0.1–1
4GLS[4] (Daresbury, UK)	0.6	50–1000	100	0.1–1
MARS[5] (BINP, Russia)	5.3	50	1	0.04

[1][Gruner *et al.* (2002)]
[2][Ben-Zvi *et al.* (2001)]
[3][Kamiya (2003)]
[4][Poole *et al.* (2002)]
[5][Kayran *et al.* (1998)]

A femtosecond electron bunch of an ERL is a powerful source of terahertz (THz) synchrotron radiation, because the electrons emit radiation coherently for a wavelength longer than the bunch length. The power of this coherent radiation is proportional to the square of the number of electrons in the bunch $P_{coh} \propto N_e^2$, while usual incoherent synchrotron radiation obeys the scaling $P_{incoh} \propto N_e$. Strong coherent radiation in the terahertz region has been demonstrated in the IR-demo, where electron bunches of 500 fs FWHM, 100 pC and 37.4 MHz are recirculated. The radiation power from the IR-demo exceeds conventional THz sources (storage rings and black body thermal sources) by 6 orders of magnitude [Carr *et al.* (2002)].

We have seen that the ERL is a universal technology applicable to future radiation devices, such as high power FELs, X-ray light sources and terahertz sources. To realize such ERL radiation sources, a couple of R&D issues of accelerator technologies must be solved: (1) superconducting cavity development for high average current and high accelerator gradient, (2) instability phenomena emerging in ERLs, (3) electron sources for low emittance and high average current. From the stability analyses and benchmark experiments at CEBAF and the IR-demo, it has been found that higher order modes (HOM) excited in superconducting cavities are a major source of instability, but stable ERL operation is possible for an average current up to 200–300 mA with TESLA superconducting cavities (1.3 GHz, 9-cell) driven at an accelerating gradient of 15–25 MV/m. For electron sources, two options have been suggested: an RF gun with a K_2CsSb photo-cathode and a DC gun with a GaAs photo-cathode. In the RF photo-cathode gun, thermal design fitting for heavy heat loads from continuous wave operation is critical. Suppression of emittance growth from a gun (~ 500 kV) through a pre-accelerator (~ 10 MeV) is a key issue in developing a DC photo-cathode gun.

Bibliography

Abo-Bakr, M., Feikes, J., Holldack, K., Wüstefeld, G. and Hübers, H.-W. (2002). Steady-state far-infrared coherent synchrotron radiation detected at BESSY II, *Phys. Rev. Lett.* **88**, 254801.

Akhiezer, A. I. and Polovin, R. V. (1956). *Sov Phys. JETP* **30**, p. 915.

Anderson, J. D., Jr (1989) *Modern Compressible Flow with Historical Perspective 2nd Edition*, Chap. 5, McGraw Hill.

Anderson, S. C. and Rosenzweig, J. B. (2000). Nonequilibrium transverse motion and emittance growth in ultrarelativistic space-charge dominated beams, *PRSTAB* **3**, 094201.

Ando, A., Amano, S., Hashimoto, S., Kinosita, H., Miyamoto, S., Mochizuki, T., Niibe, M., Shoji, Y., Terasawa, M., Watanabe, T. and Kumagai, N.(1998). Isochronous storage ring of the New SUBARU project, *J. Synchrotr. Radiat.* **5**, pp. 342–344.

Ando, A. and Takayama, K. (1983). *IEEE Tran. Nucl. Sci.* **NS-30**, pp. 2604–2606.

Andreev, A. A., Zhidkov, A. G., Sasaki, A. *et al.* (2002). Effect of plasma inhomogeneity on ion acceleration when an ultra-intense laser pulse interacts with a foil target, *Plasma Phys. Control. Fusion* **44**, p. 1243.

Arthur, J. *et al.* (1998). Linac coherent light source (LCLS) design study report, *SLAC-R-521*, Apr. 1998.

Auston, D. H., Cheung, K. P. and Smith, P. R. (1984). Picosecond photoconducting Hertzian dipoles, *Appl. Phys. Lett.* **45**, pp. 284–286.

Balcou, Ph., Grillon, G., Moustaizis, S. *et al.* Neutron generation by laser irradiation of CD_4 clusters, *AIP Conf. Proc.* **611**, Second Int. Conf. on Superstrong Field in Plasmas, Lontano, M., Mourou, G., Svelto, O. and Tajima, T. eds., p. 244.

Bane, K., Krinsky, S. and Murphy, J. B. (1994). *Nucl. Instrum. and Res. in Phys. A* **346**, p. 571

Bane, K. L. F., Mosnier, A., Novokhatsky, A. and Yokoya, K. (1998). Calculations of the short range longitudinal wakefields in the NLC linac, *Proc. of the 1998 International Computational Accelerator Physics Conf.*

Banerjee, S., Valenzuela, A. R., Shah, R. C., Maksimchuk, A. and Umstadter, D.

(2002). High harmonic generation in relativistic laser-plasma interaction, *Phys. Plasmas* **9**, pp. 2393–2398.

Bauer, T., Howorth, J., Korte, F., Momma, C., Rizvi, N., Saviot, F. and Salin, F. (2002). Development of an industrial femtosecond laser micromachining system, presented at *SPIE-LASE*.

Beg, F. H., Bell, A. R., Dangor, A. E. *et al.* (1997). A study of picosecond laser solid interactions up to 10^{19} W/cm^2, *Phys. Plasmas* **4**, p. 447.

Benson, S. V. *et al.* (2001). A 10 kW IRFEL design for Jefferson Lab, *Proc. Particle Acc. Conf.* p. 249.

Ben-Zvi, I. *et al.* (2001). Photoinjected energy recovering linac upgrade for the National Synchrotron Light Source, *Proc. Particle Acc. Conf.* p. 350; Murphy J. B. *et al.* (2001). Beam dynamics for a photoinjected energy recovery linac at the NSLS, *Proc. Particle Acc. Conf.* p. 465.

Bertrand, P., Ghizzo, A., Karttunen, S. J., Pattikangas, T. J. H., Salomaa, R. R. E. and Shoucri, M. (1995). Two-stage electron acceleration by simultaneous stimulated Raman backward and forward scattering, *Phys. Plasmas* **2**, 8, pp. 3115–3129.

Blum, E.B., Dienes, M. and Murphy, J.B. eds. (1996). *Proc. of A Workshop on the Production, Measurement and Applications of Short Bunches of Electrons and Positrons in Linacs and Storage Rings*, September 28–30, 1995, Brookhaven National Laboratory, NY, USA, AIP CP **367**.

BNL web site (2003). http://www.bnl.gov/atf/publications/slice.htm

Bohn, C. L. (1997). Recirculating accelerator driver for a high-power free-electron laser: A design overview, *Proc. of the 1997 IEEE Particle Accelerator Conf.*, IEEE Cat. No. 97CH36167, p. 909.

Bolton, P. R. *et al.* (2002). Photoinjector design for the LCLs, *Nucl. Instr. & Meth. A* **483**, pp. 296–300.

Boscolo, M. *et al.* (2002). Beam dynamics study of a RF bunch compressor for high britghness injectors, *Proc. of EPAC 2002 Conf.*

Boussard, D. (1975). *CERN LABII/RF/INT/75-2*, review: LaClare, J. L. (1980). *Proc. of 11th Int. Conf. on High Energy Accelerators*, Geneva, p. 526.

Boyce, J. R. *et al.* (2002). CAARI 2002 17th *Int. Conf. on the Application of Accelerators to Research and Industry*, Denton TX, USA, AIP Conf. Proc. **680**, pp.325–328.

Brau, C. A. (1990). Free-Electron Lasers, Academic, San Diego.

Brau, C. A. (1992). The Vanderbilt University Free-Electron Laser Center, *Nucl. Instrum. Methods Phys. Res. A* **318**, p. 38.

Bruck, H. (1968). Circular particle accelerators, LA-TR-72-10 Rev. (Translated into English at Los Alamos Scientific Laboratory).

Bulanov, S. V., Esirkepov, T. Zh., Califano, F. *et al.* (2000). Generation of collimated beams of relativistic ions in laser-plasma interactions, *JETP Letters* **71**, p. 407.

Bulanov, S. V. *et al.* (1992). Nonlinear depletion of ultrashort and relativistically strong laser pulses in an underdense plasma, *Phys. Fluids B* **4**, p. 1935.

Bulanov, S. V. *et al.* (2001) *Reviews of Plasma Physics*, **22**, Shafranov, V. D. ed.,

Kluwer Academic/Plenum Publishers, New York, p. 227.

Bulanov, S. V. *et al.* (1996). Controlled wake field acceleration via laser pulse shaping, *IEEE Trans. Plasma Science* **24**, p. 393.

Bulanov, S. V., Inovenkov, I. N., Naumova, N. M. and Sakharov, A. S. (1990). Excitation of a relativistic Langmuir wave and electron acceleration through the action of an electromagnetic pulse on a collisionless plasma, *Sov. J. Plasma Phys.* **16**, p. 444.

Bulanov, S. V., Kirsanov, V. I. and Sakharov, A. S. (1989). Excitation of ultrarelativistic plasma waves by pulse of electromagnetic radiation, *JETP Lett.* **50**, p. 176.

Bulanov, S. V., Kirsanov, V. I. and Sakharov, A. S. (1991). Limiting electric-field of the wakefield plasma-wave, *JETP Lett.* **53**, p. 565.

Bulanov, S. V., Naumova, N., Pegoraro, F. and Sakai, J. (1998). Particle injection into the wave acceleration phase due to nonlinear wake wave breaking, *Phys. Rev. E* **58**, p. R5257.

Bulanov, S. V., Pegoraro, F., Pukhov, A. M. and Sakharov, A. S. (1997a). Transverse-wake wave breaking, *Phys. Rev. Lett.* **78**, p. 4205.

Bulanov, S. V. and Sakharov, A. S. (1991). Induced focusing of electromagnetic wave in a wake plasma wave, *JETP Lett.* **54**, p. 203.

Bulanov, S. V., Vshivkov, V. A., Dudnikova, G. I., Esirkepov, T. Z., Califano, F., Kamenets, F. F., Liseikina, T. V., Naumova, N. M. and Pegoraro, F. (1999a). *Plasma Phys. Rep.* **25**, p. 701.

Bulanov, S. V., Vshivkov, V. A., Dudnikova, G. I. Naumova, N. M., Pegoraro, F. and Pogorelsky, I. V. (1997b). Laser acceleration of charged particles in inhomogeneous plasmas. I, *Plasma Phys. Rep.* **23**, p. 660.

Bulanov, S. V., Califano, F., Dudnikova, G. I., Vshivkov, V. A., Liseikina, T. V., Naumova, N. M., Pegoraro, F., Sakai, J.-I. and Sakharov, A. S. (1999b). Laser acceleration of charged particles in inhomogeneous plasmas II: Particle injection into the acceleration phase due to nonlinear wake wave-breaking, *Plasma Phys. Rep.* **25**, p. 468.

Burke, D. L. *et al.* (1997). Positron production in multiphoton light-by-light scattering, *Phys. Rev. Lett.* **79** , pp. 1626–1629.

Campbell, P. M., Johnson, R. R. and Mayer, F. J. *et al.* (1977). Fast-ion generation by ion-acoustic turbulence in spherical laser plasmas, *Phys. Rev. Lett.* **39**, p. 274.

Carlsten, B. E. (1989). New photoelectric injector design for the Los Alamos National Laboratory XUV FEL accelerator, *Nucl. Instr. & Meth. A* **285**, pp. 313–319.

Carr, G. L. *et al.* (2002). High-power terahertz radiation from relativistic electrons, *Nature* **420** p. 153.

Carr, G. L., Kramer, S. L., Murphy, J. B., Lobo, R. P. S. M. and Tanner, D. B. (2001). Observation of coherent synchrotron radiation from the NSLS VUV ring, *Nucl. Instr. and Methods A* **463**, pp. 387–392.

Castleman, A. W. and Keesee, R. G. (1988). Gas-phase clusters: Spanning the state of matter, *Science* **241**, pp. 36–42.

Chambaret *et al.* (1996). Generation of 25 TW 32 fs pulses at 10 Hz, *Opt. Lett.*

21, p. 1921.

Chang, Z., Rundquist, A., Wang, H., Murnane, M. M. and Kapteyn, H. C. (1997). Generation of coherent soft X rays at 2.7 nm using high harmonics, *Phys. Rev. Lett.* **79**, pp. 2967–2970.

Chen, S.-Y., Maksimchuk, A., Esarey, E. and Umstadter, D. (2000). Observation of phase-matched relativistic harmonic generation, *Phys. Rev. Lett.* **84**, pp. 5528–5531.

Chen, S.-Y., Maksimchuk, A. and Umstadter, D. (1998). Experimental observation of relativistic nonlinear Thomson scattering, *Nature* **396**, pp. 653–655.

Cheriaux *et al.* (1996). Aberration-free stretcher design for ultrashort pulse amplification, *Opt. Lett.* **21**, p. 414.

Clark, E. L., Krushelnick, K., Zepf, M. *et al.* (2000). Energetic heavy-ion and proton generation from ultraintense laser-plasma interactions with solids, *Phys. Rev. Lett.* **85**, p. 1654.

Courant E. D. and Snider, H. S. (1958). Theory of the alternating-gradient synchrotron, *Ann. of Phys.* **4**, pp. 1–48.

Cowan, T. E., *et al.* (1999). Photonuclear fission from high energy electrons from ultraintense laser-solid interactions, *Phys. Rev. Lett.* **84**, pp. 903–906.

Cowan, T. E. *et al.* (1999). High energy electrons, nuclear phenomena and heating in petawatt laser-solid experiments, *Laser and Particle Beams* **17**, pp. 773–783.

Crosson, E. R. *et al.* (1995). Sub-picosecond FEL micropulse length and electron bunch measurements, *Nucl. Instrum. Methods A* **358**, p. 216.

Dawson, J. M. (1959). Nonlinear electron oscillations in a cold plasma, *Phys. Rev.* **133**, p. 383.

Decoste, R. and Ripin, B. H. (1978). High-energy ion expansion in laser-plasma interactions, *Phys. Rev. Lett.* **40**, p. 34.

Derbenev, Ya. S., Rossbach, J., Saldin, E. L. and Shiltsev, V. D.(1995). Microbunch radiative tail-head interaction, TESLA-FEL 95-05.

Ditmire, T., Donnelley, T., Falcone, R. W. and Perry, M. D. (1995). Strong X-ray emission from high-temperature plasmas produced by intense irradiation of clusters, *Phys. Rev. Lett.* **75**, p. 3122.

Ditmire, T., Tisch, J. W. G., Springate, E., Mason, M. B., Hay, N., Smith, R. A., Marangos and Hutchinson, M. H. R. (1996). High-energy ions produced in explosions of superheated atomic clusters, *Nature* **386**, p. 54.

Ditmire, T., Zweibeck, J., Yanovsky, V. D., Cowan, T. E., Hays and Wharton K. B. (1999). Nuclear fusion from explosions of femtosecond laser-heated deuterium clusters, *Nature* **398**, p. 489.

Dodd, E., Kim J. K. and Umstadter, D. (1997). Ultrashort-pulse relativistic electron gun/accelerator, *Advanced Accelerator Concepts*, Amer. Inst. of Conf. Proc. No. **398**, Chattopadhyay, S. ed., AIP Press, New York.

Dodd, E., Kim, J. K. and Umstadter, D. (1999). Electron injection by dephasing electrons with laser fields, *Advanced Accelerator Concepts: Eighth Workshop*, Lawson, W., Bellamy, C. and Brosius, D. eds., AIP Conference Proceedings, AIP Press, New York, **472**, p. 886.

Douglas, D., Benson, S. V. *et al.* (2002), A 10 kW IRFEL design for Jefferson

Lab, *Proc.: 2001 Particle Accelerator Conf.*, Lucas, P. W. and Bebber, S. eds., IEEE, Piscataway, NJ.

Dowell, D.H., Haynard, T.D. and Vetter, A.M. (1995). Magnetic pulse compression using a third harmonic RF linearizer , *Proc. of PAC'95.* pp. 992-994.

Durfee, C. G., Lynch, J. and Milchberg, H. M. (1995). Development of a plasma wave-guide for high-intensity laser-pulses, *Phys. Rev. E* **51**, p. 2368.

Ehler, A. W. (1975). High-energy ions from a CO_2 laser-produced plasma, *J. Appl. Phys.* **46**, p. 2464.

Elsaesser, T., Fujimoto, J. G., Wiersma, D. A. and Zinth, W., eds. (1998). *Ultrafast Phenomena XI* **63**, Springer-Verlag, Berlin.

Emma, P. (1998). Bunch Compressor Optics for the New TESLA Parameters, LCLS-TN-01-1.DAPNIA/SEA, SEA-98-54.

Emma, P. (2001). X-Band RF Harmonic Compensation for Linear Bunch Compression in the LCLS, LCLS-TN-01-1.

Emma, P (2002). Presented at *Joint ICFA Advanced Accelerator and Beam Dynamics Workshop: The Physics & Applications of High Brightness Electron Beams*, Sardinia.

Esarey, E., Hafizi, B., Hubbard, R. and Ting, A. (1998). Trapping and acceleration in self-modulated laser wakefields, *Phys. Rev. Lett.* **80**, 25, pp. 5552–5555.

Esarey, E., Hubbard, R. F., Leemans, W. P., Ting, A. and Sprangle, P. (1997). Electron injection into plasma wake fields by colliding laser pulses, *Phys. Rev. Lett.* **79**, 14, pp. 2682–2685.

Esarey, E. and Pilloff, M. (1995). Trapping and acceleration in nonlinear plasma waves, *Phys. Plasmas* **2**, p. 1432.

Esarey, E., Ride, S. K. and Sprangle, P. (1993). Nonlinear Thomson scattering of intense laser-pulses from beams and plasmas, *Phys. Rev. E* **48**, p. 3003.

Esarey, E., Sprangle, P., Krall, J. and Ting, A. (1996). Overview of plasma-based accelerator concepts, *IEEE Trans. Plasma Sci.* **24**, 2, pp. 252–288.

Esarey, E., Schroeder, C. B., Leemans, W. P. and Hafizi, B. (1999). Laser-induced electron trapping in plasma-based accelerators, *Phys. Plasmas* **6**, 5, pp. 2262–2268.

Esirkepov, T. Zh. (2001) Exact charge conservation scheme for particle-in-cell simulation with an arbitrary form-factor, *Comput. Phys. Comm.* **135**, p. 144.

Esirkepov, T. Zh., Sentoku, Y., Mima K. *et al.* (2000). Ion acceleration by superintense laser pulses in plasmas, *JETP Lett.* **70**, p. 82.

Falcoz *et al.* (1995). Self-starting self mode-locked femtosecond diode-pumped Cr:LISAF laser, *Opt. Lett.* **20**, p. 1874.

Faure, J., Malka, V., Marques, J.-R., Amiranoff, F., Courtois, C., Najmudin, Z., Krushelnick, K., Salvati, M., Dangor, A. E., Solodov, A., Mora, P., Adam, J.-C. and Heron, A. (2000). Interaction of an ultra-intense laser pulse with a nonuniform preformed plasma, *Phys. Plasmas* **7**, p. 3009.

Ferrario, M. (2002). Recent advances and novel ideas on for high brightness electron beam production based on RF photo-injectors, *Proc. of the ICFA Workshop on The Physics and Applications of High Brightness Electron Beams*, Chia Laguna (Italy), June 2002.

Ferrario, M. *et al.* (1996). Multi-bunch energy spread induced by beam-loading in a standing wave structure, *Particle Accelerators* **52**, 1.

Ferré *et al.* (2000). Multijoule ultra short femtosecond pulse laser at 10 Hz repetition rate, *CLEO Europe*.

Fews, A. P., Norreys, P. A., Beg, F.N. *et al.* (1994). Plasma ion emission from high intensity picosecond laser pulse interactions with solid targets, *Phys. Rev. Lett.* **73**, p. 1801.

Fraser, J. S., Sheffield, R. L. and Gray, E. R. (1986). A new high-brightness electron injector for free electron lasers driven by RF linacs, *Nucl. Instr. Meth. A* **250**, pp. 71–76.

Fraser, J. S., Sheffield, R. L., Gray, E. R. and Rodenz, G. W. (1985). High-brightness photoemitter injector for electron accelerators, *IEEE Trans. Nucl. Sci.* **32**, 5, pp. 1791–1793.

Fubiani, G., Esarey, E., Schroeder, C. B. and Leemans, W. P. (2003). Laser triggered trapping of electrons in plasmas waves by the collision of a drive pulse with a single injection pulse (in preparation).

Fukuda, M., Hirose, T., Kurihara, Y., Ohashi, A., Okugi, T., Omori, T., Sakai, I., Urakawa, J., Washio, M. and Yamazaki, I. (2003a). *Proc. of Int. Workshop on Quantum Aspects of Beam Physics.* Hiroshima Univ., Japan, Jan. 7–11.

Fukuda, M., Aoki, T., Dobashi, K., Hirose, T., Kurihara, Y., Okugi, T., Omori, T., Sakai, I., Urakawa, J. and Washio, M. (2003b). Polarimetry of short-pulse gamma rays produced through inverse compton scattering of circularly polarized laser beams, *Phys. Rev. Lett.* **91**, 164801.

Gahn, C. *et al.* (2001). Generation of MeV electrons and positrons with femtosecond pulses from a table-top laser system, *Phys. Plasmas* **9**, pp. 987–999.

Gallardo, J. C. and Palmer, R. B. (1990a). *Proc. Workshop Prospects for a 1Å FEL*, BNL 52273 1990, Upton NY, p. 136.

Gallardo, J. C. and Palmer, R. B. (1990b). Preliminary study of gun emittance correction, *IEEE J. of Quantum Electronics* **26**, 8, p. 1328.

Gao, J. (1991). Nonlinear repairing in phase space emittance recovering techniques, *Nucl. Instr. & Meth. A* **304**, pp. 353–356.

Gitomer, S. J., Jones, R. D., Begay, F. *et al.* (1986). Fast ions and hot electrons in the laser plasma interaction, *Phys. Fluids* **29**, p. 2679.

Glotin, F. *et al.* (1993). Infrared subpicosecond laser pulses with a free-electron laser, *Phys. Rev. Lett.* **71**, 2587.

Gordon, G. Tzeng, K. C., Clayton, C. E., Dangor, A. E., Malka, V., Marsh, K. A., Modena, A., Mori, W. B., Muggli, P., Najmudin, Z., Neely, D., Danson, C. and Joshi, C. (1998). Observation of electron energies beyond the linear dephasing limit from a laser-excited relativistic plasma wave, *Phys. Rev. Lett.* **80**, p. 2133.

Gotts, N. G., Lethbridge, P. G. and Stace, A. J. (1992). Observation of Coulomb explosion in doubly charged atomic and molecular clusters, *J. Chem. Phys.* **96**, p. 408.

Gruner, Sol M. *et al.* (2002). Energy recovery linacs as synchrotron radiation sources, *Rev. Sci. Instrum.* **73**, p. 1402; Gruner, Sol M. *et al.* (2001). Study for a proposed phase I energy recovery Linac (ERL) synchrotron light source at Cornell University, *CHESS Tech. Memo* 01-003 / JLAB-ACT-01-04.

Gryaznykh, D. A., Kandiev, Ya. Z. and Lykov, V. A. (1998). Estimates of electron–positron pair production in the interaction of high-power laser radiation with high-Z targets, *JETP Lett.* **67**, pp.257–262.

Haissinski, J. (1973). *Nuove Cimento* **18B** (1) p. 72.

Hajima, R. *et al.* (2003). First demonstration of energy-recovery operation in the JAERI superconducting linac for a high-power free-electron laser, *Nucl. Instrum. Methods A* **507**, p. 115.

Hajima, R. and Nagai, R. (2003). Generation of a self-chirped few-cycle optical pulse in a FEL oscillator, *Phys. Rev. Lett.* **91**, 024801.

Hatchett, S. P., Brown, C.G., Cowan, T.E. *et al.* (2000). Electron, photon, and ion beams from the relativistic interaction of petawatt laser pulses with solid targets, *Phys. Plasmas* **7**, p. 2076.

He, F., Lau, Y., Umstadter, D. and Kowalczyk, R. (2003). Backscattering of an intense laser beam by an electron, *Phys. Rev. Lett.* **90**, 055002.

Hemker, R. G., Hafz, N. M. and Uesaka, M. (2002). Computer simulations of a single-laser double-gas-jet wakefield accelerator concept, *Phys. Rev. ST* **5**, 041301.

Hemker, R. G., Tzeng, K. C., Mori, W. B., Clayton, C. E. and Katsouleas, T. (1998). Computer simulations of cathodeless, high-brightness electron-beam production by multiple laser beams in plasma, *Phys. Rev. E* **57**, 5, pp. 5920–5928.

Herring, C. and Nichols, M. H. (1949). Thermionic emission, *Rev. Mod. Phy.* **21**, pp. 185–270.

Hironaka, Y., Tange, T., Inoue, T., Fujimoto, Y., Nakamura, K. G., Kondo, K. and Yoshida, M. (1999). Picosecond pulsed X-ray diffraction from a pulsed laser heated Si(111), *Jpn. J. Appl. Phys.* **38** 8, pp. 4950–4951.

Hirose, T., Dobashi, K., Kurihara, Y., Muto T., Omori, T., Okugi, T., Sakai, I., Urakawa J. and Washio, M. (2000). Polarized positron source for the linear collider, JLC, *Nucl. Instrum. Methods Phys. Res. A.* **455**, p. 15.

Hosokai, T., Kinoshita, K., Zhidkov, A., Nakamura, K., Watanabe, T., Ueda, T., Kotaki, H., Kando, M., Nakajima, K. and Uesaka, M. (2003). Effect of a laser prepulse on a narrow-cone ejection of MeV electrons from a gas jet irradiated by an ultrashort laser pulse, *Phys. Rev. E* **67**, 036407.

JLab Web Site (2003). http://www.jlab.org/FEL/.

Johnson, L. C. and Chu, T. K. (1974). Measurements of electron-density evolution and beam self-focusing in a laser-produced plasma, *Phys. Rev. Lett.* **32**, p. 517.

Kamiya, Y. (2003). A future plan of KEK-PF, *Proc. Annual Meeting of Japanese Society for Synchrotron Radiation Research.*

Kamiya, Y., Kumita, T., Hirose, T., Kashiwagi, S., Washio, M., Omori, T., Urakawa, J., Yokoya K. and Siddens, D.P. (2003). *Proc. of Joint 28th Advanced Beam Dynamics on Quantum Aspect of Beam Physics*, Hiroshima University, Japan, Jan. 7–11.

Kawabata, A. and Kubo, R. (1966). Electronic properties of fine metallic particle. II. Plasma response absorption, *J. Phys. Soc. Jpn.* **21**, p. 1765.

228   *Femtosecond Beam Science*

Kayran, D. A. *et al.* (1998). MARS—a project of the diffraction limited fourth
generation X-ray source, *Proc. 1st Asian Particle Acc. Conf.* p. 704.
Keil, E. and Schnell, W. (1969). Concerning longitudinal stability in the ISR,
CERN Report, ISR-TH-RF/69-48.
Khachatryan, A. G. (1998). Ion motion and finite temperature effect on relativistic
strong plasma waves, *Phys. Rev. A* **58**, p. 7799.
Kim, J. K., Dodd, E. and Umstadter, D. (1997). All-optical femtosecond electron
acceleration, *Applications of High Field and Short Wavelength Sources VII*,
1997 OSA Technical Digest Series, **7**, Optical Society of America, Wash-
ington DC, p. 121.
Kim, K.-J. (1989). Rf and space-charge effects in laser-driven rf electron guns,
Nucl. Instr. & Meth. A **275**, p. 201.
See for example: Kim, K. J., Bisognano, J. J., Garren, A. A., Halbach, K. and
Peterson, J.M. (1985). *Nucl. Instrum. and Methods in Phys. Res. A*, **239**,
pp. 54–61
Kim, K. J., Chattopadhyay S. and Shank, C. V. (1994). Generation of femtosecond X-rays by 90-degrees Thomson scattering, *Nucl. Instrum. Methods A*
341, p. 351.
Kimura, W. D. (2002). First demonstration of high-trapping efficiency and nar-
row energy spread in a laser-driven accelerator, *AIP Conf. Proc. No. 647*,
Advanced Accelerator Concepts, Jun. 23–28, Mandalay Beach, CA, Clayton, C. E. and Muggli, P. eds., American Institute of Physics, New York,
pp. 269–277.
Kimura, W. D., Campbell, L. P, Dilley, C. E., Gottschalk, S. C., Quimby, D. C.,
van Steenbergen, A., Babzien, M., Ben-Zvi, I., Gallardo, J. C.,
Kusche, K. P., Pogorelsky, I. V., Skaritka, J., Yakimenko, V., Cline, D. B.,
He, P., Liu, Y., Steinhaue, L. C. and Pantell, R. H. (2001). Detailed exper-
imental results for laser acceleration staging, *Phys. Rev. ST Accel. Beams*
4, 101301.
Kimura, W. D. *et al.* (2001a). First staging of two laser accelerators, *Phys. Rev.
Lett.* **86**, p. 4041.
Kinoshita, K., Ohkubo, T., Yoshii, K., Harano, H. and Uesaka, M. (2003). Time-
resolved X-ray diffraction for visualization of atomic motion, *Int. J. of Appl.
Electromagnetics and Mechanics* **14**, pp. 233–236.
Kishimoto, Y., Masaki, T. and Tajima, T. (2002). High energy ions and nuclear
fusion in laser-cluster interaction, *Phys. Plasmas* **9**, p. 589.
Kishimoto, Y., Masaki, T. and Tajima, T. (a) Laser-cluster interaction for nuclear
fusion, *AIP Conf. Proc.* **611**, Second Int. Conf. on Superstrong Field in
Plasmas, Lontano, M., Mourou, G., Svelto, O. and T. Tajima eds., p. 264.
Kishimoto, Y., Masaki, T. and Tajima, T. (b) Cluster dynamics in strong fields
and its application to fusion science, *AIP Conf. Proc.* **634**, Science of Su-
perstrong Field Interactions, Nakajima. K. and Deguchi, M. eds., p. 147.
Kmetec, J. D., Gordon III, C. L., Macklin, J. J., Lemoff, B. E., Brown, G. S. and
Harris, S. E. (1992). MeV X-ray generation with a femtosecond laser, *Phys.
Rev. Lett.* **68**, p. 1527.
Knippels, G. M. H. *et al.* (1995) Intense far-infrared free-electron laser pulses with

a length of six optical cycles, *Phys. Rev. Lett.* **75**, p. 1755.

Kobayashi, T., Uesaka, M., Katsumura, Y., Muroya, Y., Watanabe, T., Ueda, T., Yoshii, K., Nakajima, K., Zhu, X. and Kando, M. (2002). High-charge S-band photocathode RF-gun and linac system for radiation reseach, *J. Nucl. Sci. and Tech.* **39**, pp. 6–14.

Kramer, S. L. and Podobedov, B. (2002). *Proc. 2002 EPAC*, pp. 1523–1525.

Krafft, G. A. (1997). Use of Jefferson Lab's high average power FEL as a Thomson backscatter X-ray source, *Proc. of the 1997 IEEE Particle Accelerator Conf.*, IEEE Cat. No. 97CH36167, p. 739.

Kroll, N. M., Morton, P. L. and Rosenbluth, M. N.,(1981). Free-electron lasers with variable parameter wigglers, *IEEE J. Quant. Electron.* **QE-17**, p. 1436.

Kruer, W. L and Wilks, S. C. (1992). Kinetic simulations of ultra-intense laser plasma interactions, *Plasma Phys. Controlled Fusion* **34**, p. 2061.

Kubo, K., Akemoto, M., Anderson, S., Aoki, T., Araki, S., Bane, K. L. F., Blum, P., Corlett, J., Dobashi, K., Emma, P., Frisch, J., Fukuda, M., Guo, Z., Hasegawa, K., Hayano, H., Higo, T., Higurashi, A., Honda, Y., Iimura, T., Imai, T., Jobe, K., Kamada, S., Karataev, P., Kashiwagi, S., Kim, E., Kobuki, T., Kotseroglou, T., Kurihara, Y., Kuriki, M., Kuroda, R., Kuroda, S., Lee, T., Luo, X., McCormick, D. J., McKee, B., Mimashi, T., Minty, M., Muto, T., Naito, T., Naumenko, G., Nelson, J., Nguyen, M. N., Oide, K., Okugi, T., Omori, T., Oshima, T., Pei, G., Potylitsyn, A., Qin, Q., Raubenheimer, T., Ross, M., Sakai, H., Sakai, I., Schmidt, F., Slaton, T., Smith, H., Smith, S., Smith, T., Suzuki, T., Takano, M., Takeda, S., Terunuma, N., Toge, N., Turner, J., Urakawa, J., Vogel, V., Woodley, M., Yocky, J., Young, A. and Zimmermann, F. (2002). Extremely low vertical-emittance beam in the accelerator test facility at KEK, *Phys. Rev. Lett.* **88**, 194801.

Landau, L. D. and Liftshitz, E. M. (1975). The Classical Theory of Fields, *Pergamon Press,* Fourth Revised English edition, p. 163, formula (63.8).

Larsen, J. T. and Lane, S. M. (1994). HYADES—a plasma hydrodynamics code for dense plasma studies, *J. of Quant. Spectrosc. Radiat. Transfer* **51**, pp. 179–186.

Lawson, W. S., Rambo, P. W. and Larson, D. J. (1997). One-dimensional simulations of ultrashort intense laser pulses on solid-density targets, *Phys. Plasmas* **4**, p. 788.

LCLS: Service, R. F. (2002). Battle to become the next-generation X-ray source *Science* **298** p. 1356; Cho, A., The ultimate bright idea *Science* **296** p. 1008; LCLS Conceptual Design Report, SLAC-R-593/UC-414 (2002).

Le Blanc, S. P., Downer, M. C., Wagner, R., Chen, S.-Y., Maksimchuk, A., Mourou, G. and Umstadter, D. (1996). Temporal characterization of a self-modulated laser wakefield, *Phys. Rev. Lett.* **77**, p. 5381.

Lee, S.Y. (1999). Accelerator Physics, World Scientific.

Leemans, W. P. and Esarey, E. (1998). Summary report: Working group 2 on 'plasma based acceleration concepts', *Proc. of Advanced Accelerator Concepts 8th Workshop*, Baltimore, MD, Jul. 5–11, 1998, Lawson, W., Bellamy,

C, and Brosius, D., American Institute of Physics, New York, 1999, p. 174.

Leemans, W. P., Catravas, P., Esarey, E., Geddes, C. G. R., Toth, C., Trines, R., Schroeder, C. B., Shadwick, B. A., van Tilborg, J. and Faure, J. (2002). Electron-yield enhancement in a laser-wakefield accelerator driven by asymmetric laser pulses accelerator driven by asymmetric laser pulses, *Phys. Rev. Lett.* **89**, 174802.

Leemans, W. P., Rodgers, D., Catravas, P. E., Geddes, C. G. R., Fubiani, G., Esarey, E., Shadwick, B. A., Donahue, R. and Smith, A. (2001). Gamma-neutron activation experiments using laser wakefield accelerators, *Phys. Plasmas* **8**, 5, pp. 2510–2516.

L'Huillier, A. and Balcou, Ph. (1993). High-order harmonic generation in rare gases, *Phys. Rev. Lett.* **70**, pp. 774–777.

Li, R. (1999). The impact of coherent synchrotron radiation on the beam transport of short bunches, *Proc. the 1999 Particle Acc. Conf.* p. 118.

Liang, E. P., Wilks, S C. and Tabak, M. (1998). Pair production by ultraintense lasers, *Phys. Rev. Lett.* **81**, pp. 4887–4890.

Limborg, C. (1998). Freund, A. K., Freund, H. P. and Howells, M. R., eds. (1998). Time Structrue of X-ray Sources and its Applications, San Diego, CA *SPIE*.

Litvinenko, V. N. (1996). On a possibility to suppress microwave instability in storage rings using strong longitudinal focusing, *Proc. of ICFA Workshop*, Arcidosso, Italy, AIP CP **395**, p. 275 (1997).

Litvinenko, V. N. (2002). CSR via strong longitudinal focusing, *Proc. of Workshop Coherent Synchrotron Radiation in Storage Rings,* Napa, CA, USA (http://www-als.lbl.gov/LSWorkshop/VLitvinenko.pdf).

Litvinenko, V. N., Burnham, B., Emamian, M., Hower, N., Madey, J. M. J., Morcombe, P., OShea, P. G., Park, S. H., Sachtschale, R., Straub, K. D., Swift, G., Wang, P., Wu, Y., Canon, R. S., Howell, C. R., Roberson, N. R., Schreiber, E. C., Spraker, M., Tornow, W., Weller, H. R., Pinayev, I. V., Gavrilov, N. G., Fedotov, M. G., Kulipanov, G. N., Kurkin, G. Y., Mikhailov, S. F., Popik, V. M., Skrinsky, A. N., Vinokurov, N. A., Norum, B. E., Lumpkin, A. and Yang B. (1997). Gamma-ray production in a storage ring free electron laser, *Phys. Rev. Lett.* **78** pp. 4569–4572.

Litvinenko, V. N. and Shevchenko, O. A. (2003). Stability criteria for femtosecond electron bunches in storage rings, *Phys. Rev. Lett.* (in preparation).

Litvinenko, V. N., Shevchenko, O. A., Mikhailov, S. F. and Wu, Y. K. (2001). Project for generation of femtosecond X-ray beams from the Duke Storage Ring, *Proc. of 2001 Particle Accelerator Conf.*, Chicago, IL, USA.

Litvinenko, V. N., Shevchenko, O. A., Mikhailov, S. F. and Wu, Y. K. (2002). Novel femtosecond hard X-ray source at the Duke storage ring, *Proc. of the 23rd Int. Free Electron Laser Conf.*, Darmstadt, Germany, *Special Edition of Nucl. Instrum. and Res. in Phys. A* **483**, Brunken, M., Genz, H. and Richter, A. eds., p. II-63.

Litvinenko, V. N. and Wu, Y. (1998). On possibility of suppression of microwave instability and production of femtosecond pulses of radiation in storage rings, *Proc. of APAC'98,* Tsukuba, p. 831 (http://accelconf.web.cern.ch/AccelConf/a98/proc.html).

De Loos, M. J. (2002). *Proc. 2002 Eourp. Part. Accel. Conf.*, Paris, June 3–7.

Mackinnon, A. J., Sentoku, Y., Patel, P. K. *et al.* (2002). Enhancement of proton acceleration by hot-electron recirculation in thin foils irradiated by ultraintense laser pulses, *Phys. Rev. Lett.* **88**, 215006.

Macklin, J. J., Kmetec, J. D. and Gordon III, C. L. (1993). High-order harmonic generation using intense femtosecond pulses, *Phys. Rev. Lett.* **70**, pp. 766–769.

Maine, P., Strickland, D., Bado, P., Pessot, M. and Mourou, G. (1988). Generation of ultrahigh peak power pulses by chirped pulse amplification, *IEEE J. Quantum Electron.* **24**, p. 398.

Maksimchuk, A., Gu, S., Flippo K. *et al.* (2000). Forward ion acceleration in thin films driven by a high-intensity laser, *Phys. Rev. Lett.* **84**, p. 4108.

Malka, G. *et al.* (2002a). Relativistic electron generation in interactions of a 30 TW laser pulse with a thin foil target, *Phys. Rev. E* **66**, 066402.

Malka, V., Faure, J., Marques, J. R., Amiranoff, F., Rousseau, P., Ranc, S., Chambaret, J. P., Najmudin, Z., Walton, B., Mora, P. and Solodov, A. (2001). Characterization of electron beams produced by ultrashort (30 fs) laser pulses, *Phys. Plasmas* **8**, p. 2605.

Malka, V., Fritzler, S., Lefebvre, E., Aleonard, M.-N., Burgy, F., Chambaret, J.-P., Chemin, J.-F., Krushelnick, K., Malka, G., Mangles, S. P. D., Najmudin, Z., Pittman, M., Rousseau, J.-P., Scheurer, J.-N., Walton, B. and Dangor, A. E. (2002b). Electron acceleration by a wake field forced by an intense ultrashort laser pulse, *Science* **298**, p. 1596.

Marshall, T. C., Wang, C., and Hirshfield, J. L. (2002). Femtosecond planar electron beam source for micron-scale dielectric wake field accelerator, *Phys. Rev. ST Accel. Beams* **4**, 121301.

Matsumoto, H. *et al.* (2002). The C-band (5712-MHz) linac for the Spring-8 Compact SASE Source (SCSS), *Proc. Linac Conf.*

Mini, S. M., ed. (1998). *Materials Research Society, Warrendale, PA.*

Modena, A., Najmudin, Z., Dangor, A. E., Clayton, C. E., Marsh, K. A., Joshi, C., Malka, V., Darrow, C. B., Danson, C., Neely, D. and Walsh, F. N. (1995). Electron acceleration from the breaking of relativistic plasma waves, *Nature* **377**, p. 606.

Moore, C. I., Ting, A., Krushelnick, K., Esarey, E., Hubbard, R. F., Hafizi, B., Burris, H. R., Manka, C. and Sprangle, P. (1997). Electron trapping in self-modulated laser wakefields by raman backscatter, *Phys. Rev. Lett.* **79**, 20, pp. 3909–3912.

Moore, C. I., Ting, A., McNaught, S. J., Qiu, J., Burris, H. R. and Sprangle, P. (1999). A laser-accelerator injector based on laser ionization and pondero-motive acceleration of electrons, *Phys. Rev. Lett.* **82**, p. 1688.

Mori, W. B., Decker, C. D., Hinkel, D. E. and T. Katsouleas (1994). Raman forward scattering of short-pulse high-intensity lasers, *Phys. Rev. Lett.* **72**, p. 1482.

Murakami, Y., Kitagawa, Y., Sentoku, Y. *et al.* (2001). Observation of proton rear emission and possible gigagauss scale magnetic fields from ultra-intense laser illuminated plastic target, *Phys. Plasmas* **8**, p. 4138.

Murakami, H., Ohtake, H. Sarukura, N., Hara, T., Kondo, Y., Hosokai, T., Iijima, H., Kinoshita, K., Watanabe, T. and Uesaka, M. (2003). *Jpn. J. Appl. Phys.* (in preparation).

Murphy, J. B. and Kramer, S. L. (2000). First observation of simultaneous alpha buckets in a quasi-isochronous storage ring, *Phys. Rev. Lett.* **84**, pp. 5516–5519.

Murphy, J. B. and Krinsky, S. (1994). *Nucl. Instrum. and Res. in Phys.* A **346**, p. 571.

Musumeci, P. (2002). Velocity bunching: experiment at neptune photo-injector, *Proc. of the ICFA Workshop on The Physics and Applications of High Brightness Electron Beams*, Chia Laguna (Italy), June 2002.

Musumeci, P. and Rosenzweig, J. B. (2003). Private communication.

Nakajima, K. (1996). Challenge to a tabletop high-energy laser wake-field accelerator, *Phys. Plasmas* **3**, p. 2169.

Nakajima, K., Fisher, D., Kawakubo, T., Nakanishi1, H., Ogata1, A., Kato, Y., Kitagawa, Y., Kodama, R., Mima, K., Shiraga, H., Suzuki, K., Yamakawa, K., Zhang, T., Sakawa, Y., Shoji, T., Nishida, Y., Yugami, N., Downer, M. and Tajima, T. (1995). Observation of ultrahigh gradient electron acceleration by a self-modulated intense short laser pulse, *Phys. Rev. Lett.* **74**, p. 4428.

Nakajima, K., Koga, J. and Nakagawa, K. (2001). Relativistic self-focusing of ultra-intense laser pulses and ion acceleration, *AIP Conf. Proc.* **569**, p. 97.

Nakashima, K. and Takabe, H. (2002). Numerical study of pair creation by ultraintense lasers, *Phys. Plasmas* **9**, pp. 1505–1512.

Nakazato, T., Oyamada, M., Niimura, N., Urasawa, S., Konno, O., Kagaya, A., Kato, R., Kamiyama, T., Torizuka, Y., Nanba, T., Kondo, Y., Shibata, Y., Ishi, K., Ohsaka, T. and Ikezawa, M., (1989). Observation of coherent synchrotron radiation, *Phys. Rev. Lett.* **63**, pp. 1245–1248.

Neil, G. R., Bohn, C. L., Benson, S. V., Biallas, G., Douglas, D., Dylla, H. F., Evans, R., Fugitt, J., Grippo, A., Gubeli, J., Hill, R., Jordan, K., Li, R., Merminga, L., Piot, P., Preble, J., Shinn, M., Siggins, T., Walker, R. and Yunn, B. (2000). Sustained kilowatt lasing in a free-electron laser with same-cell energy recovery, *Phys. Rev. Lett.* **84** p. 662.

Nemoto, K. *et al.* (2001). Laser-triggered ion acceleration and table top isotope production, *Appl. Phys. Lett.* **78** , pp. 595–597.

Ohkubo, T., Kinoshita, K., Hosokai, T., Kanegae, Y., Zhidkov, A. and Uesaka, M. (2002). Fundamental study for time-resolved imaging by laser plasma X-rays, *17th Int. Conf. on the Application of Accelerators in Research and Industry*, CAARI2002.

Ohtake, H., Murakami, H., Yano, T., Ono, S., Sarukura, N., Takahashi, T., Suzuki, Y., Nishijima, G. and Watanabe, K. (2003). Anomalous power and spectrum dependence of terahertz radiation from femtosecond-laser-irradiated indiumarsenide in high magnetic fields up to 14 T, *Appl. Phys. Lett.* **82**, pp 1164–1166.

Ohtake, H., Ono, S., Sakai, M., Liu, Z., Tsukamoto, T. and Sarukura, N. (2000). Saturation of THz-radiation power from femtosecond-laser irradiated InAs in a high magnetic field, *Appl. Phys. Lett.* **76**, pp. 1398–1400.

Oishi, Y. *et al.* (2001). Production of relativistic electrons by irradiation of 43-fs-laser pulses on copper film, *Appl. Phys. Lett.* **79**, pp. 1234–1236.

Okugi, T., Kurihara, Y., Chiba, M., Endo, A., Hamatsu, R., Hirose, T., Kumita, T., Omori, T., Takeuchi, T. and Yoshioka, M. (1996). Proposed method to produce a highly polarized e(+) beam for future linear colliders, *Jpn. J. Appl. Phys.* **35**, p. 3677–3680

Omori, T., Aoki, T., Dobashi, K., Hirose, T., Kurihara, Y., Okugi, T., Sakai, I., Tsunemi, A., Urakawa, J., Washio, M. and Yokoya, K. (2003). Design of a polarized positron source for linear colliders, *Nucl. Instrum. Methods Phys. Res.* **A500**, p. 232.

Palik, E. D. and Furdyna, F. K. (1970). Infrared and microwave magnetoplasma effects in semiconductors *Rep. Prog. Phys.* **33**, pp. 1193–1322.

Palmer, R. B. (1972). Interaction of relativistic particles and free electromagnetic waves in the presence of a static helical magnet, *J. Appl. Phys.* **43**, p. 3014.

Pearlman, J. S. and Morse, R. L. (1978). Maximum expansion velocities of laser-produced plasmas, *Phys. Rev. Lett.* **40**, p. 1652.

Pessot, M., Maine, P. and Mourou, G. (1987). 1000 times expansion/compression of optical pulses for CPA, *Opt. Commun.* **62**, pp. 419–421.

Piot, Ph. *et al.* (2002). Sub-ps compression by velocity bunching in a photo-injector, *TESLA Note* TESLA-FEL-02-07.

Podobedov, B. *et al.* (2001). *Proc. 2001 IEEE PAC*, pp. 1921–1923.

Pogorelsky, I. V. (1998). Ultra-bright X-ray and gamma sources by Compton backscattering of CO_2 laser beams, *Nucl. Instrum. Methods A* **411**, p. 172–187.

Pogorelsky, I. V., Ben-Zvi, I., Hirose, T., Kashiwagi, S., Yakimenko, V., Kusche, K., Siddens, P., Skaritka, J., Kumita, T., Tsunemi, A., Omori, T., Urakawa, J., Washio, M., Yokoya, K., Okugi, T., Liu, Y., He, P. and Cline, D. (2000). Demonstration of 8×10^{18} photons/second peaked at 1.8Åin a relativistic Thomson scattering experiment, *Phys. Rev. ST-AB* **3**, 090702/1-8.

Poole, M. W. *et al.* (2002). 4GLS: An advanced multi-source low energy photon facility for the UK, *Proc. European Particle Acc. Conf.* p. 733.

Preston, S. G., Sanpera, A., Zepf, M., Blyth, W. J., Smith, C. G., Wark, J. S., Key, M. H., Burnett, K., Nakai, M., Neely, D. and Offenberger, A. A. (1996). High-order harmonics of 248.6-nm KrF laser from helium and neon ions, *Phys. Rev. A* **53**, pp. 31–34.

Pukhov A. (2001). Three-dimensional simulations of ion acceleration from a foil irradiated by a short-pulse laser, *Phys. Rev. Lett.* **86**, p. 3562.

Pukhov, A. and Meyer-ter-Vehn, J. (2002). Laser wake field acceleration: the highly non-linear broken-wave regime, *Appl. Phys. B* **74**, p. 355.

Qiu, X., Batchelor, K., Ben-Zvi, I. and Wang, X.-J. (1996). Demonstration of emittance compensation through the measurement of the slice emittance of a 10-ps electron bunch, *Phys. Rev. Let.* **76**, p. 3723

Rau, B., Tajima, T. and Hojo, H. (1997). Coherent electron acceleration by sub-cycle laser pulses, *Phys. Rev. Lett.* **78**, p. 3310.

Rischel, C., Rousse, A., Uschmann, I., Albouy, P. A., Geindre, J. P., Audebert, P., Gauthier, J. C., Forster, E., Martin, J. L. and Antonetti, A. (1997). Femtosecond time-resolved X-ray diffraction from laser-heated organic films, *Nature* **390** 6659, pp. 490–492.

Rosenzweig, J. B. (1999). *Proc. 1999 Particle Accelerator Conf.*, NY, March 29–April 2, p. 2045.

Rosenzweig, J. B. and Colby, E. (1995). *AIP Conf. Proc.*, **335**, p. 724.

Rosenzweig, J., Travish, G. and Tremaine, A. (1995). Coherent transition radiation diagnosis of electron beam microbunching, *Nucl. Inst. Meth. Phys. Res. A* **365**, pp. 255–259.

Rose-Petruck, C. *et al.* (1999). Picosecond-milliangstrom lattice dynamics measured by ultrafast X-ray diffraction, *Nature* **398** 6725, pp. 310–312.

Rossbach, J. (1996) A VUV free electron laser at the TESLA test facility at DESY, *Nucl. Instrum. Methods A* **375**, p. 269.

Ruggiero, A. and Vaccaro, V. (1968). CERN/ISR-RF/68-33.

Sacherer, F. J. (1974). *Proc. 9th Int. Conf. on High Energy Accel.*, Stanford, pp. 347–351.

Sakabe, S., Michizuki, T., Yabe T. *et al.* (1982). Velocity distributions of multi-ion species in an expanding plasma produced by a 1.05-μm laser, *Phys. Rev. A* **26**, p. 2159.

Sakai, I., Aoki, T., Dobashi, K., Fukuda, M., Higurashi, A., Hirose, T., Iinuma, T., Kurihara, Y., Okugi, T., Urakawa, J. and Washio, M. (2002). Production of high brightness gamma rays through backscattering of laser photons on high-energy electrons, *KEK Preprint* 2002-101, to be published in *Phys. Rev. ST-AB*.

Saldin, E. L., Schneidmiller, E. A. and Yurkov, M. V. (1997). On the coherent radiation of an electron bunch moving in an arc of a circle, *Nucl. Instr. Meth. A* **398**, 373;

Salin, F. and Squier, J. (1992). Gain guiding in solid state lasers, *Opt. Lett.* Vol 17, no. 19, pp. 1352–1354.

Salin, Squier, Mourou and Vaillancourt (1991). Multikilohertz Ti:Al2O3 amplifier for high power femtosecond pulses, *Opt. Lett.* **16**, p. 1964.

Sands, M. (1970). SLAC-121/UC-28.

Santala, M. I. L. *et al.* (2001). Production of radioactive nuclides by energetic protons generated from intense laser-plasma interactions, *Appl. Phys. Lett.* **78** , pp. 19–21.

Sarachik, E. S. and Schappert, G. T. (1970). Classical theory of the scattering of intense laser radiation by free electrons, *Phys. Rev. D* **1**, pp. 2738–2753.

Sarukura, N., Ohtake, H., Izumida, S. and Liu, Z. (1998). High average-power THz-radiation from femtosecond laser-irradiated InAs in a magnetic field and its elliptical polarization characteristics, *J. Appl. Phys.* **84**, pp. 654–656.

Schiff, L. I. (1946). Production of particle energies beyond 200 MeV, *Rev. Sci. Instr.* **17**, p. 6.

Schoenlein, R. W. *et al.* (2000). Generation of femtosecond pulses of synchrotron radiation, *Science* **287**, p. 2237.

Schottky, W. (1914). The expulsion of electrons from heated wires under retarding potentials, *Ann. Physik* **44**, pp. 1011–1032.

Schreiber, S. (2002). *Proc. 2002 European Particle Accelerator Conf.*, Paris France, June 3–7, p. 1804.

Schroeder, C. B., Lee, P. B., Wurtele, J. S., Esarey, E. and Leemans, W. P. (1999). Generation of ultrashort electron bunches by colliding laser pulses, *Phys. Rev. E* **59**, 5, pp. 6037–6047.

Seltzer, S. M. (1991). Electron-photon Monte Carlo calculations: The ETRAN code, *Appl. Radiation and Isotopes* **42**, pp. 917–941.

Serafini, L. (1992). *AIP Proc. 3rd Workshop on Advanced Accel. Concepts*, Port Jefferson NY, June 14–20.

Serafini, L. (2002). Prospective of coherent X-ray source in Italy, *Proc. of Linac-2002 Conf.*

Serafini, L. *et al.* (1992). RF gun emittance correction using unsymmetrical RF cavities, *Nucl. Instr. Meth. A* **318**, pp. 275–281.

Serafini, L. *et al.* (1992). Neutralization of the emittance blowup induced by rf time dependent forces in rf guns, *Nucl. Instr. Meth. A* **318**, pp. 301–307.

Serafini, L. *et al.* (2001). Ultra-short electron bunch generation with a rectilinear compressor, *IEEE Cat.* No. 01CH37268C, p. 2242.

Serafini, L. and Ferrario, M. (2001). Velocity bunching in photoinjectors, *AIP CP* **581**, 87.

Serafini, L. and Rosenzweig, J. B. (1997). Envelope analysis of intense relativistic quasilaminar beams in rf photoinjectors: A theory of emittance compensation. *Phys. Rev. E* **55**, 6, pp. 7565–7590.

Shoji, Y., Ando, A., Tanaka, H. and Takao, M. (1997). Fundamental restriction to the isochronus ring FEL, *Nuclr. Instr. Methods A* **390**, pp. 417–418.

Shoji, Y., Tanaka, H., Takao, M. and Soutome, K. (1996). Longitudinal radiation excitation in an electron storage ring, *Phys. Rev. E* **54**, R4556.

Smedley, J. (2001). Ph.D. thesis, Stony Brook University, Stony Brook NY, 2001.

Soutome, K. *et al.* (1999). *Proc. 1999 EPAC*, pp. 1008–1010.

Srinivasan-Rao, T. (1997) *Proc. 1997 Particle Accelerator Conf.*, Vancouver BC May 12–16, p. 2790.

Srinivasan-Rao, T., Fischer, J. and Tsang, T. (1991). Photoemission studies on metals using picosecond ultraviolet laser pulses *J. Appl. Phys.* **69**, pp. 3291–3296.

Srinivasan-Rao, T., Fischer, J. and Tsang, T. (1991) Picosecond field assisted photoemission from yttrium microstructures, *J. Opt. Soc. of Am. B* **8**, 2, pp. 294–299.

Steinhauer, L. C. and Kimura, W. D. (1999). Longitudinal space charge debunching and compensation in high frequency accelerators, *Phys. Rev. ST Accel. Beams* **2**, 081301.

Sze, S. M. (1981). Physics of Semiconductor Devices, Wiley, New York.

Tajima, T. and Dawson, J. (1979). Laser electron accelerator, *Phys. Rev. Lett.* **43**, p. 267.

Tajima, T., Kishimoto, Y. and Downer, M. (1999). Optical properties of cluster plasma, *Phys. Plasmas* **6**, p. 3759.

Uesaka, M., Hosokai, T., Kinoshita, K. and Zhidkov, A. (2003). Generation of femtosecond electron bunches and hard-X-rays by ultra-intense laser wake field acceleration in a gas jet, *Proc. of PAC2003*.

Uesaka, M., Kinoshita, K., Watanabe, T., Sugahara, J., Ueda, T., Yoshii, K., Kobayashi, T., Hafz, N., Nakajima, K., Sakai, F., Kando, M., Dewa, H., Kotaki, H. and Kondo, S. (2000). Experimental verification of laser photocathode RF gun as an injector for a laser plasma accelerator, *J. Plasma Sci.* **28**, 1133–1142.

Uesaka, M., Kinoshita, K., Watanabe, T., Ueda, T., Yoshii, K., Harano, H., Sugawara, J., Nakajima, K., Ogata, A., Sakai F., Dewa, H., Kando, M., Kotaki, H. and Kondo, S. (1999). *Proc. of Advanced Accelerator Concepts: AIP Conf. Proc.* **472**, AIP, New York, p. 229.

Uesaka, M., Tauchi, K., Kozawa, T., Kobayashi, T., Ueda, T. and Miya, K. (1994). Generation of a subpicosecond relativistic electron single bunch at the S-band linear-accelerator, *Phys. Rev. E* **50**, 3068–3076.

Uesaka, M., Ueda, T., Kozawa, T. and Kobayashi, T. (1998). Precise measurement of a subpicosecond electron single bunch by the femtosecond streak camera, *Nucl. Instrum. Methods*, **406** 371–379.

Uesaka, M., Watanabe, T., Kobayashi, T., Ueda, T., Yoshii, K., Li, X., Muroya, Y., Sugahara, J., Kinoshita, K., Hafz, N. and Okuda, H. (2001). Hundreds- and tens-femtosecond time-resolved pump-and-probe analysis system, *Radiat. Phys. and Chem.* **60**, 303–306.

Umstadter, D. (2001). Review of physics and applications of relativistic plasmas driven by
ultra-intense lasers, *Phys. Plasmas* **8**, pp. 1774–1785.

Umstadter, D., Chen, S.-Y., Maksimchuk, A., Mourou, G. and Wagner, R. (1996). Nonlinear optics in relativistic plasmas and laser wake field acceleration of electrons, *Science* **273**, p. 472.

Umstadter, D. *et al.* (2001). *Proc. of the 2001 Particle Accelerator Conf.*, PAC 2001, Chicago, Illinois U.S.A., Lucas, P. and Webber, S. eds., IEEE, New Jersey, p. 117.

Umstadter, D., Kim, J. K. and Dodd, E. (1996). Laser injection of ultrashort electron pulses into wakefield plasma waves, *Phys. Rev. Lett.* **76**, 12, pp. 2073–2076.

van Steenbergen, A., Gallardo J., Sandweiss, J. and Fang, J.-M. (1996). Observation of energy gain at the BNL inverse free-electron-laser accelerator, *Phys. Rev. Lett.* **77,** p. 2690.

Verluise *et al.* (2000). Amplitude and phase control of ultrashort pulses by use of an acousto-optic programmable dispersive filter: pulse compression and shaping, *Opt. Lett.* **25**, p. 75.

Wada, Y., Kubota, T., Ogata, A. *et al.* (2002). Ion acceleration from a thin foil with 1 TW laser, *Proc. High Field Physics Workshop*, Himeji.

Wang, X. J. and Chang, X. Y. (2002). Femto-seconds kilo-ampere electron beam generation, *Proc. of FEL 2002 Conf.*

Wang, X. J. *et al.* (1996). *Phys. Rev. E* **54**, p. R3121.

Wang, X. J. *et al.* (1998). *Proc. Linac' 98 conference*, Chicago, August 23–28, p. 866.

Wang, X. J., Srinivasan-Rao, T., Batchelor, K., Ben-Zvi, I. and Fischer, J. (1995) Measurements on photoelectrons from a magnesium cathode in a microwave electron gun, *Nucl. Instr. & Meth. A* **356**, pp. 159–166.

Watanabe, K., Awaji, S., Motokawa, M., Mikami, Y., Sakuraba, J. and Watazawa, K. (1998). 15 T cryocooled Nb_3Sn superconducting magnet with a 52 mm room temperature bore, *J. Jpn. Appl. Phys.* **37**, pp. L1148–1150.

Weiss, C., Wallenshtein, R. and Beigang, R. (2000). Magnetic-field-enhanced generation of terahertz radiation in semiconductor surfaces, *Appl. Phys. Lett.* **77**, pp. 4160–4162.

Whitham, G. B. (1974). Linear and Nonlinear Waves, Wiley, New York.

Wickens L. M., Allen, J. E. and Rumsby, P. T. (1978.) Ion emission from laser-produced plasmas with two electron temperatures, *Phys. Rev. Lett.* **41**, p. 243.

Wiedemann, H. (1976). PEP-220.

Wiik, B. H. (1997). The TESLA project: an accelerator facility for basic science, *Nucl. Instrum. Methods A* **398**, p. 1; TESLA Technical Design Report (2001).

Wilks, S. C., Langdon, A. B., Cowan, T. E. *et al.* (2001). Energetic proton generation in ultra-intense laser-solid interactions, *Phys. Plasmas* **8**, p. 542.

Yang, J. *et al.* (2002). Low-emittance electron-beam generation with laser pulse shaping in photocathode radio-frequency gun. *J. Appl. Phys.* **92**, pp. 1608–1612.

Yokoya, K., CAIN2.23, http://www-acc-theory.kek.jp/members/cain/default.html.

Yorozu, M., Yang, J., Okada, Y., Yanagida, T., Sakai, F., Takasago, K., Ito, S. and Endo, A. (2002). Fluctuation of femtosecond X-ray pulses generated by a laser-Compton scheme, *Appl. Phys. B* **74**, p. 327.

Zel'dovich, Ya. B. and Raizer, Yu. P. (1967). *Physics of shock waves and high-temperature hydrodynamics phenomena*, **1**, p. 342; bf 2 p. 515, Academic Press, New York.

Zhang, P., Saleh, N., Sheng, Z. M. and Umstadter, D. (2003a). Laser-energy transfer and enhancement of plasma waves and electron beams by interfering high-intensity laser pulses, *Phys. Rev. Lett.* **91**, p. 225001.

Zhang, P., Saleh, N., Chen, S., Sheng, Z. M. and Umstadter, D. (2003b). An optical trap for relativistic plasma, *Phys. Plasmas* **10**, p. 2093.

Zhang, X.-C., Lin, Y., Hewitt, T. D., Sangsiri, T., Kingsley, L. E. and Weiner, M. (1993). Magnetic switching of THz beams, *Appl. Phys. Lett.* **62**, pp. 2003–2005.

Zhidkov, A., Koga, J., Kinoshita, K. and Uesaka, M. (2003). *Phys. Rev. Lett.* (in press).

Zhidkov, A. and Sasaki, A. (1999). Subpicosecond pulse laser absorption by an overdense plasma with variable ionization, *Phys. Rev. E* **59**, pp. 7085–7095.

Zhidkov A. and Sasaki A. (2000). Effect of field ionization on interaction of an intense subpicosecond laser pulse with foils, *Phys. Plasmas* **7**, p. 1341.

Zhidkov, A., Sasaki, A., Tajima T., Auguste T. *et al.* (1999). Direct spectroscopic observation of multiple-charged-ion acceleration by an intense femtosecond-pulse laser, *Phys. Rev. E* **60**, p. 3273.

Zhidkov, A., Sasaki, A. and Tajima, T. (2000a). Emission of MeV multiple-charged ions from metallic foils irradiated with an ultrashort laser pulse, *Phys. Rev. E* **61**, p. R2224; Energetic-multiple-charged-ion sources on short-laser-pulse irradiated foils, *Rev. Sci. Instrum.* **72**, p. 931.

Zhidkov, A., Sasaki, A., Utsumi, T., Fukumoto, I., Tajima, T., Saito, F., Hironaka, Y., Nakamura, K. G., Kondo, K. and Yoshida, M. (2000b). Prepulse effects on the interaction of intense femtosecond laser pulses with high-Z solids, *Phys. Rev. E* **62**, pp. 7232–7240.

Zhidkov, A., Uesaka, M., Sasaki, A. and Daido, N. (2002). Ion acceleration in a solitary wave by an intense picosecond pulse, *Phys. Rev. Lett.* **89**, 215002.

Zhou, F., Ben-Zvi, I., Babzin, M., Chang, X. Y., Doyuran, A., Malone, R., Wang, X. J. and Yakimenko, V. (2002). Experimental characterization of emittance growth induced by the nonuniform transverse laser distribution in a photoinjector, *Phys. Rev. ST Accel. Beams* 5, 094203.

Zholents, A. A. and Zolotorev, M. S. (1996). *Phys. Rev. Lett.* **76**, pp. 912–915.

Chapter 3

Diagnosis and Synchronization

3.1 Pulse Shape Diagnostics

3.1.1 *Streak camera*

K. SUZUKI

System Division, Hamamatsu Photonics K.K.

812,Joko-cho,Hamamatsu City,

SHIZUOKA 431-3196, JAPAN

Streak cameras are used in the study of ultrafast light phenomena and provide intensity vs. temporal information. We describe here the principle of an electronic streak camera (Hamamatsu, FESCA 200, see Fig. 3.1), a new circuit design and measured results demonstrating a temporal resolution of ~ 200 fs.

The heart of the streak camera described here is a specially designed electromagnetic streak tube. In order to reduce the temporal dispersion of photoelectrons, a high voltage pulse of −5 kV is applied to the streak tube's photocathode, on top of −9 kV DC. Using this technique, an electric field of 8.75 kV/mm near the photocathode is achieved. A sweep deflection circuit, consisting of an avalanche pulse driver, Gaussian filter and push-pull transformer, produces a high sweep speed of 8.76×10^8 m/s at the output phosphor screen.

The temporal resolution and dynamic range were measured using a mode-locked Ti:Sapphire laser, which produces a pulse width of 100 fs at a repetition rate of 80 MHz. A temporal resolution of ~ 200 fs at a dynamic range of 10 to 20 was achieved. The photoelectron number was estimated to be in the range of 100.

Fig. 3.1   External appearance of femtosecond streak camera.

3.1.1.1   *Principle of the streak camera*

The streak camera consists of a streak tube, an electronic circuit and optics for input and output coupling. Figure 3.2 shows a diagram of the principle of the streak camera. The light to be measured is projected onto a slit that is then lens relayed to the photocathode. The light is converted to electrons by the photocathode, which are then accelerated and swept by the deflection plates. A high speed ramp voltage is applied to the deflection plates, synchronized with the incoming light signal, and swept in a descending direction at the streak tube's output phosphor screen. The photoelectrons strike the phosphor screen and are converted back into photons to form an optical image. This optical image is then amplified by an external image intensifier and is read by a digital CCD camera interfaced to a PC.

3.1.1.2   *Consistent characteristic impedance matched deflection circuit*

The femtosecond streak tube is based on an electromagnetic focus design in which an electric field of up to ~ 6 kV/mm is achieved near the photocathode. Figure 3.3 shows a cross section of the streak tube and electromagnetic focusing coil. The distance between the photocathode and the accelerating mesh electrode is 1.6 mm. The focusing coil produces a magnetic flux

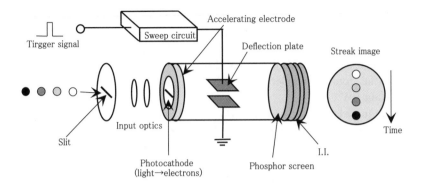

Fig. 3.2 Principle of streak camera.

density of ~ 200 G at the center of the streak tube. A meandering-type traveling wave deflector, which has a characteristic impedance of 100 Ω and bandwidth up to 900 MHz, is designed to transmit the high speed deflection voltage. The deflector is terminated with a condenser-coupled resistance of 100 Ω.

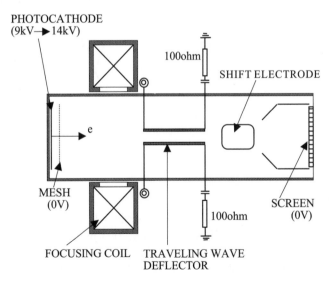

Fig. 3.3 Cross section of the streak tube and focusing coil.

In order to avoid arcing and background noise near the photocathode, a pulse of -5 kV and ~ 1 μs width is applied on top of -9 kV DC and synchronized with the streak sweep. The photoelectrons are accelerated by a high electric field of -14 kV/1.6 mm (-8.75 kV/mm). The electric potential from the accelerating mesh electrode to the phosphor screen is a constant 0 V.

An impedance matched deflection circuit, which consists of an avalanche pulse driver, Gaussian filter, coaxial switch and push-pull transformer, is shown in Fig. 3.4. The deflection ramp voltage from the push-pull transformer is applied to the meander-type traveling wave deflector using a coaxial cable with an impedance of 100 Ω. DC biasing for the streak sweep's park position is applied to the deflector pair via a condenser coupling.

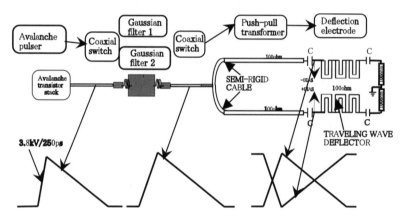

Fig. 3.4   Schematic diagram of deflection circuit.

3.1.1.3   *Measurement example*

(a)   Experimental setup

The femtosecond streak camera was evaluated using a mode-locked Ti:Sapphire laser (Spectra-Physics, Tsunami) that produces pulses of 100-fs duration with a repetition rate of 80.00 MHz at a fundamental wavelength of 800 nm. Figure 3.5 shows the setup for dynamic characteristics tests. The fundamental laser beam was split into two beams. One was directed to the input of the streak camera, the other beam triggered an optical pin diode (Hamamatsu, C1808-03) for sweep synchronization.

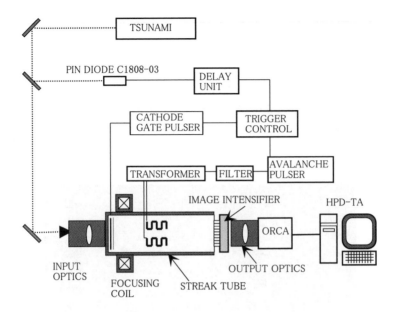

Fig. 3.5   Experimental setup for measurement of dynamic characteristics.

Input optics (Hamamatsu, A1976-01) with a temporal dispersion of less than 75 fs were used to relay the slit image to the photocathode. The output streak image was read using a lens-coupled digital CCD camera (Hamamatsu, ORCA 100) interfaced to a PC and analyzed using dedicated analysis software (Hamamatsu, HPD-TA).

(b)   Temporal resolution
The transit time spread t_{ts} of the photoelectrons is determined by the initial velocity distribution and accelerating electric field $E(V/m)$, calculated to be

$$t_{ts} = 2.34 \times 10^{-6}(\Delta\varepsilon)^{1/2}/E \qquad (3.1)$$
$$= 2.34 \times 10^{-6}(0.3)^{1/2}/8.75 \times 10^6 \ [\text{s}] = 0.15 \times 10^{12} \ \text{s},$$

where E is the acceleration electric field and $\Delta\varepsilon$ is the energy distribution of the photoelectrons ($\Delta\varepsilon \sim 0.3$ eV at $\lambda = 800$ nm).

The line spread w (60 μm) in the non-sweep mode also limits the temporal resolution. The temporal spread t_f is given by

$$t_f = w/v \tag{3.2}$$
$$= 60 \times 10^{-6}/8.76 \times 10^8 \ [\text{s}] = 0.068 \times 10^{-12} \ \text{s}\,,$$

where v is the sweep speed (8.76×10^8 m/s).

The overall temporal resolution t_r of the streak camera is given by

$$t_r = (t_{ts}^2 + t_f^2)^{1/2} \tag{3.3}$$
$$= 0.16 \times 10^{-12} \ \text{s}.$$

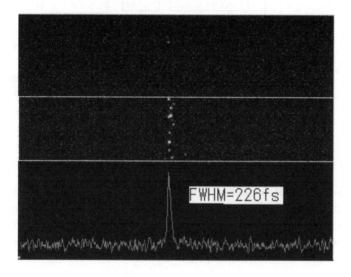

Fig. 3.6   Streak image and intensity profile of Ti:Sapphire laser.

The streak image and associated intensity profile are shown in Fig. 3.6. The horizontal and vertical axes represent time and transverse spatial respectively. The full width at half maximum (FWHM) is measured to be ~ 226 fs.

(c) Dynamic range

Dynamic range is defined as the ratio of the number of photoelectrons where 20% broadening of the measured pulse width occurs, to the minimum detectable number of photoelectrons. Figure 3.7 shows a plot of temporal resolution vs. the number of photoelectrons. A dynamic range of 10–20 is achieved at a temporal resolution of 200 fs.

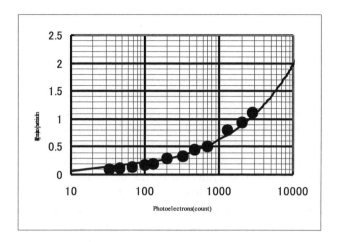

Fig. 3.7   Streak image and intensity profile of Ti:Sapphire laser. The plot of the photo-electron number depends on the temporal resolution.

3.1.2   *Coherent radiation interferometer*

C. SETTAKORN
H. WIEDEMANN
Applied Physics Department
and Synchrotron Radiation Laboratory,
Stanford University, Stanford,
CA 94309-0210, USA

During the past decade it became possible to produce picosecond and sub-picosecond electron pulses [Kung *et al.* (1994); Wiedemann *et al.* (1997)] and observe coherent radiation derived from such pulses [Nakazato *et al.* (1989); Happek *et al.* (1991)]. To further develop this ability, methods and instrumentation to measure such short pulses are required. In the subpicosecond regime the effectiveness of streak cameras reached its limits and new techniques had to be developed. A frequency domain technique has been proposed by Barry [Barry (1991)] and its efficacy was demonstrated by Lihn [Lihn *et al.* (1996); Lihn (1996)]. He was able for the first time to successfully measure the duration of ultrashort electron bunches of the order of 100 fs rms.

3.1.2.1 *Technique*

The technique is based on the autocorrelation of coherent transition radiation (CTR) derived from ultrashort bunches and then passed through a Michelson interferometer. The technique requires the processing of a radiation pulse that exactly resembles the particle distribution of the electron pulse. Therefore, very fast conversion of the electron pulse to electromagnetic radiation is required. Transition radiation (TR) is specially well suited because the emission process is very fast, much faster than the duration of realistic electron pulses as evidenced by the broad TR spectrum reaching up to hard X-rays. This, for example, would not be true for undulator radiation, where the radiation pulse is greatly stretched compared to the electron pulse. Only for very fast emission processes can we expect the radiation pulse to be a true replication of the particle distribution.

An electron bunch can emit coherent radiation at wavelengths comparable to and longer than the bunch length. An electron beam, passing through a metal interface, emits TR and its coherent radiation field spectrum is determined by folding the single-electron radiation spectrum with the Fourier transform of the particle distribution. Because of the uniformity of the single-electron TR spectrum, the CTR intensity spectrum is determined solely by the particle distribution or the bunch form-factor $f(\omega) = \left| \int e^{ik\mathbf{n} \cdot \mathbf{r}} S(\mathbf{r}) \mathrm{d}^3 r \right|^2$, where $S(\mathbf{r})$ is the 3-D particle distribution, k the wavenumber and $\mathbf{n}$ the unit vector in the direction of observation and is given by $I_{\mathrm{CTR}}(\omega) = N_e^2 I_{\mathrm{TR}}(\omega) \left| \int e^{ik\mathbf{n} \cdot \mathbf{r}} S(\mathbf{r}) \mathrm{d}^3 r \right|^2$, where N_e is the number of electrons per bunch and $I_{\mathrm{TR}}(\omega)$ the single-electron TR spectrum. When the transverse distribution can be neglected, $I_{\mathrm{CTR}}(\omega) \propto N_e^2 \left| \int e^{ikz} S_1(z) \mathrm{d}z \right|^2 = N_e^2 \left| S_1(\omega) \right|^2$, where $S_1(z)$ is the longitudinal particle distribution and $S_1(\omega)$ is its Fourier transform. The measurement of the radiation spectrum therefore is sufficient to determine the pulse length of the electron bunch. Of course, in the frequency domain measurement, phase information is lost and it is not possible to recover more information about the pulse shape beyond the effective pulse length. It has been reported that the application of the Kramers–Kronig relationship can be used to reconstruct the particle distribution. Further studies show [Settakorn (2001)] that, for example, a time reversal ambiguity cannot be resolved. The radiation spectrum does depend on the particle distribution and is therefore different for a Gaussian or rectangular distribution. A precise measurement of the spectrum can be used to distinguish between both distributions, although other, more pathological distributions, could create similar spectra. Any temporal pulse asymmetry is fully symmetrized in this frequency domain method and cannot be resolved anymore. In spite of these shortcomings, the described method is a simple and effective way of controlling sub-picosecond electron bunch facilities in support of efforts to achieve ever shorter electron pulses or to set up efficient beam conditions for the production of ultrashort radiation pulses.

Since the CTR intensity spectrum is proportional to the square of the particle distribution spectrum $|S_1(\omega)|^2$, a Michelson interferogram represents the autocorrelation of the particle distribution, and the bunch length can be obtained directly from the interferogram. This method is based on optical principles and works therefore for any bunch length for which the corresponding coherent electromagnetic radiation spectrum can be extracted and suitable optical elements are available. The spectrum from subpicosecond electron bunches extends from the microwave into the far infrared regime and appropriate optical materials must be employed in the experimental setup. For an exact measurement of the electron pulse duration any possible modification of the radiation spectrum must be avoided as much as possible and, where this is not possible, corrections to the results must be applied. At the long wavelength end of the spectrum significant suppression of radiation must be expected due to cutoff in metallic beam pipes [Novdick and Saxon (1954)]. This creates a practical upper limit by this method for bunch lengths of the order of a few ps. Experimental investigations have been pursued at the SUNSHINE facility [Kung *et al.* (1994); Wiedemann *et al.* (1997)], where much shorter and intense electron bunches at 120 fs rms can be produced and where the coherent radiation spectrum reaches from microwaves down well into the far infrared regime.

A schematic diagram of the bunch length measurement based on the CTR autocorrelation technique is shown in Fig. 3.8. Backward CTR is generated from electron bunches impinging at 45° on an Al foil. The radiation exits from the evacuated electron beam environment through a high density polyethylene window and enters a Michelson interferometer.

The far infrared Michelson interferometer consists mainly of a beam splitter, a fixed mirror and a movable mirror, arranged as shown in Fig. 3.8. The radiation field entering the interferometer is split into two parts by the beam splitter, both traveling in orthogonal directions, to be reflected back

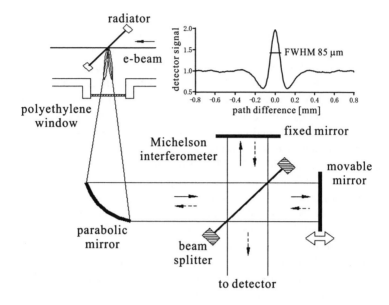

Fig. 3.8 Michelson interferometer setup (schematic) for bunch length measurements with a typical interferogram.

by mirrors. After reflection, parts of the two radiation pulses are combined again and absorbed by an intensity detector. At an optical path difference δ between both spectrometer arms, the combined radiation pulse is the addition of the electric field pulse from the fixed arm, $E_f(t) = TRE(t)$, and that from the movable arm delayed in time by δ/c, $E_m(t) = RTE(t + \delta/c)$. Here $R = R(\omega)$ and $T = T(\omega)$ are the reflection and transmission coefficients of the beam splitter, respectively. The intensity measured at the detector is then,

$$I_d(\delta) \propto \int \left| TRE(t) + RTE(t + \frac{\delta}{c}) \right|^2 dt \qquad (3.4)$$

or

$$I_d(\delta) - I_{d0} \propto 2\,|RT|^2 \int E(t)E^*(t + \frac{\delta}{c})dt\,, \qquad (3.5)$$

where $I_{d0} \propto 2\,|RT|^2 \int |E(t)|^2 dt$ is independent of the path difference δ. Equation (3.5) represents the autocorrelation of the radiation pulse called the *interferogram*, of which an ideal sample for a Gaussian particle distribution is shown in Fig. 3.9 while a real measurement is shown in the inset of Fig. 3.8.

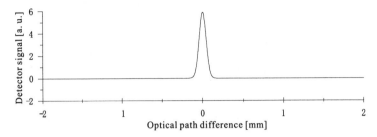

Fig. 3.9 Simulation of an ideal interferogram of a Gaussian pulse with a bunch length of $\sigma_z = 30\ \mu$m.

3.1.2.2 *Michelson interferometer*

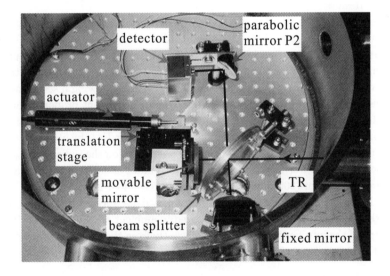

Fig. 3.10 In-vacuum Michelson interferometer.

An experimental setup of a Michelson interferometer is shown in Fig. 3.10, which can be operated in ambient air or under ultrahigh vacuum conditions. CTR is extracted from the radiator and guided through an evacuated pipe to the Michelson interferometer. An Al-coated off-axis parabolic mirror P1 (not shown) located in this pipe is used to convert the divergent radiation from the radiator into a quasi-parallel beam, which is directed toward the Kapton foil beam splitter in the interferometer. The recombined radiation from the beam splitter is collected by a second off-axis parabolic mirror P2 and directed to a pyroelectric detector. The movable mirror is mounted on a translation stage and its movement is executed by an actuator controlled by a PC operating in a LABVIEW environment. An interferogram is obtained by recording the radiation intensity from the detector while moving the movable mirror in steps of 1–10 μm. All mirrors in this setup have a diameter of 1 inch.

As long as the optical path difference δ is longer than the pulse length, the detector signal is constant and the interferogram is zero. Only when the pulses from both mirrors overlap will there be a signal in the interferogram. The width of the signal in terms of optical path difference is the radiation and thereby the electron pulse length (see inset of Fig. 3.8).

(a)   Results

The width of the interferogram main peak is equal to the bunch length only under ideal conditions. We need to discuss the more relevant effects that can impact on this bunch length measurement. Since the measurement occurs in the frequency domain, all effects that modify the radiation spectrum must be avoided or corrected for. For the following discussions it is useful to introduce into Eq. (3.5) a frequency dependent proportionality factor $F(\omega) = F_{\mathrm{bs}}(\omega)\, F_{\mathrm{det}}(\omega)\, F_{\mathrm{amb}}(\omega)\, F_{\mathrm{source}}(\omega)$. Significant spectral modifications occur in the beam splitter F_{bs}, the detector F_{det}, by absorption in humid ambient air F_{amb} and from source size effects on the radiator F_{source}. The impact of these factors has been studied and evaluated in considerable detail [Settakorn (2001)] and will be briefly recounted here.

Figure 3.11 includes an interferogram and its spectrum is measured with a Michelson interferometer operated in ambient air, Fig. 3.11(a). The shape of the interferogram differs significantly from the expected one shown in Fig. 3.9 and we will discuss each of the undesired features.

(b)   Beam splitter

The width of the central peak is essentially the pulse length. However, the valleys close to the main peak in Fig. 3.11(a) indicate interference effects, which originate from the beam splitter. To minimize excessive absorption in the beam splitter, thin Kapton or Mylar foils are used as beam splitters.

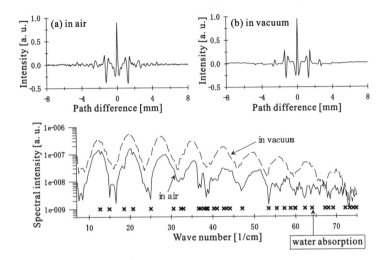

Fig. 3.11 Comparison of interferograms and radiation spectra obtained from an in-air and an in-vacuum Michelson interferometer.

The spectral efficiency of the beam splitter $F_{bs}(\omega)$ is dominated by thin film interference effects [Lihn *et al.* (1996); Lihn (1996)] due to reflections on the front and back surfaces; $F_{bs}(\omega) = |R(\omega)T(\omega)|^2$. The efficiency becomes zero at certain frequencies, where destructive interference occurs. In Fig. 3.12 the spectral efficiency is shown for foils of different thicknesses. A thin beam splitter seems to cover the widest spectral range with good efficiency. However, the thinner the beam splitter, the more low frequency components are suppressed. This suppression generates the interference effects in the interferogram, evident by the valleys next to the main peak causing this peak to narrow and leading to an underestimate of the pulse length. A detailed study of the beam splitter thickness effect has been performed by Lihn [Lihn (1996)], with the results showing that a thick beam splitter is preferable to keep the width of the interferogram main peak equal to that of the electron pulse. Too thick a beam splitter, however, leads to strong absorption. As a rule, the minimum allowable thickness for which no correction is required must be at least equal to half the electron pulse length. For thinner beam splitters corrections must be applied [Lihn (1996)]. This ability to correct is limited to beam splitter thicknesses equal to at least 20% of the bunch length. For thinner beam splitters no correction is possible. It is therefore prudent to use a Kapton beam splitter that is at least as thick as half the bunch length to be measured, although the

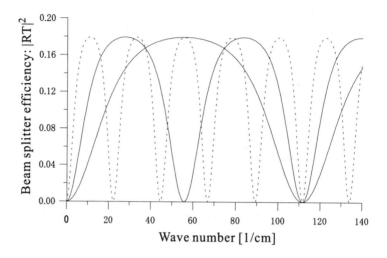

Fig. 3.12   Beam splitter efficiency $|R(\omega)T(\omega)|^2$ of Kapton foils of different thicknesses: 25.4 μm (solid line), 50.8 μm (dash-dootted line), and 127 μm (dotted line)

absorption in the beam splitter may be noticeable or significant.

(c)   Pyroelectric detector

The two signatures at about ± 1.2 mm from the main peak are detector artifacts. This effect becomes clearer in in-vacuum measurements, as shown in Fig. 3.11(b). We notice periodic features both in the interferogram and spectrum, as shown on a linear scale in Fig. 3.13, indicating an interference process.

The periodic signatures in the interferogram also suggest that multiple reflections and interferences occur. The periodicity of the interference pattern is not consistent with a 25-μm Kapton beam splitter but is associated with the pyroelectric detector used in this experiment [Settakorn (2001)]. The Molectron P1-65 detector used here is made of a 100-μm thick LiTaO$_3$ pyroelectric crystal coated front and back with different metals to enhance efficiency, but also allowing multiple reflections of the radiation within the crystal. Measurements performed with a different thickness of the detector crystal, 25-μm thick, verified this explanation [†] [Settakorn (2001)]. A pyroelectric detector like the P1-65 therefore does not exhibit a smooth detection spectrum. However, since we are interested only in the width of the main interferogram peak, it can be used as an efficient and convenient

---

[†]Test sample courtesy of Molectron Detactor, Inc

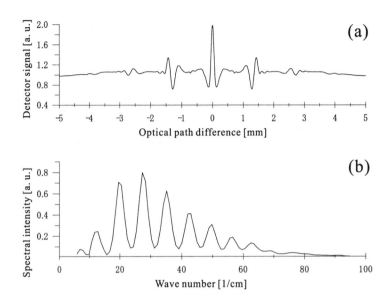

Fig. 3.13 (a) Interferogram and (b) radiation spectrum taken from the in-vacuum Michelson interferometer.

room temperature detector for bunch length measurements.

(d) Humidity in air

The noise-like variations along the interferogram in Fig. 3.11 are caused by many narrow water absorption lines. Known absorption lines are indicated by cross marks in the spectral graph. Because the spectral resolution is the inverse of the interferogram length, these absorption lines can be detected only if a long interferogram is recorded. Water absorption lines are very narrow and therefore are expected to have little impact on the overall radiation spectrum and the determination of the bunch length.

Figure 3.11 shows the comparison of in-air and in-vacuum interferograms and radiation spectra. The small oscillations in the interferogram have vanished in the in-vacuum interferogram since the radiation is free from water absorption lines. Comparing interferograms taken for the same beam both in-air and in-vacuum we note a significant increase in the width of the central peak for the in-air measurement compared to the in-vacuum measurement. The observed FWHM of the main peak, after applying the appropriate corrections, are 126 ± 3 μm in vacuum and 158 ± 5 μm in air, a difference of 32 μm or 25%. This broadening can be explained by dispersion in water vapor of humid air [Settakorn (2001)]. The refractive index of humid air is not constant over the far infrared spectrum of interest and therefore different radiation frequencies propagate with different velocities. Consequently, the radiation pulse spreads as it travels through humid air. It is worth noting that the pulse spreading depends on the radiation spectral range. For example, a long bunch of the order of 1 ps generating mainly coherent mm radiation will not be altered by dispersion since the variation of the group velocity within the radiation spectral range of interest is negligible. A much shorter bunch, on the other hand, generates a broader spectrum reaching into the far infrared regime and is subjected to larger pulse spreading when it passes through humid air. An error of 25%, as obtained in the example above, is significant where absolute bunch length measurements are important and an in-vacuum Michelson interferometer should be employed. When the knowledge of the absolute bunch length is of lesser importance and bunch length measurements are done mostly to develop or optimize an ultrashort bunch electron pulse source a much more convenient in-air interferometer can be used.

(e) Source size

The radiation spectrum can be further modified by a finite source area. Recalling the 3-D form factor $f(\omega) = \left| \int e^{i k \mathbf{n} \cdot \mathbf{r}} S(\mathbf{r}) \mathrm{d}^3 r \right|^2$, we may evaluate this integral for, say, a Gaussian distribution to get $f(\omega) = e^{-(k\sigma_\rho \sin\theta)^2} e^{-(k\sigma_z \cos\theta)^2}$, where $k = \omega/c$, σ_ρ is the radial extent of a round Gaussian beam and θ is the observation angle. Transverse effects can be neglected as long as $\sigma_\rho \sin\theta \ll \sigma_z \cos\theta$. As the source radius and/or the observation or radiation collection angle increases, high frequency radiation becomes suppressed, thus altering the radiation spectrum and giving an incorrect longer pulse length. To investigate effects of beam size on the bunch length measurements, interferograms for different electron beams sizes were recorded using the in-vacuum Michelson interferometer and shown in Fig. 3.14 [Settakorn (2001)]. The FWHM of each interferogram gives a measured value σ_m associated with each beam size. The measured bunch length scales up or down with the source size and should approach the real bunch length when the beam size is small enough that the contribution of the transverse distribution becomes negligible compared to that of the longitudinal distribution. The horizontal error bars are due mostly to elliptical source shapes and resemble the minimum and maximum source diameter.

The theoretical estimate of the bunch length as a function of beam size

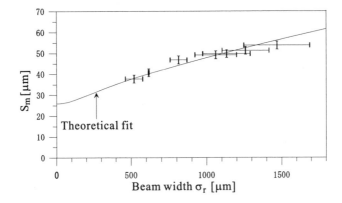

Fig. 3.14   Measured bunch length σ_m as a function of beam cross section at the radiator.

is shown in Fig. 3.14 by the solid line. The calculation was performed for the transition radiation generated by a 26-MeV beam, collected over an acceptance angle of ±160 mrad from the experimental conditions. The trend in the experimental results and the calculations suggest an actual bunch length of $\sigma_z \approx 26$ μm, however the extrapolation to reach this result is large. For a more precise bunch length measurement, it is imperative to focus the beam to a small size at the radiator so that transverse effects are negligible. In a particular case, bunch length measurements as a function of beam size may need to be performed to separate longitudinal and transverse contributions.

3.1.2.3   *Bunch length measurements with coherent diffraction radiation*

Diffraction radiation (DR) is emitted when a charged particle travels in the neighborhood of some inhomogeneity. The theory of DR was developed in the late 1950's [Dnestrikovskii and Kostomorov (1959); Bootovskii and Voskresenskii (1966)] for charged particles passing through simple structures including circular apertures and slits. In 1995, the first observation of coherent DR (CDR) generated from a 150-MeV beam passing through a circular opening or iris was reported [Shibata *et al.* (1995)].

A more systematic study was conducted to quantify the observable radiation characteristics and its relationship with the pulse length [Settakorn (2001)]. Backward CDR from a circular opening can be observed by rotating the iris 45° with respect to the beam trajectory. Since the perturbation of radiation generation on the electron beam is relatively small, it is then possible to generate diffraction radiation at several experimental stations as the beam travels along the beam line. This capability makes coherent diffraction radiation (CDR) of great interest for nondestructive bunch length measurements and control using the autocorrelation technique.

(a)   DR from a circular aperture

Radiation emitted from an electron moving with velocity v and passing through a circular aperture of radius r in an ideal conducting screen is essentially TR with a correction factor $D(\omega)$ to account for the size of the aperture [Dnestrikovskii and Kostomorov (1959); Bootovskii and Voskresenskii (1966); Shibata *et al.* (1995)],

$$I_{DR}(\omega) = I_{TR}(\omega) D(\omega)$$
$$= I_{TR}(\omega) \left[J_0 \left(\frac{\omega r}{c} \sin \theta \right) \left(\frac{\omega r}{v\gamma} \right) K_1 \left(\frac{\omega r}{v\gamma} \right) \right]^2, \qquad (3.6)$$

where θ is the observation angle with respect to the beam axis, J_0 is the Bessel function of order zero and K_1 is the modified Bessel function of the first order. The radiated intensity approaches that of pure TR as the aperture radius vanishes $(r \rightarrow 0)$.

An experimental test setup is shown schematically in Fig. 3.15 where the DR is generated by a 26-MeV electron beam passing through a circular aperture in a 1.5-mm thick aluminum plate. Three aperture sizes, 1.5 mm, 3 mm and 5 mm diameter, are available. The aperture size can be selected by moving the plate vertically to center the selected aperture on the beam trajectory. The Al plate is tilted by 45° with respect to the beam path and the backward DR, emitted at 90°, exits through a 19-mm diameter and 1.25-mm-thick polyethylene window. Also available on the Al plate is a fluorescent screen which can be selected to monitor the beam position and spot size. The beam profile is observed through a CCD camera (not being shown in the diagram). Similar to CTR, the intensity of coherent DR also scales with the square of the number of electrons per bunch and includes pulse length information. In Fig. 3.16 the spectra measured in a Michelson interferometer are shown for different aperture diameters, 1.5 mm, 3.0 mm and 5.0 mm, together with regular TR. The low frequency suppression in the spectrum is due to thin film interference effects in the 25-μm Kapton beam splitter. High frequency suppression is caused by the spectral distribution of the diffraction radiation expressed in Eq. (3.6) and becomes increasingly

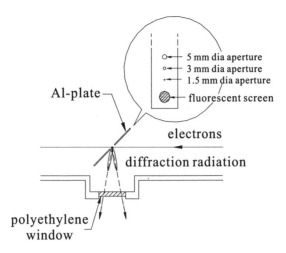

Fig. 3.15 Schematic diagram of the setup to generate diffraction radiation. The target can be moved in the direction normal to the plane of the figure to select different apertures.

severe as the aperture size increases.

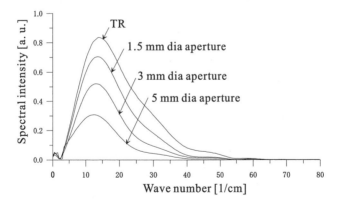

Fig. 3.16 Spectral distribution of coherent transition radiation (CTR) and coherent diffraction radiation (CDR) generated at SUNSHINE.

(b) Bunch length measurements with CDR
The possibility of using CDR for nondestructive bunch length measurements has been shown before. A closer study of the process, however,

indicates the need for careful adjustment of the experimental parameters to obtain an accurate representation of the actual bunch length [Settakorn (2001)]. Detailed experimental studies have been conducted at SUNSHINE to determine the required experimental procedures. Different from TR, the spectral distribution of DR is now frequency dependent and must be folded with the form factor of the particle distribution for correct bunch length measurements.

As mentioned before, the high frequency part of the radiation spectrum can be suppressed if too large an aperture is used. CDR is emitted at wavelengths $\lambda \gg r/\gamma$, where r is the radius of the aperture. However, the radiation spectrum does not resemble the particle distribution if the aperture of the radiator is too large. In this case, the suppression of higher frequency components in Eq. (3.6) of the coherent radiation results in a broader interferogram and a seemingly longer bunch length. To investigate this, the apparent pulse length has been measured with radiation generated from circular apertures of different diameters, ranging from 0 to 5 mm. As expected, the *measured* bunch length increases with aperture in agreement with theory (Eq. (3.6)). Based on this agreement, we define experimental conditions that must be met to perform bunch length measurements that do not require extensive corrections. As a practical rule of thumb and accepting a bunch length measurement error of no more than 5%, the aperture radius should not exceed a limit given by

$$r \, [\mathrm{mm}] \lesssim 0.06 \, \sigma_\tau \, [\mathrm{ps}] \, \gamma \,, \tag{3.7}$$

where σ_τ is the expected bunch length. This puts severe limits on the radiator aperture for low beam energies, but is quite easily satisfied for high energy beams ($\gamma \gg 1$).

3.1.2.4   *Pulse shape reconstruction procedure*

T. WATANABE
M. UESAKA
Nuclear Engineering Research Laboratory, University of Tokyo
2-22 Shirane-shirakata, Tokai, Naka,
IBARAKI 319-1188, JAPAN

(a)   Bunch form factor

Here we recall the theory of coherent radiation and explain the reconstruction procedure. When a bunch of N electrons emits radiation, e.g. transition radiation, by passing through an interface, the resulting total electric field at the observation point is the superposition of that emitted from each electron (see Fig. 3.17). If the observation point is far from the interface,

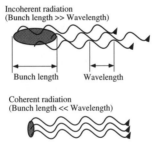

Incoherent radiation
(Bunch length >> Wavelength)

Bunch length   Wavelength

Coherent radiation
(Bunch length << Wavelength)

Fig. 3.17   Coherent radiation and incoherent radiation. When the bunch length is longer than the wavelength of the radiation, the total radiation shows incoherency. When the bunch length is shorter than the wavelength, the radiation becomes coherent.

the total intensity at wavelength λ, using the far-field approximation, can be expressed as

$$\left(E_n \sum_{k=1}^{N} e^{-in\varphi_k} \right)^2 = E_n^2 \left(\sum_{k=1}^{N} e^{-i2n\varphi_k} + \sum_{k\neq q} e^{-in(\varphi_k+\varphi_q)} \right) , \qquad (3.8)$$

since the phase of the electric field emitted by the k-th electron in the bunch at time t is $\varphi_k - \omega t$ and the electric field of the k-th electron in the frequency domain contains $e^{-in\varphi_k}$. Here it is assumed that the energy dispersion and transverse beam divergence are not considered, although the peak energy and the transverse beam distribution are included. Let the radiation emitted by a single electron be

$$I_n = \left(E_n e^{-in\varphi_k}\right)^2. \tag{3.9}$$

Then,

$$\left(E_n \sum_{k=1}^{N} e^{-in\varphi_k}\right)^2 = I_n \sum_{k=1}^{N} + I_n \sum_{k \neq q} e^{-in(\varphi_q - \varphi_k)}$$

$$= I_n N + I_n \sum_{k \neq q}^{N} \cos[n(\varphi_k - \varphi_q)], \tag{3.10}$$

where the first term in Eq. (3.10) represents the intensity of the N independent sources, while the second part takes into account the phase relations between different electrons. Hence it can be understood that the first term is that for incoherent radiation and the second for coherent radiation. For completely coherent radiation, $\varphi_k - \varphi_q = 0$, then

$$\left(E_n \sum_{k=1}^{N} e^{-in\varphi_k}\right)^2 = I_n N + I_n N(N-1)$$

$$= N^2 I_n. \tag{3.11}$$

The total electric field is proportional to N^2. By introducing the bunch form factor $f(\nu)$, the total radiation intensity can be rewritten as

$$\left(E_n \sum_{k=1}^{N} e^{-in\varphi_k}\right)^2 = I_n N + I_n N(N-1)f(\nu)$$

$$\simeq N^2 f(\nu). \tag{3.12}$$

From Eqs. (3.10) and (3.12);

$$f(\nu) = [1/N(N-1)] \sum_{k \neq q} \cos n(\varphi_k - \varphi_q). \tag{3.13}$$

When it can be assumed that the distribution of each electron in the bunch is symmetric to some reference angle, and each electron is independent from one another, the bunch form factor is reduced to the simple equation

$$f(\nu) = \left(\int \cos(n\varphi) S(\varphi) d\varphi\right)^2, \tag{3.14}$$

where $S(\varphi)d\varphi$ is the probability that the electron exists from φ through $\varphi + d\varphi$. When the distribution has a Gaussian function the probability becomes

$$S(\varphi) = \frac{1}{\sigma\pi} \exp(-\frac{\varphi^2}{\sigma^2}) \qquad (3.15)$$

and the bunch form factor $f(\nu)$ is eventually deduced to be [Novdick and Saxon (1954)]

$$f(\nu) = \exp(-2\pi^2\sigma^2\nu^2). \qquad (3.16)$$

The quantity $f(\nu)$ is given by the Fourier transform of the distribution of the function $S(x)$ of the electrons in the bunch:

$$f(\nu) = \left| \int S(r) \exp\left[i2\pi(\mathbf{n} \cdot \mathbf{r})\nu\right] dr \right|^2, \qquad (3.17)$$

where $\mathbf{n}$ is the unit vector directed from the center of the bunch to the observation point and $\mathbf{r}$ is the position vector of the electron relative to the bunch center (see Fig. 3.18). $f(\nu)$ can be divided into two parts, the

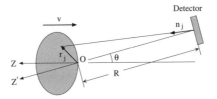

Fig. 3.18 Coordinates associated with coherent radiation spectrum produced by asymmetric electron pulse. The origin of the coordinate system is located at the front of the bunch. θ is the polar angle between the detector and bunch velocity, $\mathbf{r}_j$ locates the j-th electron, $\mathbf{n}_j$ is a unit vector pointing to the j-th electron and R, the distance from the detector to the bunch, is much larger than the extent of the bunch.

longitudinal bunch form factor $f_L(\nu)$ and the transverse bunch form factor $f_T(\nu)$, as follows:

$$f(\nu) = f_L(\nu) \cdot f_T(\nu), \qquad (3.18)$$

where

$$f_L(\nu) = \left| \int_{-\infty}^{\infty} h(z) \exp(i2\pi \cos\theta/\nu) dz \right|^2, \qquad (3.19)$$

$$f_T(\nu) = \left| 2\pi \int_0^\infty g(\rho) J_0(2\pi\rho \sin\theta\nu)\rho\,\mathrm{d}\rho \right|^2 , \qquad (3.20)$$

and $h(z)$ and $g(\rho)$ are, respectively, the longitudinal (z) and transverse (ρ) distribution functions of the electron bunch. With regards beam dynamics, $f_L(\nu)$ and $f_T(\nu)$ can be regarded as a projection of the horizontal axis of longitudinal and transverse phase space, respectively. The vertical axes, i.e. the energy divergence for longitudinal phase space and the beam divergence for transverse space, are not considered, as mentioned at the beginning of the present section. Here we also assume, for simplicity, that the transverse beam distribution is circular. $f_T(\nu)$ is obtained by measuring the transverse distribution of the electron bunch. When one observes the transition radiation in the on-axis or nearly on-axis direction, i.e. $\theta \gg 1$, then $\cos\theta$ and $\sin\theta$ can be unity and zero, respectively. Hence Eqs. (3.19) and (3.20) can be approximately rewritten as

$$f_L(\nu) = \left| \int_{-\infty}^\infty h(z)\exp(i2\pi\nu z)\mathrm{d}z \right|^2$$
$$= |F_L(\nu)|^2 , \qquad (3.21)$$

$$f_T(\nu) = \left| 2\pi \int_0^\infty g(\rho) J_0(0)\rho\,\mathrm{d}\rho \right|^2 , \qquad (3.22)$$

where $F_L(\nu)$ is the Fourier transform of the longitudinal bunch distribution $h(z)$, $g(\rho)$ is the transverse bunch distribution and $J_0(0)$ isthe zeroth-order Bessel function.

From $f_L(\nu)$, there are two ways to get the longitudinal bunch distribution; one is for a symmetric bunch shape and the other for an asymmetric bunch shape. The analysis of the asymmetric bunch shape by using Kramers–Kronig relations has been developed mainly by Lai and Sievers *et al.* [Lai and Sievers (1997); Happek *et al.* (1991); Lai and Sievers (1994); Lai *et al.* (1994); Lai and Sievers (1995a)]. We examine the details of this in the next section. Here we shall focus on the basic equation for the last step of the reconstruction. For both symmetric and asymmetric analysis, the inverse Fourier transform is applied as follows:

$$h(z) = \frac{1}{2\pi} \int_{-\infty}^{+\infty} F_L(\omega)\exp[-i\omega t]\,\mathrm{d}\omega , \qquad (3.23)$$

$$F_L(\omega) = \sqrt{f(\omega)} , \qquad (3.24)$$

where $h(z)$ is the longitudinal bunch form factor, namely the final objective of this procedure.

(b) Interference spectroscopy

An interferogram is obtained from a Michelson interferometer by measuring the detector signal as a function of the path difference in the interferometer's two arms, as shown in Fig. 3.19. The intensity of the recombined radiation

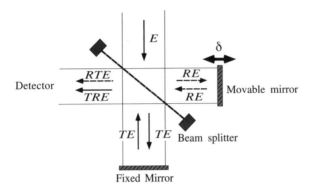

Fig. 3.19 Michelson interferometer.

intensity at the detector can be expressed in the time domain with an additional time delay δ/c for the movable arm by

$$I(\delta) \propto \int_{-\infty}^{+\infty} |TRE(t) + RTE\left(t + \frac{\delta}{c}\right)|^2 dt \qquad (3.25)$$

$$= 2|RT|^2 Re \int_{-\infty}^{+\infty} E(t)E^*\left(t + \frac{\delta}{c}\right) dt$$

$$+ 2|RT|^2 \int_{-\infty}^{+\infty} |E(t)|^2 dt. \qquad (3.26)$$

Alternatively, in the frequency domain an extra phase difference $e^{-i\omega\delta/c}$ is added to the radiation from the movable arm, and the intensity at the detector can be expressed by

$$I(\delta) \propto \int_{-\infty}^{+\infty} |TRE(\omega) + RTE(\omega)e^{-i\omega\delta/c}|^2 dt \qquad (3.27)$$

$$= 2Re \int_{-\infty}^{+\infty} |RT|^2 E(\omega)e^{-i\omega\delta/c} d\omega$$

$$+2|RT|^2 \int_{-\infty}^{+\infty} |RT|^2 |E(\omega)|^2 d\omega \,, \tag{3.28}$$

where δ is the optical path difference and c the speed of light. Equations (3.26) and (3.28) are related by the Fourier transform

$$E(\omega) = \frac{1}{\sqrt{2\pi}} \int_{-\infty}^{+\infty} E(t) e^{i\omega t} dt \,. \tag{3.29}$$

The baseline is defined as the intensity at $\delta \to \pm\infty$,

$$I_\infty \propto 2|RT|^2 \int_{-\infty}^{+\infty} |E(t)|^2 dt \tag{3.30}$$

$$= 2 \int_{-\infty}^{+\infty} |RT|^2 |E(\omega)|^2 d\omega \,. \tag{3.31}$$

By definition, the interferogram can be written

$$S(\delta) = I(\delta) - I_\infty \tag{3.32}$$

$$\propto 2|RT|^2 Re \int_{-\infty}^{+\infty} E(t) E^* \left(t + \frac{\delta}{c} \right) dt \tag{3.33}$$

$$= 2Re \int_{-\infty}^{+\infty} |RT|^2 |E(\omega)|^2 e^{-i\omega\delta/c} d\omega \,. \tag{3.34}$$

Therefore, the interferogram $S(\delta)$ is the autocorrelation of the incident radiation [Bell (1972)], and its Fourier transform is the power spectrum of the radiation [Williamson (1952)]. Solving for $|E(\omega)|^2$ in Eq. (3.34) yields

$$|E(\omega)|^2 \propto \frac{1}{4\pi c |RT|^2} \int_{-\infty}^{+\infty} S(\delta) e^{i\omega\delta/c} d\delta \,, \tag{3.35}$$

where $|E(\omega)|^2 = |E(-\omega)|^2$ is used since $E(t)$ is a real function. Using Eq. (3.8), the relation $I_{total}(\lambda) \propto |E(2\pi c/\lambda)|^2$, the bunch form factor can be obtained, giving

$$f(\lambda) \propto \frac{1}{N-1} \left[\frac{\int_{-\infty}^{+\infty} S(\delta) e^{i2\pi\delta/\lambda} d\delta}{4\pi c |RT|^2 N I_e(\lambda)} - 1 \right] \,. \tag{3.36}$$

Hence the interferogram contains the frequency spectrum of the coherent transition radiation and can be used to derive the bunch duration.

The beam splitter we used for the far infrared regime, a Mylar foil, does not provide constant and equal reflectance and transmittance for all frequencies. This departure from an ideal beam splitter is caused by the interference of light reflected from both surfaces of the beam splitter, which is equivalent to thin film interference in optics [Hecht and Zajac (1974)]. The total amplitude reflection coefficient for a Mylar foil of thickness t and refractive index n mounted at a $45°$ angle to the direction of the incoming radiation is given by [Chantry (1971)]

$$R = -r\frac{1 - e^{i\phi}}{1 - r^2 e^{i\phi}}, \qquad (3.37)$$

where r is the amplitude reflection coefficient of the air-to-Mylar interface at an incident angle of $45°$ and ϕ is defined as $4\pi t\sigma\sqrt{(2n^2 - 1)/2}$ for wavenumber $\sigma = 1/\lambda$ [Bell (1972)]. The total amplitude transmission coefficient for the same condition is

$$T = (1 - r^2)\frac{e^{i\phi/2}}{1 - r^2 e^{i\phi}}. \qquad (3.38)$$

No absorption in the foil is assumed and the refractive index is assumed to be constant ($n = 1.85$) over all frequencies [Bell (1972)]. The phase difference between R and T is $\pi/2$ at any frequency. With this, energy conservation in the interferometer can be proved [Chantry (1971)].

Unlike an ideal beam splitter, the efficiency of the beam splitter, defined as $|RT|^2$, is not constant over all frequencies and becomes zero at certain frequencies where radiation reflected from both surfaces of the beam splitter interferes destructively (see Fig. 3.12). It is worth noting that equations in the frequency domain, such as Eq. (3.28), are still valid, although equations in the time domain, such as Eq. (3.26), are no longer valid and need to be replaced by appropriate convolution integrals.

(c) Calibration of interferometer

In the diagnosis of coherent radiation by spectroscopy, it is important to estimate the absolute value of the radiation. The intensity of the radiation observed by the detector has already been suppressed by absorption in ambient air and the optics, the interference effect and other mechanisms, all of which are functions of wavelength λ. Therefore, one has to calibrate a response function in order to estimate the absolute value of the intensity. For this purpose, a high pressure mercury arc is used as a reference light source for far infrared radiation. Although a mercury arc is not an exact source of black-body radiation, we use Plank's law for black bodies to estimate the absolute value of the radiation intensity emitted from the mercury arc. The radiation power per unit solid angle, per unit wavenumber, per unit area, P_b [W/(sr·m)], is expressed as

$$P_b = 2hc^2\nu^3 \left\{ \exp\left(\frac{hc\nu}{kT}\right) - 1 \right\}^{-1} \cos\theta, \qquad (3.39)$$

where h is the Plank constant, c is the velocity of light, k is the Boltzmann constant, T is the temperature and θ is the angle against the normal vector of the unit surface. Since the radiation is emitted almost perpendicular to the black body, we let $\cos\theta$ be unity.

The radiation efficiency of a high pressure mercury arc $P = 2 \sim 3$ is larger than that of a low pressure mercury arc. The low wavelength component of the radiation, of wavelength smaller than 100 μm, is radiated from the wall of the quartz, while the longer wavelength component is dominated by the radiation from the mercury arc, the radiation spectrum of which can be regarded as that of a black body with ~ 4000 K [Kondo *et al.* (1993); Lichtenberg and Sesnic (1967)].

Now we consider detailed calibration in a real experiment. The output voltage of the detector V_b for black-body radiation is calculated using Plank's law,

$$V_b = RK \iint P_b \mathrm{d}S \mathrm{d}\Omega$$

$$= RK P_b S_b \Omega_b, \qquad (3.40)$$

where S_b is the emission area, ω_b [sr] is the solid angle, R [V/W] isthe sensitivity of the detector and K is the response function including the transmission of the filter, reflectivity of the mirrors and the beam splitter efficiency.

Using the radiation power P_T [W·m/sr] per unit solid angle and unit wavenumber, the output of the detector V_T can also be expressed as

$$V_T = RK \int P_T \mathrm{d}\Omega. \qquad (3.41)$$

Therefore the intensity measured in the experiment is estimated to be

$$\int P_T d\Omega = \frac{V_T}{V_b} P_b S_b \Omega_b. \qquad (3.42)$$

It is worth noting that the sensitivity of the detector R [V/W] and the response function K, including the sensitivity of each channel of the polychromator and the beam splitter efficiency $|RT|^2$, are clearly calibrated.

It is easy to get the number of photons per unit wavenumber by multiplying Eq. (3.41) by $1/hc\nu$.

(d)   Kramers–Kronig relations

It is well known that the real bunch distribution of any kind of accelerator is not symmetric due to the acceleration mechanism and experimental perturbation. Although it is not necessary to account for the asymmetry of the bunch in most cases, to obtain information on the asymmetry of the bunch is challenging.

In order to obtain the most information about the bunch shape from the measured spectral data, we redefine the integral in Eq. (3.19) so that [Lai and Sievers (1997); Happek *et al.* (1991); Lai and Sievers (1994); Lai *et al.* (1994); Lai and Sievers (1995a)]

$$
\begin{aligned}
F(\omega) &= \int_0^\infty h(z) e^{i\frac{\omega}{c}z} \mathrm{d}z \\
&= \rho(\omega) e^{i\phi(\omega)} \,,
\end{aligned}
\tag{3.43}
$$

where $h(z)e^{i\frac{\omega}{c}z}$ is the complex form factor amplitude and

$$
f(\omega) = F(\omega)F^*(\omega) = \rho^2(\omega) \,.
\tag{3.44}
$$

A measurement of $F(\omega)$ over the entire frequency intervals gives directly the magnitude of the form factor amplitude $\rho(\omega)$.

According to Eq. (3.36) the electric field for the coherent part of the emission spectrum can be written as

$$
\frac{E_{total}(\omega)}{\sqrt{N(N-1)}} = F(\omega)E(\omega) \,,
\tag{3.45}
$$

so that the effective electric field is linearly related by $F(\omega)$ to the electric field produced by an individual particle. The integration in Eq. (3.43) is only over positive z since the effective electric field cannot reach the detector before that of the first particle located at $z = 0$, a consequence of causality. It follows that $F(\omega)$ can be analytically continued into the upper half complex plane by virtue of the factor $\exp[-z\mathrm{Im}(\omega/c)]$, where z and $\mathrm{Im}(\omega/c)$ are positive.

Note the formal similarity between Eq. (3.43), which involves an integral over space, and the corresponding expression for the complex degree of coherence, which involves an integral of similar form over positive frequencies [Wolf (1962)], or with the input–output response function analysis used in optics to obtain the complex reflectivity at an interface, which involves an integral of similar form over positive time [Wooten (1972)].

In analogy with these earlier studies, we write

$$\ln S(\omega) = \ln \rho(\omega) + i\phi(\omega). \tag{3.46}$$

The dispersion relation for Eq. (3.46) has the same form as for the real and imaginary parts of $\ln[r(\omega)]$ [Wooten (1972)] so that [Toll (1956); Stern (1963)]

$$\phi_m(\omega) + \phi_{Blaschke}(\omega) = -\frac{2\omega}{\pi} P \int_0^\infty \frac{\ln \rho(x)}{x^2 - \omega^2} dx + \sum_j \arg\left(\frac{\omega - \omega_j}{\omega - \omega_j^*}\right), \tag{3.47}$$

where ϕ_m is the minimal phase and ω_j identifies the zeros of $S(\omega)$ in the upper half of the complex frequency plane. It is known that the minimum phase evaluated from the first term in Eq. (3.47) is a good approximation of the actual phase in cases where the form factor has no nearby zeros in the upper half of the complex frequency plane. The minimum phase $\phi_m(\omega)$ is obtained from the resulting Kramers–Kronig relation. The singularity at $x = \omega$ can be removed by adding to the first term of Eq. (3.47)

$$-\frac{2\omega}{\pi} P \int_0^\infty \frac{\ln \rho(x)}{x^2 - \omega^2} dx = 0. \tag{3.48}$$

The final expression for calculating the minimum phase is

$$\phi_m(\omega) = \mp \frac{2\omega}{\pi} P \int_0^\infty \frac{\ln \rho(x)/\rho(\omega)}{x^2 - \omega^2} dx. \tag{3.49}$$

Once $f(\omega)$ is experimentally determined over the entire frequency spectrum then, by Eq. (3.44), so is $\rho(\omega)$. With the aid of Eq. (3.49), $\rho(\omega)$ can be used to find the frequency dependent phase $\phi_m(\omega)$ to complete the determination of the frequency dependence of the complex form factor amplitude. The normalized density distribution function can now be obtained from the inverse Fourier transform of Eq. (3.43), namely,

$$h(z) = \frac{1}{\pi c} \int_0^\infty \rho(\omega) \cos\left[\phi_m(\omega) - \frac{\omega z}{c}\right]. \tag{3.50}$$

Note that the difference between Eq. (3.50) and Eq. (3.23) is the frequency dependent phase factor $\phi_m(\omega)$. The information about the bunch asymmetry is contained in the nonlinear contribution to the phase factor.

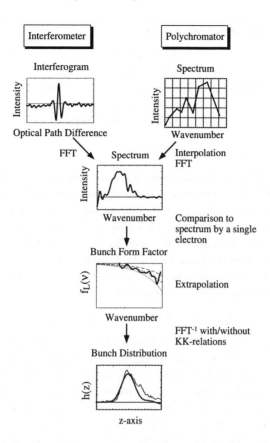

Fig. 3.20 Reconstruction procedures from interferogram to longitudinal bunch distribution.

A schematic of the reconstruction procedures from experimental results to longitudinal bunch distribution is shown in Fig. 3.20.

3.1.3   *Far infrared polychromator*

T. WATANABE

M. UESAKA

Nuclear Engineering Research Laboratory, University of Tokyo

2-22 Shirane-shirakata, Tokai, Naka,

IBARAKI 319-1188, JAPAN

3.1.3.1   *Single-shot measurement*

Before proceeding to the technique of coherent radiation spectroscopy, we shall recall the basic principle of the frequency domain technique [Barry (1991)]. Transition radiation (TR) emitted from an electron pulse is coherent when the wavelength of the radiation is longer than the pulse length. The threshold wavelength usually stays in the far infrared or longer wavelength region, since the pulse length from existing accelerators is typically from a few μm to tens of mm. The enhancement of the spectral intensity due to the coherent effect is expressed as

$$I_{total}(\lambda) = N[1 + (N - 1)f(\lambda)]I_0(\lambda),   \qquad (3.51)$$

where $I_0(\lambda)$ and $I_{total}(\lambda)$ are the spectral intensities of a single electron and the total electrons at wavelength λ, respectively, N is the number of electrons and $f(\lambda)$ is the bunch form factor. The total intensity $I_{total}(\lambda)$ and the number of electrons N can be measured by experiment, while $I_0(\lambda)$ can be estimated from theory [Jackson (1975)]. Accordingly, one can obtain the bunch distribution by applying a Fourier transform to the bunch form factor $f(\lambda)$ with the Kramers–Kronig relations [Lai and Sievers (1995a)]. The total spectral intensity $I_{total}(\lambda)$ can be deduced from the observed interferogram by means of the Wiener–Khinchine theorem; the interferogram taken by a Michelson interferometer corresponds to the first-order correlation function. Thus, the bunch length can be evaluated by interferometry. The scheme has an intrinsic limitation; only the averaged pulse duration is obtained. The bunch shape of a practical electron beam, on the other hand, fluctuates from shot to shot due to the timing jitter between the accelerating RF phase and the electron emission at the gun. Figure 3.21 indicates a typical five-shot bunch distribution of a sub-picosecond pulse taken by a femtosecond streak camera (Hamamatsu Photonics Co., Ltd.). It is apparent that those fluctuations cannot be observed by interferometry.

Another way to obtain the spectral intensity $I_{total}(\lambda)$ is to observe the

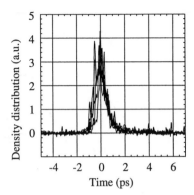

Fig. 3.21   Typical 5-shot bunch distribution taken by a femtosecond streak camera.

spectrum of radiation directly by a spectrometer. A multi-channel spectrometer, hereafter called a *polychromator*, is capable of detecting a spectrum by a single shot. Not only can a single-shot measurement be performed, but also monitoring for beam tuning can be done.

3.1.3.2   *10-channel polychromator*

Grating spectrometers come in many shapes and sizes, varying in wavelength and other characteristics. Any regular periodic structure can be used to resolve light into its constituent colors provided the period is small enough. A grating for a spectrometer can be, for example, a series of finely spaced grooves on a glass substrate [O'Shea, D. C. (1985)]. In many instances, the surface of the grating is metalized to produce a reflective surface. The 10-channel polychromator, which was developed by Kondo *et al.* [Kondo *et al.* (1993)], consists of two sets of reflection gratings and 10 InSb detectors, as shown in Fig. 3.22. Partially coherent radiation incident on a grating is diffracted over different angles.

For a concave grating, in particular, as one can see in Fig. 3.22, the diverging light from S can be imaged onto a particular point P after diffraction at the surface of the grating. In addition, if the source point S is put on the circle of diameter equal to the radius of curvature, the imaging point P is found on the same circle. Such a circle is called the *Rowland circle*. Using the notation used in Fig. 3.22, the relation between the incident and

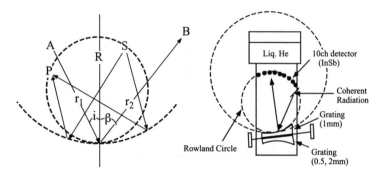

Fig. 3.22 Schematics of a grating on the Rowland circle and the 10-channel polychromator. On the left the light source S, the concave grating and the detector plane are located on a circle of diameter equal to the radius of curvature R of the grading. On the right the 10-channel polychromator consists of a concave grating and 10 InSb detectors. Incident light is imaged on the circle and diffracted by the grating.

diffracted angles is expressed as

$$d(\sin i + \sin \beta) = m\lambda. \tag{3.52}$$

Let the distance from the source A and the observation point B to the center of the circle of the concave grating be r_1 and r_2, respectively. Eq. (3.52) is then

$$\cos i \left(\frac{\cos i}{r_1} - \frac{1}{R} \right) + \cos \beta \left(\frac{\cos \beta}{r_2} - \frac{1}{R} \right) = 0. \tag{3.53}$$

If the light source A is located on the Rowland circle, i.e. $r_1 = R\cos i$, the position of the observation point is $r_2 = R\cos \beta$, namely on the same circle. On the right of Fig. 3.22, the 10-channel detectors are on the Rowland circle and the incident coherent radiation is focused to a point on the same circle [Kondo *et al.* (1993)].

3.1.3.3 *Bunch length measurement*

A diagnostic experiment with a polychromator was performed and the results were compared with those for a femtosecond streak camera. The polychromator detects a discrete power spectrum of the CTR, as shown in Fig. 3.23. The output signal in the figure contains the response function of the optics and the sensitivity of the detector as a function of frequency. Making use of a standard mercury arc, the response function and the sensitivity of the detector are calibrated. The discreteness of the experimental

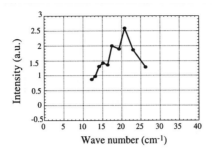

Fig. 3.23   Output of a polychromator for a picosecond electron pulse.

data, which depends on the grating pitch (1 mm) and the configuration of the measurement bins installed in the polychromator, is interpolated in order to obtain a quasi-continuous spectrum. According to the procedure of the analysis, the longitudinal bunch form factor is calculated from the spectrum. Figure 3.24 represents the bunch form factor; the solid line is

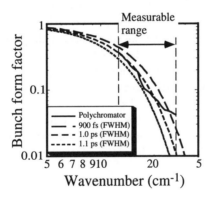

Fig. 3.24   Longitudinal bunch form factor for a subpicosecond electron pulse.

the experimental data with the interpolation. In the figure, the longitudinal bunch form factors are limited due to the measurement range determined by the grating and the detectors. Therefore, it is necessary to introduce a theoretical extrapolation function assuming a Gaussian distribution, as indicated by the dashed lines in Fig. 3.24. Here the extrapolation was applied with a Gaussian function of 1.0 ps FWHM.

Thus the longitudinal bunch distributions are reconstructed from the

extrapolated bunch form factors via a Fourier transform with the aid of
the Kramers–Kronig relations, as shown in Fig. 3.25. The solid line is

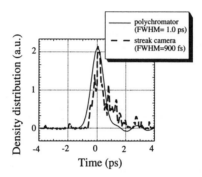

Fig. 3.25   Bunch distributions obtained by a streak camera (dashed) and a polychromator (solid).

the longitudinal bunch distribution obtained by a polychromator and the
dashed line is a typical result from a streak camera. Since the pulse distribution changes for each shot, the pulse length was measured to be from
750 fs to 1.2 ps. Consequently we conclude that the temporal resolution of
the polychromator was better than the fluctuation of the practical pulse. It
should be noted, however, that more careful investigations on avoiding erroneous measurement should be carried out. For example, the lack of spectral
information should be improved by increasing the number of bins. Once
precise measurements can be done by the polychromator, an ideal measurement, i.e. a non-destructive single-shot measurement, becomes possible via
the combination of the polychromator and coherent diffraction radiation
(CDR).

3.1.4   *Fluctuation*

3.1.4.1   *Theory*

M. ZOLOTOREV

Accelerator and Fusion Research Division

Lawrence Berkeley National Laboratory (LBNL)

1 Cyclotron Road, Berkeley,

CA 94720, USA

G. V. STUPAKOV

Stanford Linear Accelerator Center (SLAC)

PO BOX 4349, M/S 26 Stanford,

CA 94309 USA

The measurement of the longitudinal beam profile of a relativistic beam of charged particles is an important diagnostic tool in modern accelerators and is used both for routine monitoring and dedicated studies of beam physics. For bunch lengths in the range of picoseconds, such measurements can be performed by means of a streak camera. Shorter bunches usually require some kind of special technique. Several methods have been proposed that have the capability of measuring ultrashort bunches [Kung *et al.* (1994); Shintake (1996)]. In this Section we describe a method first proposed by Zolotorev and Stupakov [Zolotorev and Stupakov (1996); Zolotorev and Stupakov (1998); Krzywinski *et al.* (1997)] that can obtain the properties of a bunch of charged particles through a measurement of the fluctuations of incoherent emission from the bunch. Emission can be produced by any kind of incoherent radiation, such as that generated in a bend or wiggler, transition or Cerenkov radiation, Smith-Purcell radiation, etc.

Each particle of the bunch radiates an electromagnetic pulse with an electric field given by a function $e(t)$. If the longitudinal position of the k-th particle within the bunch is marked by a time variable t_k, the total radiated field $E(t)$ of all particles is

$$E(t) = \sum_{k=1}^{N} e(t - t_k), \qquad (3.54)$$

where N is the total number of particles in the bunch. We assume that t_k is random, with a probability of finding t_k between t and $t + dt$ equal to

$f(t)dt$, where $f(t)$ is the bunch distribution function (normalized so that $\int_{-\infty}^{\infty} f(t)dt = 1$). For a Gaussian distribution, $f(t) = (2\pi\sigma_t^2)^{-1/2}e^{-t^2/2\sigma_t^2}$, where σ_t is the bunch length in units of time. We also assume that the positions of the different particles in the bunch, t_k and t_i for $k \neq i$, are uncorrelated, $\langle t_k t_i \rangle = \langle t_k \rangle \langle t_i \rangle$, with angular brackets denoting an average. The spectral properties of the radiation are related to the Fourier transform $\hat{E}(\omega)$ of the radiated field by

$$\hat{E}(\omega) = \int_{-\infty}^{\infty} E(t)e^{i\omega t}dt = \hat{e}(\omega)\sum_{k=1}^{N} e^{i\omega t_k}, \qquad (3.55)$$

where $\hat{e}(\omega) = \int_{-\infty}^{\infty} e(t)e^{i\omega t}dt$. In an experiment, one is interested in the spectrum of the radiation $P(\omega)$, which is proportional to $|\hat{E}(\omega)|^2$ (for ease of notation, we take $P(\omega) = |\hat{E}(\omega)|^2$):

$$P(\omega) = |\hat{e}(\omega)|^2 \sum_{k,l=1}^{N} e^{i\omega(t_k - t_l)}. \qquad (3.56)$$

Averaging this equation, we find

$$\langle P(\omega) \rangle = |\hat{e}(\omega)|^2 \sum_{k,l=1}^{N} \int_{-\infty}^{\infty}\int_{-\infty}^{\infty} dt_k dt_l f(t_k)f(t_l)e^{i\omega(t_k - t_l)}$$

$$= |\hat{e}(\omega)|^2(N + N^2|\hat{f}(\omega)|^2), \qquad (3.57)$$

where $\hat{f}(\omega) = \int_{-\infty}^{\infty} f(t)e^{i\omega t}dt$ is the Fourier transform of the distribution function (for the Gaussian distribution mentioned above, $\hat{f}(\omega) = e^{-\omega^2/2\sigma_t^2}$) and we set $N - 1 \approx N$. The first term in Eq. (3.57) is the incoherent radiation, proportional to the number of particles in the bunch. The second term is the coherent radiation that scales quadratically with N. The coherent radiation term carries information about the distribution function of the beam, but only at relatively low frequencies, of the order of $\omega \lesssim \sigma_t^{-1}$, where $\hat{f}(\omega)$ is not zero. At high frequencies, where $N|\hat{f}(\omega)|^2 \ll 1$, the coherent radiation is negligible in comparison with the incoherent radiation.

However, the original (not averaged) expression for the spectral power, Eq. (3.56), shows that the properties of the radiation, even at high frequencies, carry information about the distribution function. Indeed, each term $e^{i\omega(t_k - t_l)}$ considered separately oscillates as a function of frequency, with a period $\Delta\omega = 2\pi/(t_k - t_l) \sim 2\pi/\sigma_t$. Because of the random distribution of particles in the bunch, the sum in Eq. (3.56) fluctuates randomly as a function of frequency ω (see Fig. 3.26) and the statistical properties of these fluctuations depends on properties of the distribution function of the bunch.

To obtain a quantitative characteristic of the fluctuation, we first calculate the average value of the product $P(\omega)P(\omega')$,

$$\langle P(\omega)P(\omega')\rangle = |\hat{e}(\omega)|^2|\hat{e}(\omega')|^2 \sum_{k,l,m,n=1}^{N} \langle e^{i\omega(t_k-t_l)+i\omega'(t_m-t_n)}\rangle. \quad (3.58)$$

Assuming that $N|f(\omega)|^2, N|f(\omega')|^2 \ll 1$ (which means that we can neglect the coherent radiation), it is straightforward to find

$$\langle P(\omega)P(\omega')\rangle = N^2|\hat{e}(\omega)|^2|\hat{e}(\omega')|^2(1 + |\hat{f}(\omega-\omega')|^2), \quad (3.59)$$

where the contribution to the final result comes from terms with $k = l$, $m = n$, $k \neq m$ and $k = n$, $l = m$, $k \neq l$. It is convenient to define the second order correlation function $g(\omega-\omega') = \langle P(\omega)P(\omega')\rangle/2\langle P(\omega)\rangle\langle P(\omega')\rangle$. Then, from Eqs. (3.56) and (3.59) if follows that

$$g(\Omega) = \frac{1}{2}\left(1 + |\hat{f}(\Omega)|^2\right). \quad (3.60)$$

Note that $\hat{f}(0) = 1$ and $\hat{f}(\infty) = 0$, correspondingly $g(0) = 1$ and $g(\infty) = 1/2$.

Another useful quantity is the Fourier transform $G(\tau)$ of the spectrum [Zolotorev and Stupakov (1998)],

$$G(\tau) = \int_{-\infty}^{\infty} P(\omega)e^{i\omega\tau}d\omega. \quad (3.61)$$

Analogous to the derivation of Eq. (3.59), one can show that the variance $D(\tau) = \langle |G(\tau) - \langle G(\tau)\rangle|^2\rangle$ is proportional to the convolution of the distribution function f as

$$D(\tau) = A\int_{-\infty}^{\infty} f(t)f(t-\tau)dt. \quad (3.62)$$

3.1.4.2 *Discussion*

The above consideration assumes a filament beam, with a negligibly small transverse size. In this case, the fluctuation of the spectral intensity $P(\omega)$ is 100% and the autocorrelation function g in Eq. (3.60) varies from 1 (at $\Omega = 0$) to 1/2 (at $\Omega \to \infty$). This regime requires that the transverse beam size be smaller than the transverse coherence size of the radiation, which is of order of λ/σ_θ, where the reduced wavelength $\lambda = c/\omega$ and σ_θ is the angular spread of the radiation. In the opposite case, the fluctuation is less pronounced.

We also used a classical description of the radiation process and neglected quantum effects. This approach is justified if the quantum fluctuations of the number of radiated photons are negligible. We can easily derive the condition for the beam intensity when this requirement is met. If the source of the radiation is synchrotron light, the number of photons n_{ph} that reach the detector in the frequency interval $\Delta\omega$ (they are radiated from $(1/\gamma)(\omega_c/\omega)^{1/3}$ radians of the beam trajectory, where ω_c is the critical frequency for the synchrotron radiation) can be estimated to be [Sands (1970)]

$$\langle n_{ph}\rangle \approx \frac{1}{2}\alpha N_e \frac{\Delta\omega}{\omega}, \tag{3.63}$$

where α is the fine structure constant and N_e is the number of particles in the bunch. Taking $\Delta\omega \sim \sigma_t^{-1}$ and $\lambda \sim 1\ \mu\text{m}/2\pi$ (visible light) we can rewrite Eq. (3.63) as $\langle n_{ph}\rangle \approx 3\cdot 10^5 (I/I_A)$, where I is the peak current in the beam and $I_A = 17$ kA is the Alfven current. Quantum effects can be neglected when $\langle n_{ph}\rangle \gg 1$. Note that in this estimate we assumed that the beam radiation is coherent in the transverse direction (see above).

For wiggler radiation, the estimate of the number of radiated photons is

$$\langle n_{ph}\rangle \approx 4\alpha N_e M \frac{K^2}{1+K^2}\frac{\Delta\omega}{\omega}, \tag{3.64}$$

where K is the wiggler parameter and M is the number of periods in the wiggler.

If the condition $\langle n_{ph}\rangle \gg 1$ is not met, quantum fluctuations superimpose on the random fluctuations described above. However, information about the pulse shape is still present in the measured signal, although more statistics would be needed to suppress the additional noise introduced by the quantum fluctuations.

Strictly speaking, knowledge of the spectral function $|\hat{f}(\omega)|$ (or a convolution function) does not allow a unique restoration of $f(t)$, because the information about the phase of $\hat{f}(\omega)$ is lost. However, using a phase retrieval technique allows restoration of the beam profile in many practical cases [Lai and Sievers (1995b)].

Finally, we would like to emphasize that measurement of the beam profile using spectral fluctuations can be done in a single shot if the width of the measured spectrum is wide enough. In this case, averaging of different quantities can be done over the spectrum, rather than over multiple shots.

3.1.4.3 *Experiment*

Measurement of the bunch length using spectral fluctuations of incoherent radiation was reported in Refs. [Catravas *et al.* (1999); Sajaev (2000)]. Here we summarize the main results of Ref. [Sajaev (2000)].

In Ref. [Sajaev (2000)], radiation spectra were measured at the Low-Energy Undulator Test Line at the Advanced Photon Source (APS) at the Argonne National Laboratory using a high resolution spectrometer with a cooled CCD imager, with a resolution of 0.4 Å per pixel. Measurement of the single-shot spectrum is shown in Fig. 3.26. The spectrum is composed of spikes of random amplitude and frequency that have a characteristic width $\Delta\omega \sim \sigma_t^{-1}$ and intensity fluctuation of almost 100%. The shape of an individual spectrum changes randomly from shot to shot, but the average of many shots approaches the wiggler spectrum.

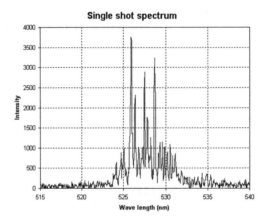

Fig. 3.26   Typical single-shot spectrum from Ref. [Sajaev (2000)].

The normalized second-order correlation function g averaged over 100 shots is plotted in Fig. 3.27. We see that g varies from unity to a value close to 0.5, in agreement with Eq. (3.60). The spike width at half maximum, according to Fig. 3.27, is about two pixels. The frequency step corresponding to one pixel is $\delta\omega = 2.4 \times 10^{11}$ rad/s. Therefore, assuming a Gaussian

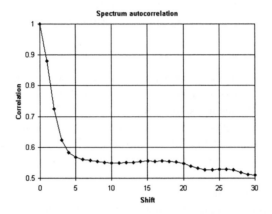

Fig. 3.27   Autocorrelation of spectrum from Ref. [Sajaev (2000)]. The horizontal axis is the number of CCD pixels; one pixel corresponds to 2.4×10^{11} rad/s.

beam, the sigma of the Gaussian distribution is $\sigma_t \sim 1/2\delta\omega \approx 2$ ps. This gives a FWHM length of the bunch of 4.5 ps.

Using the phase extraction technique [Lai and Sievers (1995b)], Sajaev [Sajaev (2000)] was able to recover the bunch shape from spectral measurements.

Acknowledgment

We thank V. Sajaev for providing us with copies of figures from Ref. [Sajaev (2000)].

3.1.4.4   *Fluctuation in time domain*

K. NAKAMURA
T. WATANABE
M. UESAKA
Nuclear Engineering Research Laboratory, University of Tokyo
2-22 Shirane-shirakata, Tokai, Naka,
IBARAKI 319-1188, JAPAN

Here we assume, for simplicity, that the electromagnetic field of radiation can be described in terms of the classical field and that electron bunches have zero emittance. We write the total electric field of radiation from an electron bunch as

$$E(t) = \sum_{k=1}^{N} e(t - t_k), \qquad (3.65)$$

where N is the number of electrons in the bunch, t_k is the arrival time of electrons at a certain position and $e(t)$ is the electric field from an electron. Each electron in the bunch radiates independently and the radiation $e(t)$ forms a wave packet, the width of which is inversely proportional to the bandwidth of the radiation. Equation (3.65) shows that the radiation from the electron bunch consists of a random superposition of wave packets. Such light is known as *thermal* or *chaotic light* [Loudon (1983)]. The intensity of the chaotic light is characterized by the coherent length τ_{coh} as

$$\tau_{coh} \propto \frac{1}{\Delta\omega}, \qquad (3.66)$$

where $\Delta\omega$ is the bandwidth of the radiation. For a random superposition of a large number of waves, the probability distribution of the field amplitude $|E(t)|$ is given by a Gaussian function. This implies that the intensity distribution $I(t) = |E(t)|^2$ is given by an exponential distribution, the standard deviation of which is the same as its average. Consider the case in which the radiation is observed within a certain interval. If the bunch length τ_b is shorter than τ_{coh}, the time-integrated intensity, then

$$I = \int_{-\infty}^{\infty} I(t)\mathrm{d}t = \int_{-\infty}^{\infty} |E(t)|^2 \mathrm{d}t, \qquad (3.67)$$

shows 100% fluctuation shot by shot, independently of τ_b and τ_{coh}. The distribution of I follows an exponential distribution [Kim (2000)],

$$f(x) = \lambda \exp(-\lambda x), \qquad (3.68)$$

where x corresponds to I. Both the expectation $E(x)$ and the standard deviation $S(x)$ of the exponential distribution are equal to $1/\lambda$, therefore the fluctuation of I, namely the coefficient of variation (hereafter $C.V.$) is given by

$$C.V. = \frac{S(x)}{E(x)} = 1. \qquad (3.69)$$

The distribution of I in the case of $\tau_b < \tau_{coh}$, is shown in Fig. 3.28(a). In this case, the bunch length cannot be deduced from the fluctuation. If the bunch length τ_b is longer than τ_{coh}, the pulse consists of m independent slices, as

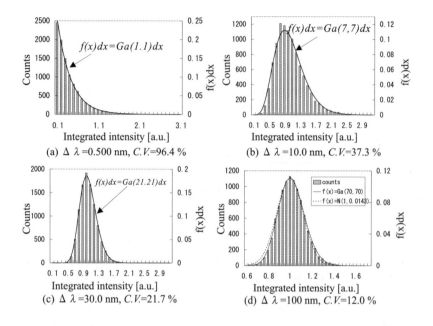

Fig. 3.28 Histograms and fluctuations of the calculated integrated intensity I over 10 000 shots with fitting curves of the gamma distribution for $\lambda_0 = 700$ nm and bunch length $\tau_b = 500$ fs. The integrated intensity I is normalized so that its average is unity. The longitudinal mode numbers are deduced from fluctuations to be about (a) 1, (b) 7, (c) 21 and (d) 69. Finally, the bunch length is deduced to be (a) 1550 fs, (b) 518 fs, (c) 509 fs and (d) 498 fs from Eq. (3.74). When the longitudinal mode number is large, the gamma distribution approaches the Gaussian distribution $N(E(x), \sigma^2)$, as shown in (d), where $E(x)$ is the expectation and σ is the standard deviation.

shown in Fig. 3.29. Since the probability distribution of the integrated intensity of each slice follows the exponential distribution independently, the resultant fluctuation in intensity is smoothed out [Kim (2000)]. The sum of m independent exponential distributions is known as the gamma distribution:

$$G_a(m, \lambda) = f(x) = \frac{\lambda^m}{\Gamma(m)} x^{m-1} \exp\left(-\lambda x\right), \qquad (3.70)$$

$$\Gamma(m) = \int_0^\infty x^{m-1} \exp\left(-x\right) dx, \qquad (3.71)$$

where the number of independent distributions accords with the number of slices in radiation intensity, and again x corresponds to I. The distribution

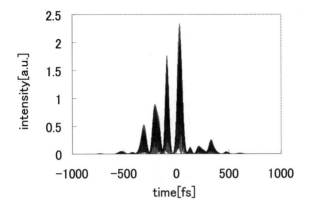

Fig. 3.29 Intensity of radiation as a function of time for $\lambda_0 = 700$ nm, bandwidth $\Delta\lambda = 15$ nm, $\tau_{coh} \simeq 50$ fs, $\tau_b = 500$ fs and $m_l = \tau_b/\tau_{coh} \simeq 10$. The temporal distribution of electrons is Gaussian.

of I in the case of $m \simeq 7$, $m \simeq 21$ and $m \simeq 70$, is shown in Fig. 3.28(b), (c) and (d). In the case of large m, the gamma distribution approaches the Gaussian distribution $N(E(x), S^2(x))$, as

$$N(E(x), S^2(x)) = f(x) = \frac{1}{\sqrt{2\pi}S(x)} \exp\left(-\frac{(x - E(x))^2}{2S^2(x)}\right) \qquad (3.72)$$

[see Fig. 3.28(d)]. The number of slices, namely the number of the longitudinal coherent mode m_l, is given by the ratio of the bunch length and the coherent length:

$$m_l = \frac{\tau_b}{\tau_{coh}}. \qquad (3.73)$$

Since the expectation and the standard deviation of the gamma distribution are m/λ and $\sqrt{m}/\lambda$, respectively, the fluctuation of the time-integrated intensity of the radiation from the electron bunch can be expressed as

$$C.V. = \frac{S(x)}{E(x)} = \frac{1}{\sqrt{m_l}} = \sqrt{\frac{\tau_{coh}}{\tau_b}} \propto \frac{1}{\sqrt{\tau_b \Delta\omega}}. \qquad (3.74)$$

Equation (3.74) shows that the bunch length τ_b can be acquired from the measured fluctuation and the bandwidth. One can control the bandwidth with a band pass filter (BPF). Note that the measurement of time-integrated intensity does not require a fast detector like a streak camera [Nakamura *et al.* (2004)].

3.1.5   *Overall comparison*

T. WATANABE

M. UESAKA

K. NAKAMURA

Nuclear Engineering Research Laboratory, University of Tokyo

2-22 Shirane-shirakata, Tokai, Naka,

IBARAKI 319-1188, JAPAN

3.1.5.1   *Theoretical discussion*

To begin with, we shall summarize theories of the four schemes presented in the previous sections.

The electric field of an electron pulse is described as

$$E(\omega) = e(\omega) \sum_{k=1}^{N} \exp(i\omega t_k) \,, \qquad (3.75)$$

where $e(\omega)$ is the spectrum of a single electron at frequency ω, and t_k is the arrival time at some reference point of the k-th electron among N electrons. Thus the first-order correlation function for different frequencies $E(\omega)$ and $E(\omega')$ is written as

$$\langle E(\omega)E^*(\omega') \rangle = e(\omega)e^*(\omega') \langle \sum_{k=1}^{N} \sum_{l=1}^{N} \exp(i\omega t_k - i\omega' t_l) \rangle \,, \qquad (3.76)$$

where the angular brackets denote an ensemble average [Zolotorev and Stupakov (1996); Saldin *et al.*] for the random variables $t_{k,l}$. The ensemble average of the phasors $e^{i\omega t_k}$ is

$$\langle \exp(i\omega t_k) \rangle = F(\omega) \,. \qquad (3.77)$$

$F(\omega)$ is related to the bunch form factor $f(\omega)$ by

$$|F(\omega)|^2 = f(\omega) \,. \qquad (3.78)$$

Equation (3.77) is an important relation between random variables and a continuous function; $F(\omega)$ is the Fourier harmonic of a continuous electron distribution. By expanding Eq. (3.76) with the aid of Eq. (3.77), the first-order correlation function can be written as

$$\langle E(\omega)E^*(\omega') \rangle = e(\omega)e^*(\omega')NF(\Delta\omega) + e(\omega)e^*(\omega')N(N-1)F(\omega)F^*(\omega') \,, \qquad (3.79)$$

where $\Delta\omega = \omega - \omega'$. The first term includes $F(\Delta\omega)$ and is proportional to the number of electrons N. The second term consists of $N(N-1)$ and two forms of $F(\omega)$. The factor F is a function of $\Delta\omega$ in the first term and of ω itself in the second term. For incoherent radiation, where $N|F(\omega)|^2 \ll 1$, the second term disappears. On the contrary, in the case of coherent radiation, where $N|F(\omega)|^2 \gg 1$, the second term is dominant, since the second term is approximately proportional to N^2 and the number of electrons N is quite large. A more general expression for the power spectrum can be obtained directly by substituting $\omega = \omega'$ into Eq. (3.79), to give

$$I(\omega) = NI_0(\omega) + N(N-1)f(\omega)I_0(\omega), \tag{3.80}$$

which is exactly the same as Eq. (3.51). Equation (3.80) indicates that the bunch distribution can be obtained only by coherent radiation, since the first term does not include the bunch information $f(\omega)$. In order to extract the bunch information from incoherent radiation, a second-order (or higher) correlation function has to be introduced. The relation between normalized first- and second-order correlation functions is known as the Siegert relation:

$$g_2(\omega, \omega') = 1 + |g_1(\omega, \omega')|^2, \tag{3.81}$$

where g_1 and g_2 are the first- and second-order correlation functions, respectively. From Eqs. (3.79) and (3.81), the second-order correlation function is written as

$$g_2(\omega, \omega') = 1 + |F(\Delta\omega)|^2 = 1 + f(\Delta\omega). \tag{3.82}$$

Here it is assumed that the wavelength of the radiation is much shorter than the pulse duration. Hence the bunch distribution can be deduced from incoherent radiation in a similar fashion as from coherent radiation.

3.1.5.2 *Experimental discussion*

The effectiveness of the diagnostic methods relies largely upon the characteristics of each accelerator. For accelerators that generate picosecond pulses, for instance, the streak camera would be the best apparatus. However, accelerators generating tens of femtoseconds pulses or shorter require an alternative scheme. The following overall evaluation was carried out for the linear accelerators at the Nuclear Engineering Research Laboratory (NERL), University of Tokyo (35 MeV with a thermionic gun, 18 MeV with a photoinjector, see Sec. 2.2.2), which will be illustrative of other accelerators.

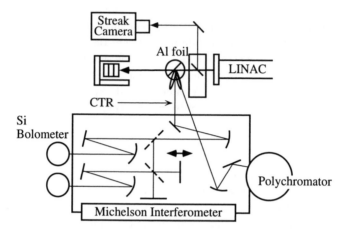

Fig. 3.30   Experimental setup for diagnostics by Michelson interferometer, 10-channel polychromator and femtosecond streak camera.

Diagnostic experiments using a femtosecond streak camera, Michelson interferometer and 10-channel polychromator were carried out at NERL [Watanabe *et al.* (2002)]. The experimental setup is illustrated in Fig. 3.30. Cherenkov radiation in visible light is sent to the streak camera and transition radiation in the far infrared region is transported to the Michelson interferometer and the 10-channel polychromator by turns. An example measurement result is shown in Fig. 3.31. The bunch durations for the streak camera, interferometer and polychromator were 1.0, 1.2 and 1.0 ps, respectively. As discussed in Ref. [Watanabe *et al.* (2002)] and Sec. 3.1.2.4, error factors have to be considered and calibrated as much as possible.

Having considered the discussions on the femtosecond streak camera (cf. Sec. 3.1.1), Michelson interferometer (cf. Sec. 3.1.2), 10-channel polychromator (cf. Sec. 3.1.3), fluctuation method (cf. Sec. 3.1.4) and a comparison with the experiment in the present section, we will summarize the characteristic features of the four diagnostic methods. For this purpose, let us first list the important diagnostic parameters:

(a) type of radiation,
(b) availability of single-shot measurement,
(c) time resolution,

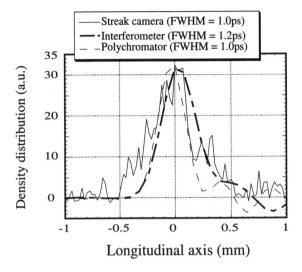

Fig. 3.31 Experimental results of femtosecond streak camera, Michelson interferometer and 10-channel polychromator.

(d) error factors,

(e) dynamic range of bunch duration,

(f) availability of timing jitter measurement.

(a) Radiation can be classified into two kinds; coherent and incoherent radiation. Coherent radiation is much brighter than incoherent radiation (see Sec. 3.1.4). Hence accelerators that cannot generate much charge require measurement with coherent radiation. (b) Single-shot measurements are quite useful for tuning the beam parameter or observing the fluctuation of a bunch distribution. (c) We can distinguish the measurement limit from the measurement error. Here we mean the time resolution to be the shortest measurable limit of the pulse length. (d) Diagnostic schemes require the observation of other quantities, such as electron charge and transverse beam profile. All the quantities observed in an experiment can include experimental errors. The assumptions of a theory can also be an error factor. Furthermore, factors that affect the measurement limit, or the absolute error, can be considered to be error factors. (e) Every diagnostic method has a measurable range of the bunch duration. If the available range is small, the bunch duration must be known precisely in advance. (f) We are interested in the timing jitter between two bunches, such as femtosecond electron pulses and femtosecond laser pulses. For this reason, the availability of timing jitter measurement is an important factor.

We shall begin our discussion with the streak camera. (a) The streak camera resolves visible light, which is incoherent radiation due to its short wavelength compared with the bunch duration. Since incoherent (optical) transition radiation and diffraction radiation are not bright enough for low energy (≤ 50 MeV) linacs, Cherenkov radiation is usually used in that case. Empirically an electron charge of hundreds of pC is enough for Cherenkov radiation. (b) One of the most powerful advantages of the streak camera is the availability of single-shot measurement. One can observe changes in a bunch distribution simultaneously. (c) The time resolution of the fastest resolving streak camera is known to be about 200 fs (see Sec. 3.1.1.2). (d) Since the streak camera directly measures the bunch distribution in the time domain, one does not need to pay attention to other beam parameters such as an electron charge, energy distribution or transverse beam emittance. As discussed in Sec. 3.1.2.2, the light intensity can distort the time resolution of the streak camera. Also, chromatic and achromatic aberrations can distort the time resolution of the measurement using radiation. (e) The dynamic range of bunch duration of the femtosecond streak camera is three orders of magnitude from 200 fs to several hundreds of ps. One can easily change the measurement range within the dynamic range to observe the practical bunch duration. (f) There is no restriction in measuring two bunches at the same time if their timing interval is not too large compared with their pulse duration.

We next consider the Michelson interferometer. (a) The Michelson interferometer is based on the interference of light and thus it detects coherent radiation, of which the wavelength should be much longer than the bunch duration. For this purpose, a far infrared detector is adopted. Since coherent radiation is bright, a liquid-He-cooled Si bolometer can detect both coherent transition and diffraction radiation without difficulty. (b) In order to acquire an interferogram from a Michelson interferometer, one has to scan the linear stage step by step. It takes at least a few minutes to get one interferogram, which corresponds to several thousands of shots. (c) The theoretical limit of the method seems to be less than 10 fs. Shorter pulses makes the measurement easier, since one can use a wide use detector for shorter wavelengths. The response functions for infrared to far infrared detectors are shown in Fig. 3.32. If the light intensity is great enough, the pyroelectric detector provides a wide frequency range (cf. Sec. 3.1.2.2). If the intensity is not large, several kinds of detectors can cover each frequency band. For pulses less than 10 fs, even an optical detector can be used. (d)

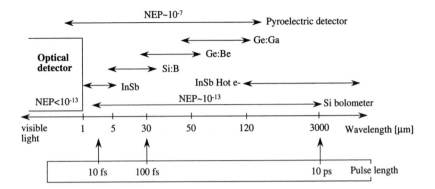

Fig. 3.32 Detectors for infrared and far infrared radiation.

An interferogram can be distorted by a misalignment of the optics. In the reconstruction procedure from the interferogram to the bunch distribution, the electron charge and transverse beam emittance have to be observed. The spectrum emitted by a single electron is estimated from theory under several assumptions of the energy distribution, the size of the radiator and so on. Chromatic and achromatic aberrations can be absolute error factors. (e) For the sake of the most reliable estimation, the interferometer detects the wavelength so that the bunch form factor is around 0.1. Hence, the dynamic range of the apparatus is of the order of 10. (f) It is impossible to measure the timing jitter, since single-shot measurement is not available.

The third apparatus is the polychromator. (a), (c), (d), (e) and (f) for the polychromator are similar to those of the Michelson interferometer; only item (b) is different. Since the polychromator can acquire a power spectrum of coherent radiation shot by shot, the bunch distribution can also be deduced by a single shot.

The last method is the fluctuation method (see Sec. 3.1.4). (a) The fluctuation method is based on a stochastic process of incoherent radiation, of which the intensity is low. Furthermore, in order to extract a narrow band from the white spectrum of Cherenkov radiation, the radiation intensity decreases. Therefore, light intensity is one of the most severe conditions for this method. If spontaneous emission is available, it is much easier to make measurements. (b) A rough evaluation of the bunch duration can be taken by a single shot. The precise estimation of the bunch distribution is, however, achieved by the stochastic analysis of the shot noise. (c) The theoretical limit of the method appears to be infinitesimal, although further investigations based on experiments need to be done. Catravas *et al.* reported that 1–5-ps pulses can be measured with an error of 35% [Catravas *et al.* (1999)]. Sajaev also reported the measurements of 4-ps pulses [Sajaev (2000)]. They used undulator radiation at BNL-ATF and APS-linac to obtain narrow band spectra with sufficient intensity simultaneously. (d) The transverse beam emittance affects the result of the longitudinal bunch duration, since the coherent volume in the three-dimensional beam is measured simultaneously in this method. (e) Though one should pay attention to the bunch duration in advance of measurements, the dynamic range is not narrow compared to measurements using coherent radiation. The dynamic range of the spectrometer in the fluctuation method is around two orders of magnitude. (f) In time domain measurements, when the bunch duration becomes short, fluctuation is enhanced, which makes measurement easier. Spectra of shorter bunch duration are also wider in the frequency domain. Hence it is easy to measure without an exceedingly high resolution spectrometer. Thus, the method becomes more powerful as the bunch duration becomes shorter.

All the characteristics that we have so far mentioned are summarized in Table 3.1 and Fig. 3.33. We conclude finally that the femtosecond streak camera is the best diagnostic apparatus within its time resolution of 200 fs, at present. Nevertheless the other three diagnostic methods, the Michelson interferometer, the polychromator and the fluctuation method, are extremely important for crosschecking results and for unlimited time resolution and low price. For practical electron bunches beyond the time resolution of the streak camera, one has to use an alternative apparatus, such as one of the other three methods. Furthermore, studies on such diagnostics reveal valuable information about the method and measurements.

Table 3.1   Characteristics of four diagnostic methods.

	Streak cam.	Interferometer	Polychromator	Fluctuation
Radiation	incoherent	coherent	coherent	incoherent
Intensity	low	high	high	low
Wavelength	visible	FIR	FIR	visible
Single shot				
Duration	available	unavailable	available	available
Distribution	available	unavailable	difficult	unavailable
Resolution	200 fs	$\sim$ 10 fs	$\sim$ 10 fs	$\sim$ 10 fs
Error	w/o	$\sim$ 20%	$\sim$ 20%	$\sim$ 35%
				[Catravas *et al.* (1999)]
Error	intensity	optical alignment		beam size
factors	aberrations	electron charge		aberrations
		beam size		
		theory		
		aberrations		
Dynamic range	$\sim 10^3$	~ 10	~ 10	$\sim 10^2$
Timing	available	unavailable	unavailable	unavailable
jitter				

3.1.6   *New trends*

M. UESAKA

K. NAKAMURA

Nuclear Engineering Research Laboratory, University of Tokyo

2-22 Shirane-shirakata, Tokai, Naka,

IBARAKI 319-1188, JAPAN

3.1.6.1   *Electro-optical method*

Recently, the application of well-established electro-optic techniques to the measurement of ultrashort electron bunches has been pursued at several laboratories [Srinivasan-Rao *et al.* (2002); Fitch *et al.* (2001); Yan *et al.* (2000)]. In this section we introduce this technique briefly as a new trend in diagnosis.

In essence, this technique utilizes birefringence induced by external electric fields derived from electron bunches at a birefringent crystal, and a synchronized short pulse laser as a probe. The conceptual experimental arrangement of a recently proposed single-shot scheme [Srinivasan-Rao *et al.*

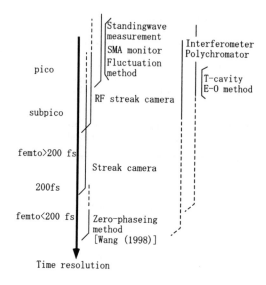

Fig. 3.33   Diagnostic schemes and available time ranges.

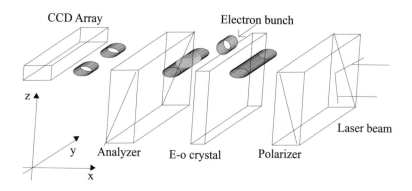

Fig. 3.34   Conceptual experimental arrangement. A linear polarized laser beam is injected into an e-o crystal as a probe of birefringence induced by an electron bunch.

(2002)] is shown in Fig. 3.34. We consider a focused electron bunch propagating above a birefringent crystal, parallel to the y-axis, as shown in Fig. 3.34. The electric field in the crystal at a distance r from the electron beam, due to the charge σdv, can be written as

$$\mathrm{d}\boldsymbol{E} = (\gamma/4\pi\varepsilon_0)\sigma dv/\varepsilon r^2 \boldsymbol{r} \,. \tag{3.83}$$

This field should be numerically evaluated with an assumed charge density profile σ. When such an electric field is applied to an anisotropic crystal, the refractive index ellipsoid of the crystal undergoes a modulation. The equation of an index ellipsoid for an anisotropic crystal in the presence of an external electric field is given by [Yariv (1989)]

$$x^2(\frac{1}{n_1^2} + r_{1j}E_j) + y^2(\frac{1}{n_2^2} + r_{2j}E_j) + z^2(\frac{1}{n_3^2} + r_{3j}E_j)$$
$$+2yz(r_{4j}E_j) + 2xz(r_{5j}E_j) - 2xy(r_{6j}E_j) = 1 \,, \tag{3.84}$$

where x, y and z are the coordinates parallel to the principal dielectric axes of the crystal when no field is present, n_1, n_2 and n_3 are the indices of refraction along x, y and z, respectively, r_{ij} are the elements of the electro-optic tensor and E_j ($j = 1, 2, 3$) are the components of the applied field along x, y and z. As can be seen from Eq. (3.84), the principal axes of the index ellipsoid are rotated by an angle ϕ in the presence of an electric field. This rotation modulates the laser intensity observed after the analyser and polarizer, and the charge density profile can be deduced from this modulation.

3.1.6.2 *T-cavity method*

Development of an electron bunch diagnostic system for a future free electron laser that uses bunches as short as tens of fs rms has been done at Stanford Linear Acclerator Center (SLAC) [Krejcik *et al.* (2003)]. The method streaks the bunch vertically using a transverse RF deflecting structure, as shown in Fig. 3.35. In this Section we give a brief introduction to the concept and recent results of measurements.

An RF transverse deflecting cavity is installed and operated at the zero-crossing phase of a field, so that a bunch is given a strong correlation between the longitudinal z-coordinate and the transverse position on the beam profile monitor (BPM). The vertical beam size at the BPM σ_y is a function of the bunch length σ_z and the deflector parameters [Emma *et al.* (2001); Akre *et al.* (2002)] as

$$\sigma_y = \sqrt{\sigma_{y0}^2 + \sigma_z^2 \beta_c \beta_p \left(\frac{2\pi e V_0}{\lambda_{rf} E_0} \sin \Delta\Psi_y \cos \varphi_{rf}\right)^2} \,. \tag{3.85}$$

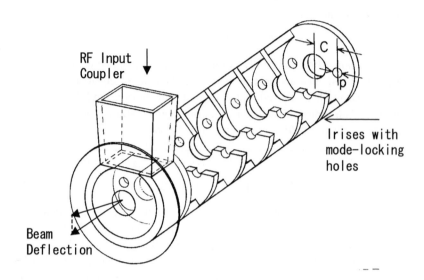

Fig. 3.35   Schematic of a SLAC S-band transverse deflecting structure. The kick is vertical in this drawing.

The bunch length is determined from fitting of the measured vertical beam sizes as a function of RF deflecting voltage and their quadratic relation

$$\sigma_y^2 = A(V_{rf} - V_{rf_{min}})^2 + \sigma_{y_0}^2 , \qquad (3.86)$$

where A gives the bunch length,

$$\sigma_z = A^{1/2} \frac{E_0 \lambda_{rf}}{R_{34} 2\pi} . \qquad (3.87)$$

A typical experimental result is given in Fig. 3.36. Fitting gives $\sigma_z = 1.2$ ps.

$$\sigma_z = 0.366 \text{ mm; } \sigma_{y0} = 0.88 \text{ mm; } \varepsilon_y/\varepsilon_{y0} = 1.07$$

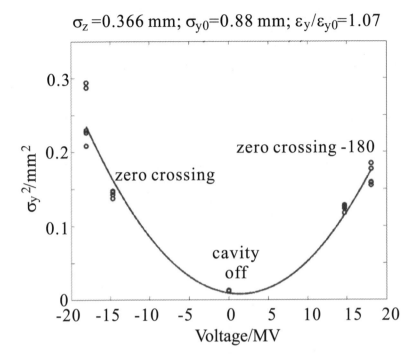

Fig. 3.36 Vertical beam size squared measured as a function of deflector voltage, yielding $\sigma_z = 1.2$ ps.

3.1.7 *Low jitter X-ray streak camera*

A. TAKAHASHI
System Division, Hamamatsu Photonics K.K.
812,Joko-cho,Hamamatsu City,
SHIZUOKA 431-3196, JAPAN

Recently, X-ray streak cameras have been used in spectroscopy and time-resolved X-ray diffraction measurements. As the X-ray pulse width decreases, sub-picosecond temporal resolution and high dynamic range are required. Commercially available X-ray streak cameras have temporal resolutions of less than 2 ps in the single-shot mode, but the data obtained from these cameras lack sufficient dynamic range due to space charge limitations of photoelectrons within the streak tube. To obtain a high signal-to-noise measurement, signal integration is required, i.e. the X-ray streak camera should operate in a signal accumulation mode. However, in the accumulation mode, the temporal resolution of the streak camera is limited by timing jitter caused by electrical noise within the deflection circuit. On the other hand, a synchroscan type streak camera can be operated in the accumulation mode with very low jitter (less than 1 ps) by means of sinusoidal sweep deflection, which phase locks to a stable laser repetition frequency up to 80 MHz [Lumpkin (1998)]. However, there are limitations in the time resolution of the synchroscan-type streak camera when operating at a laser repetition frequency less than a few kHz. In this section, we describe the principle and characteristics of the low jitter X-ray streak camera using a GaAs optical (photoconductive) switch, which has sub-picosecond temporal resolution and a high dynamic range for a accumulation frequency of a few kHz. The principle of the low jitter X-ray streak camera is shown in Fig. 3.37.

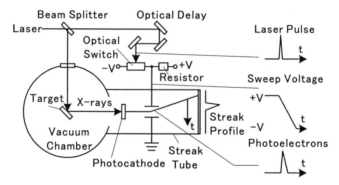

Fig. 3.37   Principle of the low jitter X-ray streak camera.

A high intensity ultrafast laser is passed through a beam splitter. One branch of the beam triggers the GaAs optical switch; the other is focused on the target that generates X-rays. The GaAs optical switch is biased at high voltage and applied to the deflection plates of the X-ray streak tube. The deflection plates are swept by a high voltage ramp waveform, which is produced by photoconduction of the GaAs optical switch. There is minimal timing jitter between the laser pulse producing the X-rays and triggering of the GaAs optical switch. Therefore, the X-ray streak camera is able to accumulate a signal with minimal reduction in temporal resolution.

The work done by a group at the University of Michigan in 1996 is shown

in Fig. 3.38 [Maksimchuk *et al.* (1996); Mourou (1996)]. It is clear from
the streak images of the X-ray emission in the figure that the accumulation
of X-ray images significantly increased the signal-to-noise ratio of the data.
These data were measured using an X-ray streak camera with a temporal
jitter of less than ±1 ps, laser energy of about 100 $\mu J/mm^2$ applied to the
GaAs optical switch and a sweep speed of 8 ps/mm.

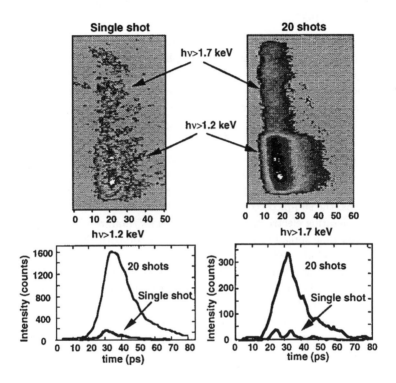

Fig. 3.38 Streaked images of broadband X-ray emission from a Ni target passing
through 25 μm (cutoff energy $h\nu > 1.2$ keV) and 50 μm (cutoff energy $h\nu > 1.7$ keV).

In 1998, the group at the European Synchrotron Radiation Facility
(ESRF) developed an X-ray streak camera with a temporal resolution of
460 fs and 640 fs for accumulation times of 900 and 54,000 shots respectively, measuring the third harmonic, 267 nm, of a Ti:Sapphire laser with
pulse width 100 fs and repetition rate 1 kHz [Scheidt and Naylor (1999)].
They achieved this by improving the streak tube design and optimizing the

laser pulse that triggers the GaAs optical switch.

The temporal resolution of the streak camera in the accumulation mode is determined by a convolution of its temporal resolution in the single-shot mode and the timing jitter caused by the laser that triggers the GaAs optical switch.

In the single-shot mode, the temporal resolution of the streak camera is limited mainly by the following two factors [Takahashi *et al.* (1994); Chang *et al.* (1997)]:

(a) Energy distribution of the photoelectrons
    Most of the transit time spread of photoelectrons in the streak tube is caused by the energy distribution of the photoelectrons in the photocathode. This time spread Δt_1 is in reverse proportion to the accelerating electric field and is defined to be

$$\Delta t_1 = 2.63 \times 10^{-6} (\Delta\varepsilon)^{1/2}/E, \qquad (3.88)$$

where E [V/m] is the acceleration electric field and $\Delta\varepsilon$ [eV] is the half width of the energy distribution of the photoelectrons.
    In order to further suppress the energy distribution of the photoelectrons, the sub-picosecond X-ray streak camera is designed so that the accelerating electric field is more than 6 kV/mm, resulting in a calculated time spread of 730 fs using an Au photocathode ($\Delta\varepsilon = 3.8$ eV) [Henke *et al.* (1981)].
    In 1997, the group at the University of Michigan estimated the time spread to be about 276 fs by means of an electric field (10 kV/mm) and a KBr photocathode ($\Delta\varepsilon = 1.1$ eV) [Chang *et al.* (1997)].

(b) Time spread caused by spatial resolution
    The time spread Δt_2 caused by the spatial resolution is determined by

$$\Delta t_2 = w/v, \qquad (3.89)$$

where w [m] is the minimum output slit width and v [m/s] is the streak sweep speed on the phosphor screen of the streak tube.
    In the case of a deflection sensitivity of 20 mm/kV and slit width of 50 μm, the time spread is estimated to be 500 fs using a sweep ramp voltage of 5 kV/ns.

The timing jitter is limited by the following four factors.

(a) Laser power stability

Laser pulse fluctuation causes the output amplitude of the GaAs optical switch to vary from shot to shot, which translates into timing jitter. Figure 3.39 shows the sensitivity to energy variation in fs/% as a function of laser energy applied to a GaAs optical switch. It is found that the GaAs optical switch goes into a saturated region with an energy variation of less than 150 fs/% at a laser energy of more than 25 μJ [Scheidt and Naylor (1999)].

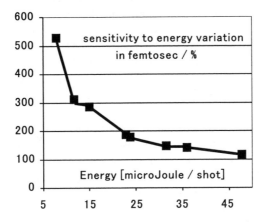

Fig. 3.39   Sensitivity to energy variation in fs/% as a function of laser energy on a GaAs optical switch.

(b) Pre-pulse

A high intensity, ultrashort laser pulse generally has an amplified spontaneous emission (ASE) buildup in the regenerative amplifier and has pre-pulses caused by leakage through the Pockels cell before the main pulse illuminates. These satellite pulses lead to an offset fluctuation of the streak ramp voltage and cause temporal jitter. To reduce the influence of these satellite pulses, a saturable absorber is placed in front of the GaAs optical switch. When the total energy in the satellite pulses is reduced to less than 10^{-4} of the main pulse energy, the influence of the satellite pulses on the timing jitter becomes negligible.

(c) Uniformity of the laser illuminating the GaAs optical switch surface

Spatial instability of the laser illuminating the GaAs optical switch surface causes temporal jitter and local stress of the GaAs optical switch. To avoid this problem a diffuser is placed in front of the GaAs optical switch.

(d) Response time of the GaAs optical switch

Timing jitter can be reduced by decreasing the response time of the GaAs optical switch. Chang *et al.* showed that the temporal resolution is reduced from 2 ps to 1.1 ps when the response time of the GaAs optical switch and deflection plate is improved from 300 ps to 150 ps [Chang (2002)].

Finally, a low jitter X-ray streak camera which has a temporal resolution of the order of 500 fs in the accumulation mode has been recently developed. However, its useful temporal resolution is limited to approximately 2 ps because of the triggering laser pulse quality. In order to reduce the temporal resolution down to the limitation of the streak tube, it is necessary to improve the laser pulse quality. This involves improving ASE build-up, pre-pulse and power instability by careful alignment and use of a saturable absorber.

In the present streak method, it is difficult to reduce the temporal resolution down to less than 500 fs because the acceleration electric field suppressing the energy distribution of the photoelectrons is limited by arcing between the photocathode and the accelerating electrode in the vacuum chamber. Further improvement in the temporal resolution will be accomplished by the development of a special photocathode with a small energy distribution of photoelectrons.

3.2   Synchronization

As well as the generation of femtosecond beams, the synchronization of pump and probe beams is crucial for ultrafast time-resolved analysis. Perfect synchronization can be established by using a terawatt laser and beamsplitter to generate pump and probe beams. A sophisticated synchronization system is not necessary in general, however, it is needed when using beams from two different beam sources, such as a laser, linac, synchrotron, etc. The precision of the synchronization is influenced by the timing jitter and drift, for short and long time scales, respectively. Unfortunately, in almost all systems currently in use, timing jitter and drift are longer than the pulse width of the beams. For instance, a 100-fs rms beam suffers from 400-fs rms jitter and 10-ps rms drift. If pump-and-probe analysis is done for a few minutes, timing jitter is critical. Furthermore, drift has to be overcome for analysis that continues for several hours. The main sources of jitter are the quality of the electronic devices in the system and thermal and quantum fluctuations. The drift sources are environmental changes, such as temperature and humidity, and successive motion of the device and the building. In this section, we introduce updated technology and achievements in the synchronization between femtosecond lasers and linacs/synchrotrons.

3.2.1 *Laser vs. linac*

H. IIJIMA
T. WATANABE
M. UESAKA
Nuclear Engineering Research Laboratory, University of Tokyo
2-22 Shirane-shirakata, Tokai, Naka,
IBARAKI 319-1188, JAPAN

3.2.1.1 *S-band linacs (thermionic and RF gun vs. active-mode-locked Ti:Sapphire laser)*

At the Nuclear Engineering Research Laboratory (NERL), University of Tokyo, Japan, the development of a timing system between a terawatt laser and an S-band linac was started in 1996. In order to evaluate the first performance of the timing system, a synchronization experiment was performed with two different systems, a Ti:Sapphire laser and a 35-MeV linac with a thermionic gun (the 35L system) and a Ti:Sapphire laser and a 18-MeV linac with RF photoinjector (18L system), as shown in Fig. 3.40. The specifications are summarized in Table 3.2.

The 35-MeV linac (35L) consisted of a thermionic injector, a subharmonic buncher (SHB), two accelerating tubes and an arc-type magnetic compressor [Uesaka *et al.* (1997)]. A single bunch was produced from the 35L for the synchronization experiment. The femtosecond-terawatt laser, synchronized with the 35L, produced the probe laser. The Ti:Sapphire oscillator of this laser generated a laser pulse train with an interval of 79.33 MHz. The interval was realized by active mode locking. The 18-MeV linac (18L) consisted of a photocathode injector, an accelerating tube and a chicane-type magnetic compressor [Uesaka *et al.* (2000)]. A YLF laser (PULRISE-100, Sumitomo Heavy Industries, Ltd.) was used as a driving laser and a Ti:Sapphire laser as the probe laser. Active mode locking was also adopted for the YLF laser. Two 7-MW klystrons supplied RF power to both the accelerating tubes of 35L, and to the photoinjector and the

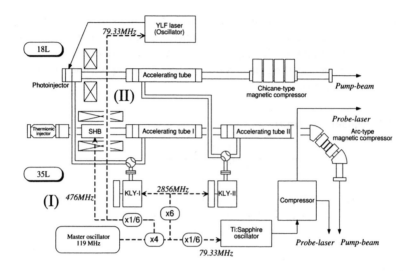

Fig. 3.40   Schematic view of 35L and 18L system. (I) Ti:Sapp. laser and 35-MeV linac with thermionic gun. (II) Ti:Sapp. laser and 18-MeV linac with RF photoinjector.

Table 3.2   Laser vs. linac synchronization system at NERL, U. of Tokyo.

	System I	System II	System III
Laser oscillator	Tsunami	PULRISE-100	Mira 900F
	(Spectra Physics)	(Sumitomo)	(Coherent)
Pulse length	100 fs	2–10 ps	100 fs
Mode locking	active (E/O)	active (E/O)	passive (Kerr lens)
Amplification	amplifier	amplifier	amplifier
	(BM Industries)	(BM Industries)	(Thales laser)
	≤300 mJ at 790 nm	≤4 mJ at 1047 nm	≤30 mJ at 790 nm
Linac max. energy	35 MeV	18 MeV	18 MeV
Pulse length	700 fs–10 ps	240 fs–10 ps	240 fs–10 ps
RF master	119 MHz	119 MHz	119 MHz
Timing stabilizer	at the fundamental		at 9th harmonic
at laser oscillator	(79.3 MHz)		(713.7 MHz)
Klystron	two, 7 MW		one, 15 MW

accelerating tube of 18L. A clock generator with a frequency of 119 MHz was used as a master clock for these systems. At first, the 4th harmonic of the master was generated for both systems. The 24th harmonic of the master was controlled to synchronize with the klystrons, and 1/6 of the 4th harmonic, i.e. 79.33 MHz, was synchronized with the oscillators. Consequently, the laser and the linac were synchronized. The timing circuit is illustrated in Fig. 3.41.

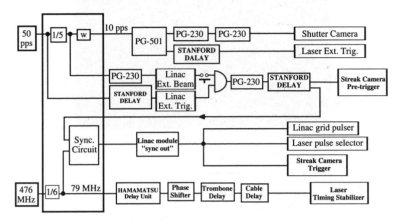

Fig. 3.41 Preliminary timing circuit.

Here the femtosecond electron and laser pulses were measured by the femtosecond streak camera (FESCA-200, Hamamatsu) simultaneously, and the timing difference between them was recorded. Since the streak camera detects visible light, the electron pulse was converted to Cherenkov radiation in air or Xe gas, depending on the electron energy beyond or below the threshold (20 MeV) in 1 atm air. This system was constructed in collaboration with the High Energy Accelerator Research Organization (KEK) [Nakajima *et al.* (1995)], JAERI [Dewa *et al.* (1998)] and Sumitomo Heavy Industries, Ltd.

The images of the laser pulse and Cherenkov radiation emitted by the electron bunch are presented in Fig. 3.42 with temporal profiles of their pulses, taken with the 35L system. In Fig. 3.42, one can see the timing difference between the two pulses. We accumulated about 100 data points and produced a histogram, shown in Fig. 3.43. If a Gaussian distribution is assumed, the timing jitter is evaluated to be 3.7 ps rms. For the 18L system, we evaluated a timing jitter of 3.5 ps rms [Uesaka *et al.* (2000)]. It was worth noting that the jitter values 3.7 ps and 3.5 ps are larger than the durations of the electron and laser pulses. From these results, the most dominant factor that worsens the total temporal resolution of the pump- and-probe analysis was not the pulse durations of the electron and laser

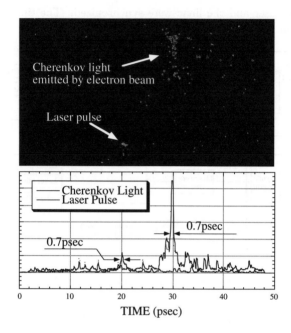

TIME (psec)

Fig. 3.42   Images of Cherenkov radiation and laser light for synchronization system I.
Cherenkov radiation is emitted by the electron pulse in air.

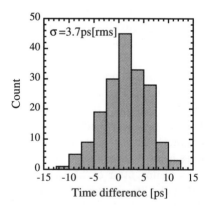

Fig. 3.43   Histogram of timing jitter for synchronization system I. The total timing jitter
is 3.7 ps rms.

pulses, but the timing jitter between them. Although the pulse durations were short enough to realize a sub-picosecond time resolution, the timing jitter was as large as picoseconds. In order to improve the timing jitter, we sought out the timing jitter sources of both the linac and the laser. The important task was to find the dominant factors that greatly affect the total timing jitter. We concluded the timing jitter sources listed below:

(a) mode locker of laser oscillator,
(b) timing stabilizer,
(c) fluctuation of the klystron output power,
(d) mutual fluctuation between the klystrons.

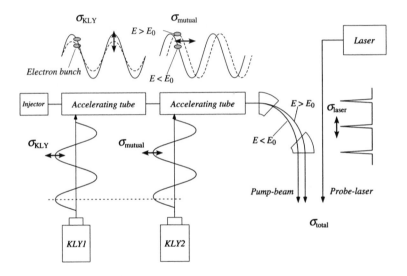

Fig. 3.44   Schematic view of jitter sources and their contributions.

The timing jitter sources and their contributions are illustrated in Fig. 3.44. Jitter sources (1) and (2) directly cause laser jitter σ_{laser}. The klystron-originating fluctuations give rise to both RF phase and power jitter. The RF phase jitter directly contributes to timing jitter in the accelerating tubes and RF gun. The RF power jitter changes the energy of the electron bunch, so that the time-of-flight via the magnetic optics becomes timing jitter. Here we use σ_{KLY} and σ_{mutual} for the two jitter caused by sources

(3) and (4), respectively. Thus, the total timing jitter σ_{total} is defined as

$$\sigma_{\text{total}}^2 = \sigma_{\text{laser}}^2 + \sigma_{\text{KLY}}^2 + \sigma_{\text{mutual}}^2 . \qquad (3.90)$$

In Sec. 3.2.1.2, an upgraded synchronization system to reduce timing jitter is introduced. In Secs. 3.2.1.3 and 3.2.1.4, we discuss contributions to and evaluation of these timing jitter sources quantitatively. Section 3.2.1.5 describes the reduced total timing jitter of the upgraded system.

3.2.1.2 *Upgraded timing system*

The upgraded system is illustrated in Fig. 3.45 [Uesaka *et al.* (2001)]. The

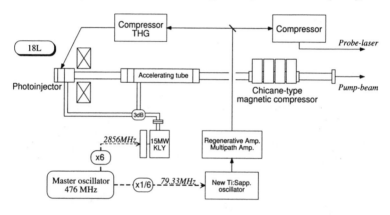

Fig. 3.45   Upgraded synchronization system of NERL, U. of Tokyo.

design was based on the work at the FELIX facility of the FOM Institute for Plasma Physics, where a timing jitter between the electron and FEL pulse of 400 fs rms was achieved [Knippels *et al.* (1998); Uesaka and Knippels]. At first, we installed a new laser oscillator (Mira 900-F, Coherent Inc.), in which a Ti:Sapphire crystal is used and mode locking is passively made by the Kerr lens effect. The new laser produces a laser power of 0.3 TW with a pulse duration of 100 fs. A diode laser (Verdi, Coherent Inc.) is used to pump it. The laser light is split: one beam is used as the driving laser of the photoinjector, the other as the probe laser. The oscillator is controlled by a new timing stabilizer (Coherent Inc., SynchroLock). The timing stabilizer can be driven by the fundamental and higher harmonic loops up to the 9th harmonic. A new 15-MW klystron (Mitsubishi, Modifed

PV-3015) was installed, leaving one 7-MW klystron. The new klystron is designed so the phase and power noises are much smaller than those of the old klystrons. Furthermore, in the new system, only one klystron supplies RF to several cavities. In this system, the mutual timing jitter mentioned in the last section disappears, i.e. $\sigma_{\text{mutual}} = 0$ fs. In addition, the master oscillator was replaced with an oscillator with a frequency of 476 MHz. The 6th harmonic is used to control the new klystron, and 1/6 of the master frequency synchronizes with the new oscillator. The new timing circuit is illustrated in Fig. 3.46.

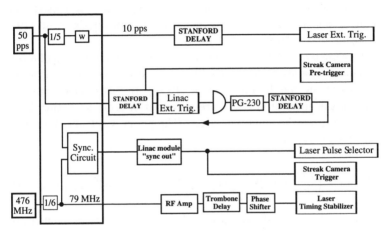

Fig. 3.46   Illustration of upgraded timing circuit.

3.2.1.3   *Timing jitter source in laser oscillators*

The precise timing of a laser pulse is predominantly determined by the oscillator or the regenerative amplifier. The optical cavity length of the oscillator fluctuates due to acoustic vibration of the mirrors. The regenerative amplifier also consists of a cavity and can generate the same kind of timing jitter as the oscillator. Mirror vibration in a cavity can enhance the timing jitter, since one mirror drift affects the timing of the laser repeatedly. In multi-pass amplifiers, all the mirrors vibrate, which also gives rise to timing jitter, however, the mirrors vibrate randomly and the jitter is canceled out and hence timing jitter is not enhanced by the vibration of mirrors in multi-pass amplifiers and similar devices. Therefore, we focus our attention on the oscillator.

Two recently developed lasers use two different kinds of mode-locking schemes of the oscillator. One scheme is active mode locking, such as that used in an oscillator. This scheme utilizes an acousto-optical (A/O) or electro-optical (E/O) crystal which work as electric shutters. The other scheme is passive mode locking, which utilizes the Kerr lens effect as described in Sec. 2.1.1.2. For active mode locking, an A/O or E/O crystal must be prepared in addition to the amplification crystal, such as the Ti:Sapphire crystal, while passive mode locking can occur in the amplification crystal itself. Therefore, the timing jitter for passive mode locking is smaller than for active mode locking [Knippels *et al.* (1998); Kobayashi *et al.* (1999)].

There are two well-known techniques for the characterization of noise in continuously operating mode-locked lasers [Takasago and Kobayashi]. One was proposed by von der Linde [Linde (1986)]. In this scheme, higher harmonics of the spectral intensity are measured in order to distinguish the phase noise from the amplitude noise. The power spectrum of an imperfectly mode-locked laser is defined by

$$P_F(\omega) = \left(\frac{2\pi}{T}\right)^2 |\tilde{f}(\omega)|^2 \sum_{\mu} [\delta(\omega_\mu) + P_A(\omega_\mu) + (2\pi\mu)^2 P_J(\omega_\mu)], \quad (3.91)$$

where $\omega_\mu = (\omega - 2\pi\mu/T)$ and μ is an integer running from minus to plus infinity. The first term corresponds to a perfect, noise free pulse train. The second term is the frequency-shifted power spectrum of the amplitude noise. The third term involves the power spectrum of a random function that describes the timing jitter. It is important to point out that the jitter term is proportional to μ^2, where μ labels the individual frequency bands. The μ^2 dependence of the third term in Eq. (3.91) is significant because it distinguishes the two different kinds of noise and therefore allows a determination of both $P_A(\omega)$ and $P_J(\omega)$, as shown in Fig. 3.47.

The other technique was proposed by Tsuchida [Tsuchida (1998)]. This method calculates the ratio between the real part and the imaginary part of the amplitude to get the phase noise in the time domain. A time domain demodulation of the pulse intensity phase provides a large dynamic range and wide frequency span. For this purpose, the conventional expression of the sinusoidal output voltage $V(t)$ of the laser is defined as

$$V_n(t) = V_{n0}[1 + \epsilon(t)] \sin\{n[2\pi f_r t + \phi(t)]\}, \quad (3.92)$$

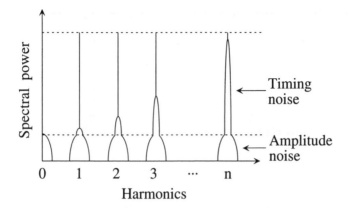

Fig. 3.47 Power spectrum of higher harmonics of mode-locked pulse train.

where V_0, f_r, ϵ and $\phi(t)$ are the nominal amplitude, frequency, amplitude and phase noise respectively. The real and imaginary parts of the amplitude of each component are given by

$$V_{re} = V_{n0}[1 + \epsilon(t)] \cos[n\phi(t)], \tag{3.93}$$

$$V_{im} = V_{n0}[1 + \epsilon(t)] \sin[n\phi(t)]. \tag{3.94}$$

The phase noise in the time domain can be extracted from the above signals by the relation

$$\phi(t) = \tan^{-1}\left(\frac{V_{im}}{V_{re}}\right). \tag{3.95}$$

The value $\phi(t)$ can be easily measured by a vector signal analyzer. In a femtosecond laser field, the measurement scheme of von der Linde is used for the timing stabilizer and that of Tsuchida is used for characterization of the oscillator.

Here, we describe the experimental evaluation of the timing jitter σ_{laser} for the new laser system. The timing jitter of the oscillator was measured by von der Linde's scheme. The laser pulse train was detected by a pin photodiode and a spectrum analyzer. The timing stabilizer, SynchroLock, can be operated with both internal and external reference triggers, and it observes the 9th harmonic of the laser pulse and references. Since the sampling frequency of the pin photodiode is 1 GHz and the laser frequency is 79.33 MHz, we could take the 14th harmonic of the power spectrum.

Femtosecond Beam Science

Figure 3.48 shows the fundamental (1st), 5th, 10th and 14th harmonics of

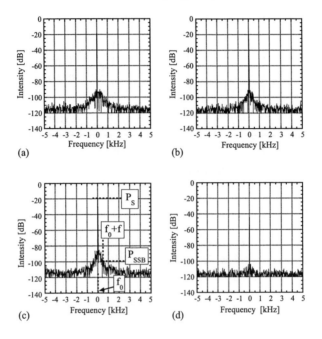

Fig. 3.48   Power spectrum of pulse trains: (a) fundamental, (b) 5th, (c) 10th and (d) 14th harmonics.

the laser power spectrum. In particular, the 10th harmonic shows enhanced noise compared to the 1st and 5th harmonics. However, the 14th harmonic did not work well, due to the limitations of the frequency response of the pin photodiode.

A function to estimate the timing jitter σ_T is derived from Eq. (3.91) as below:

$$\sigma^2 = 2 \int_{f_{\text{low}}}^{f_{\text{high}}} \frac{P_{\text{SSB}}}{P_S \cdot BW} df \tag{3.96}$$

$$= \sigma_A^2 + (2\pi n f_0)^2 \sigma_T^2, \tag{3.97}$$

where f_{high} and f_{low} are the offset frequencies, f_0 (= 79.33 MHz) is the carrier frequency, P_{SSB} is the single-sideband noise power and P_S is the power at the carrier frequency, as shown in Fig. 3.48. σ_A is the amplitude noise defined in Fig. 3.47. BW ($= 10$ Hz) is the frequency resolution of the spectrum analyzer and n is the number of harmonics. Here, σ_T is equal to σ_{laser} used in Eq. (3.90).

Timing jitter is estimated by fitting the total jitter as a function of the number of harmonics. The fitting function is

$$\sigma = a + bn^2, \qquad (3.98)$$

where a and b are fitting parameters such that

$$\sigma_A = \sqrt{a}, \qquad (3.99)$$

$$\sigma_T = \frac{\sqrt{b}}{2\pi f_0}. \qquad (3.100)$$

The fitting was carried out with two sets of data, one was from the 1st to 10th harmonics and the other from the 1st to 14th harmonics. The results gave 23.5 ps for the 1st-to-10th data set and 24.9 ps for the 1st-to-14th data set. These results are clearly inconsistent with the previous result shown in Fig. 3.43. We believe that this inconsistency can be attributed to environmental noise in our laser room. Kobayashi *et al.* measured the timing jitter of their passive mode-locked laser oscillator by using Linde's method in a super-clean room: dust class 1000; temperature ±0.1 °C; humidity ±0.1% and electromagnetically shielded. The specifications of our laser room are: dust class 10 000; temperature ±1°C; humidity below about 50% and not electromagnetically shielded. We speculate that a precise measurement needs such a highly controlled area.

In addition, we remark on one consideration for the choice of RF frequency multiplier/divider. Figure 3.49 shows measured power spectra for several input signals to a laser oscillator with a frequency of 79.33 MHz. Figure 3.49(a) shows a spectrum of an internal trigger. In comparison with the internal trigger, the jitter noise around the central frequency of an external trigger, shown in Fig. 3.49(b), is rather large. As for the spectrum in Fig. 3.49(c), there is no jitter noise from the reference synthesizer. However, as shown in the spectrum of Fig. 3.49(d), noise appears in the spectrum of the frequency divided by a 1/6-frequency divider. Actually, the Synchrolock didn't work for the cases corresponding to Fig. 3.49(b) and (d). We also measured the spectrum for a frequency of 476 MHz, but did not see any noise. Therefore, we concluded that the 1/6 RF divider gave rise to the fatal jitter noise. Figure 3.50 shows typical diagrams of RF control systems. For frequency multiplier/dividers with a multiplier of only 2, the

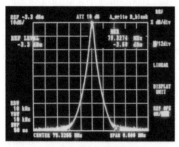

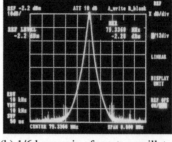

(a) Internal oscillator   (b) 1/6 harmonic of master oscillator

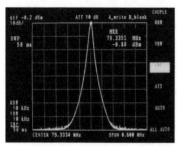

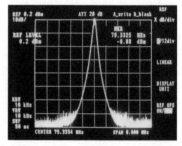

(c) Reference synthesizer   (d) 1/6 harmonic of reference synthesizer

Fig. 3.49   Power spectra of oscillators with fundamental frequency 79.33 MHz: (a) Internal oscillator of Synhcrolock. (b) 1/6 harmonic of master oscillator. (c) Reference synthesizer. (d) 1/6 harmonic of reference oscillator. Frequency span is 0.6 MHz in all spectra.

relative merits between the from-top-down and from-bottom-up diagrams have not clearly been reported so far. We have clearly found that a frequency multiplier/divider is a source of noise timing jitter. We believe that using a *timing delay* to generate a frequency of 1/3 or 1/6 multiple in our diagram gives rise to timing jitter and hence it should be avoided.

3.2.1.4   *Timing jitter source in a linac*

In a linac, the precise timing of the electron pulse is decided mainly by an RF phase in the accelerating tube. When an electron beam is generated by the injector, its energy is not relativistic and its pulse duration is a few picoseconds. The electrons are accelerated by the RF phase in the accelerating tubes to higher energies, so that the velocity of the electrons

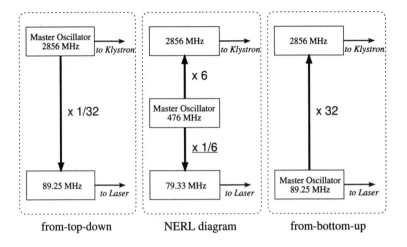

from-top-down   NERL diagram   from-bottom-up

Fig. 3.50 Typical diagrams of RF control systems.

is almost equal to the speed of light. In the process, the electrons settle at the crest of the RF phase. Once the velocity of the electrons reaches the relativistic region, the electrons are frozen to the RF crest. It is therefore found that the precise timing of the electrons at the output of the linac is decided by the phase of the accelerating RF. Only when the electrons are bent by the magnetic field is the timing jitter of the electron pulse influenced by the energy fluctuation.

Fluctuation of the amplitude in a klystron generates both phase and amplitude jitters of the RF phase in the accelerating tube. The fluctuation of the phase in the klystron directly affects the phase jitter of the RF phase in the accelerating tube, and the phase jitter directly yields the timing jitter of the electron. Moreover, if the electron is accelerated by a different phase, the voltage at the phase also differs. Both can be sources of timing jitter when the electron is bent by the arc- or chicane-type compressor shown in Fig. 3.44.

To estimate the phase jitter of the RF phase in the accelerating tube, it is necessary to measure the fluctuation of the voltage. Here we shall discuss the timing jitter caused in a klystron [Meddens *et al.* (1993); Takeshita *et al.* (1999)]. It is known that phase jitter $\Delta\phi$ is defined by

$$\Delta\phi = -\frac{2\pi L}{\lambda}\frac{eV}{m_0 c^2}(\gamma^2 - 1)^{-3/2}\frac{\Delta V}{V}, \tag{3.101}$$

where V is the voltage applied to the electron gun in the klystron, L is the length between the input and output cavities, λ is the wavelength of the klystron frequency and γ is the relativistic factor of an electron. Equation (3.101) is the most important relation between the fluctuation of the amplitude and that of the phase. In practical measurements, the phase jitter of a klystron is measured with the aid of Eq. (3.101). Since the newly installed 15-MW klystron has $L = 400$ mm and $V = 210$ kV, Eq. (3.101) becomes

$$\mathrm{d}\phi = 572\,\frac{\mathrm{d}V}{V}. \tag{3.102}$$

In experiment, the voltage was found to fluctuate within 0.2% without ripple, or 0.26% with ripple in the peak-to-peak value. Using Eq. (3.102), this corresponds to a phase jitter $\mathrm{d}\phi$ of 1.1° without ripple or 1.5° with ripple. It should be noted that the variation of the voltage, 0.2% or 0.26%, was close to the limit of measurement. The phase jitter obtained above enables us to calculate the electron beam transportation in the whole linac and evaluate the fluctuation of the time-of-flight at the end of linac. For this purpose, the particle tracking code PARMELA was used. The jitter σ_{KLY} was estimated to be 300 fs rms.

3.2.1.5   *Overall evaluation*

Measurement of the total timing jitter between the linac and the laser was carried out on the 18L system. As mentioned in the previous sections, timing jitters $\sigma_{\mathrm{KLY}} = 300$ fs and $\sigma_{\mathrm{mutual}} = 0$ fs were estimated. The laser timing jitter was assumed to be 100 fs, based on the discussion of Kobayashi [Kobayashi *et al.* (1999)]. Thus, the total timing jitter was estimated to be

$$\sigma_{\mathrm{total}} = 100^2 + 300^2 + 0^2$$
$$\simeq 320 \text{ fs}. \tag{3.103}$$

The designed and measured total jitter is summarized in Table 3.3. We have measured the improvement of the synchronization performance in the three systems:

(a) active mode-locked oscillator and two klystrons (System I),
(b) passive mode-locked oscillator with fundamental loop and new klystron (System III),
(c) passive mode-locked oscillator with 9th harmonic loop and new klystron (System III).

Table 3.3   Experimental and designed timing jitter (rms).

	System I & II (measured)		System III (designed)	(measured)
	35L	18L	18L	18L
Klystron σ_{KLY}	A few ps	A few ps	300 fs	300 fs
Mutual σ_{mutual}	A few ps	A few ps	0 fs	0 fs
Laser σ_{laser}	100 fs	2–10 ps	100 fs	100 fs
Laser mode locker	Active by A/O		Passive by Kerr lens	
Timing stabilizer	At fundamental (79.33 MHz)		At 9th harmonics (714 MHz)	
Total jitter σ_{total}	3.7 ps	3.5 ps	320 fs	330 fs

For system III, a typical image of Cherenkov radiation and laser light taken by the streak camera is shown in Fig. 3.51. Experimental histograms of the time differences are summarized in Fig. 3.52. It is possible to see the suppression of timing jitter as the system is improved. The results shown in Fig. 3.52(a) are for system I. For the next system, system II in Fig. 3.52(b), the timing jitter is clearly decreased due to the installation of the new laser and new klystron. The difference between Fig. 3.52(b) and (c) is mode locking. The timing stabilizer of system II observes the fundamental spectral intensity, while that of system III records the 9th harmonic. For system III the total timing jitter was measured to be 1.9 ps, which is much larger than the pulse duration. However, it can be seen that the dominant factor in the value 1.9 ps is drift between the laser and electron pulses. If the effect of the drift is calibrated with a slow change function, the timing jitter can be estimated to be 330 fs rms.

We found that the drift was produced by a thermal shift of the cooling water, which was supplied to the accelerating tubes, and fluctuations of room temperature. A timing jitter measurement of the laser oscillator was performed for a few seconds, while a measurement between the laser and electron pulses was done for an hour. In order to suppress the thermal drift, we introduced a stable water cooler for the accelerator tubes and RF gun, and an air conditioner. The water temperature was controlled within 0.01°C and the air temperature was controlled within 0.5°C. In addition, the transport line, of length about 50 m, guiding the laser from the laser room to the linac, causes timing drift. At first, we tried keeping the transport line in a vacuum, because laser light was being scattered by the air. However, in the vacuum the transport line was twisted by external pressure. Strain at the bellows connecting the chambers was the principle cause of

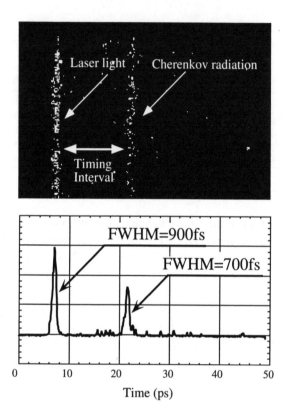

Fig. 3.51   Images of laser pulse and Cherenkov radiation emitted by an electron pulse in air.

a fluctuation of the path length of the transport line. The fluctuation of the path length occurred with fluctuation of the room temperature. Therefore, the transport line was filled with nitrogen gas and kept at atmospheric pressure. Figure 3.53 shows the result of synchronization after these modifications. The timing drift is clearly reduced, although the timing jitter is larger than that in Fig. 3.52(c) because the pulse duration was 1.6 ps FWHM. The total jitter was finally measured to be 1.6 ps rms for 2 hours and 1.4 ps rms for 1 hour between 20:30 and 21:30 as shown in Fig. 3.53.

Consequently, a new synchronization system was constructed and has already been applied to pulse radiolysis experiment.

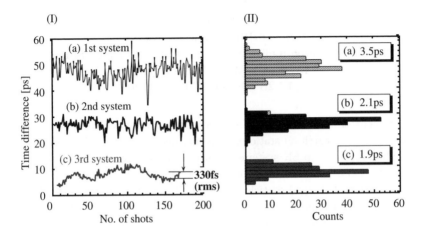

Fig. 3.52 Time difference between laser and electron pulses and corresponding histograms for (a) system I, (b) system II and (c) system III.

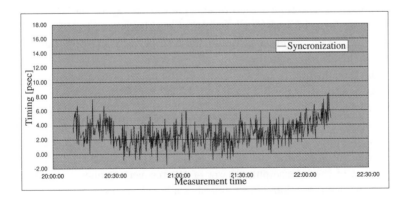

Fig. 3.53 Timing difference after modifications of temperature conditions.

3.2.2   *Laser vs. synchrotron*

Y. TANAKA

T. ISHIKAWA

SPring-8/RIKEN Harima Institute

Japan Synchrotron Radiation Research Institute (JASRI/SPring-8)

1.1.1Kouto, Mikazuki-cho, Sayo-gun,

1.1.2HYOGO 679-5198, Japan

Synchrotron radiation (SR) has been making an extensive contribution to scientific research on static phenomena, because of the wide tunability of photon energy in SR systems, from the infrared to X-ray regions, and high stability. Essentially SR sources generate pulsed radiation due to the acceleration of charged particles. Pulsation is also useful for investigations of dynamic phenomena, especially in combination with short pulse lasers. Laser–SR synchronization with a precision of less than the pulse duration is indispensable for combination experiments on ultrafast dynamics.

Synchronization has been conducted at a number of SR facilities, such as the LURE laboratory (CLIO) [Lacoursiere *et al.* (1994)], the Ultraviolet Synchrotron Orbital Radiation Facility (UVSOR) [Mizutani *et al.* (1997)], the BESSY synchrotron radtion source [Gatzke *et al.* (1995)] and third-generation SR facilities. The required precision and monitoring techniques for synchronization are dependent on the SR facilities, because they are determined by the time structure and photon energy of the SR. In this section, we focus on laser synchronization at third-generation SR facilities producing X-rays, and take SPring-8 and European Synchrotron Radiation Facility (ESRF) as examples.

The fundamental techniques and performance of laser–SR synchronization is described in Secs. 3.2.2.1 and 3.2.2.2 for SPring-8, where precise synchronization has been achieved [Tanaka *et al.* (2002); Tanaka (2001)] for time-resolved X-ray diffraction, and studies on the mixing of X-ray and optical photons. A fast X-ray shutter/chopper is an important tool for synchronization because the repetition rate of SR pulses is typically larger than that of the intense pulsed laser or the recovery frequency of the target phenomena. Section 3.2.2.3 describes a synchronous X-ray mechanical chopper that is available for investigations on macromolecules with flash photolysis at a beamline of ESRF [Wulff *et al.* (1997)]. This is followed in Sec. 3.2.2.4 by a method to obtain a time resolution higher than the SR pulse duration by using a fast X-ray streak camera [Larsson *et al.* (2002)].

Section 3.2.2.5 then closes with a discussion of the prospects for femtosecond synchronization.

3.2.2.1 *Synchronization scheme and timing monitor*

The X-ray pulses of SPring-8 have a duration of typically 40 ps FWHM in various filling patterns of the electron bunches. Thus, both a synchronization technique and a monitoring technique with a precision of less than a few tens of ps are required for achieving perfect overlap of the laser and the target X-ray pulses.

For laser–SR synchronization, it is necessary to make the laser pulses synchronize with the RF master oscillator that controls the voltage of the RF cavity for acceleration of the electron bunches in the storage ring (see Fig. 3.54(i)). Here a synchronization technique is used that has been extensively applied to synchronization between mode-locked lasers using a feedback loop to vary the cavity length by means of a piezoelectric translator (see Sec. 3.2.1). Figure 3.54(ii) shows a block diagram for a laser–SR synchronization system at BL29XUL of SPring-8 [Tanaka *et al.* (2000)]. A pico/femtosecond laser system is installed in a clean booth located just outside of the experimental hutch. The laser system is composed of a mode-locked Ti:Sapphire oscillator and a regenerative amplifier with a picosecond mask in a pulse stretcher. Trigger signals for controlling the laser timing are guided from an RF master oscillator through a 220-m long optical fiber cable with a small propagation time drift (5 ps/km/K). The mode-locked oscillator has a repetition rate determined by its cavity length. By controlling the cavity length with a phase lock electronics module, it can be locked to the 84.76-MHz signal provided from the trigger through a 1/6-frequency-divider. The mode-locked Ti:Sapphire oscillator is pumped by the second harmonic (SH) of a diode laser pumped 5WNd:YVO$_4$ laser, and produces pulses with a duration of 80 fs and average power of 700 mW at a wavelength of 800 nm. This laser beam is guided to the regenerative amplifier, which is pumped by the SH of a diode laser pumped 6-W-1-kHz Q-switched Nd:YLF laser. The regenerative amplifier is composed of a cavity with a gain medium of a Ti:Sapphire crystal and a pulse stretcher/compressor assembly. The timing of seed-in and extraction of the pulse from the cavity is controlled by high voltage (HV) pulses applied to the Pockels cells (P1 and P2). The trigger for the HV pulses and Q-switching of the pumping laser was carried out by a 84.76-MHz reference with a 1/n frequency divider ($n = 84854$) and a digital delay pulse generator. Here, the number n was

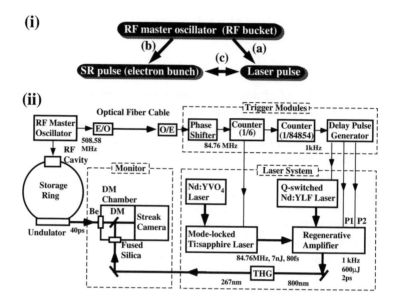

Fig. 3.54 (i) Scheme of a synchronization system with (ii) a block diagram of the synchronization and timing monitor system. The trigger of the pulse laser is obtained from the master oscillator that controls the RF cavity in the storage ring of SPring-8, through an optical fiber cable and a counter (frequency divider). A beam from a mode-locked Ti:Sapphire laser is guided to the regenerative amplifier pumped by the second harmonic of a Q-switched Nd:YLF laser. The third harmonic of the laser is reflected off a dichroic mirror onto the photocathode of a streak camera along with the SR.

determined by considering the harmonic number, 2436, of SPring-8, so that the laser pulses meet the SR pulses originating from a particular electron bunch [Tanaka *et al.* (2001)]. A set of slits in the stretcher/compressor works as a band-pass filter and consequently produces picosecond pulses with an energy of 700 μJ. The output beam of the regenerative amplifier is guided out of the laser booth into the experimental hutch through a duct with Ag steering mirrors.

For monitoring the timing of the hard X-ray SR pulses and laser pulses, an X-ray streak camera (see Sec. 3.1.7) was used, as illustrated in Fig. 3.54(ii). The synchronization of the pulses was monitored by detecting both beams simultaneously on a photocathode of the streak camera in 84.76MHz synchroscan mode. This method ensures a precise measurement of the interval between both beams without being affected by the drift of the streak trigger timing. The output beam of the regenerative amplifier was converted by β-barium borate crystals into the third harmonic with a photon energy of 4.7 eV, which is higher than the work function of gold (4.3 eV), so that the photocathode was sensitive to the laser as well as X-rays. The laser and SR beams (16.4 keV) were introduced onto the photocathode through a dichroic mirror (DM) with a surface of polished Be plate and a fused silica window.

3.2.2.2 *Performance of the synchronization at SPring-8*

(a) Timing jitter between laser and RF master oscillator

In order to investigate the effect of jitter in the trigger signal on the laser output pulse timing, we used a digital sampling oscilloscope with a fast sampling head of bandwidth 50 GHz and rise time of 7.0 ps. The 1/6 frequency divider in the trigger modules was found to generate 33 ps FWHM jitter between the RF reference (508.58 MHz) and the output with 84.76 MHz, as shown in Fig. 3.55(a). It should be noted that the jitter described here is estimated considering the 4.4-ps jitter in the sampling oscilloscope itself, which was measured by applying a RF reference to both the sampling head and the trigger input.

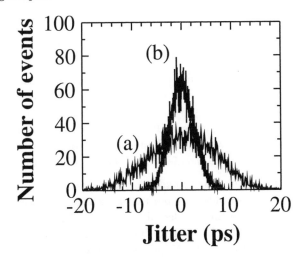

Fig. 3.55 (a) Timing jitter originating from the frequency divider. (b) Jitter between the RF reference (508.58 MHz) and the laser pulses, including the jitter of the oscilloscope and photodetector.

In spite of the 33-ps jitter on the 1/6 frequency divider output, which is used for the trigger of the laser oscillator, synchronization between the laser pulse and the RF reference (508.58 MHz) is achieved with a jitter of less than 5 ps, as seen in Fig. 3.55(b). This is examined by using a sampling oscilloscope and a picosecond photodetector with a rise time of 70 ps (10–90% of the peak). The measured jitter between the signal from the picosecond photodetector and the RF reference is estimated to be 2.4 ps including the jitter of the photodetector. This implies that the trigger jitter has little effect on the repetition rate of the laser, since the mechanical feedback to control the cavity length does not respond to fast pulse-to-pulse jitter on the trigger, and functions as a low pass filter.

(b)   Time drift between X-ray pulses and RF master oscillator

Here we describe the characteristics of synchronization between the RF master oscillator and the electron bunches. As the phase of the RF master oscillator is precisely locked to that of the RF cavity for acceleration of the electron bunches in the storage ring, the phase drift between the RF cavity and the electron bunches is discussed.

Theoretically, the phase ϕ of the RF electric field, where the electrons in the storage ring keep their energy balance, can be derived to be [Rowe (1979); Thompson and Dykes (1994)]

$$\phi = \arcsin(P_\ell/(V_0 \times I)),, \tag{3.104}$$

where P_ℓ, V_0 and I are the power loss of the electrons, the total peak voltage of the RF cavities and the ring current, respectively. According to the phase stability, $90° \leq \phi \leq 180°$ should be satisfied in Eq. (3.104), as shown in Fig. 3.56. Equation (3.104) indicates that the phase drift depends on the *undulator power* and the drop of the RF voltage due to *beam loading*, which is also related to the life time of the electrons in the storage ring.

A. *Electron energy loss by undulator radiation*

Since the laser pulses are precisely locked to the phase of the RF pulse in the storage ring, as described in Sec. 3.2.2.1, the synchronization system can also be used to show that closing undulator gaps shifts the arrival time of the SR pulses, which is due to the electron energy loss produced by the undulator radiation. The graphs in Fig. 3.57 are obtained with (a) the gaps of the 14 undulators fully opened and (b) closed, while the undulator gap of BL29XUL is fixed for monitoring the timing. The shift of the SR pulses between (a) and (b) by 42.2 ps is in good agreement with expectations for the increased power loss

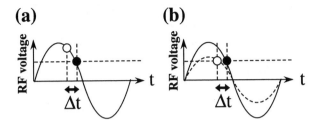

Fig. 3.56 RF voltage for acceleration and synchronized phase of electron bunches depicted as solid circles. Open circles represent the electron bunches drifted by (a) the electron energy loss and (b) the beam loading effect.

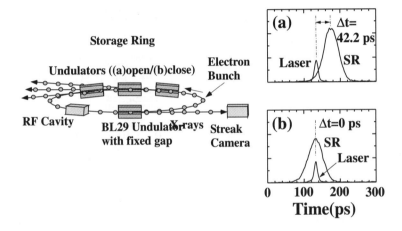

Fig. 3.57 Time drift of SR pulses due to the electron energy loss produced by the undulator radiation. Schematic illustration shows the experimental scheme to observe the drift. The graphs are obtained with (a) the undulator gaps open and (b) the 14 undulators closed.

[Tanaka *et al.* (2000)]. This estimate suggests that full power change at an undulator beamline produces a few ps shift. Practically, it is not necessary to take the effect into account in the user's operation mode under normal conditions. More precise experiments (less than a few ps) may need a feedback system to compensate the time drift of the centroid in the SR pulse.

B. *Beam loading effect*

Using a streak camera [Hara *et al.* (2000)] with a dual time base scan in which a slow time, orthogonally oriented, sweep unit is combined with the synchroscan unit, the relative phase of the electron bunches on the RF voltage for acceleration can be observed. A non-negligible drop

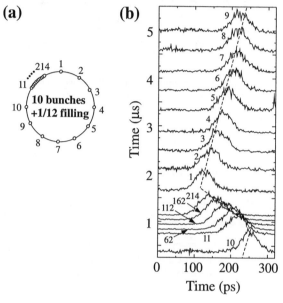

Fig. 3.58 (b) Beam loading effect in (a) an asymmetric filling pattern of electron bunches.

of the RF voltage due to beam loading appears for the asymmetric filling patterns of the electron bunches in the storage ring, as shown in Fig. 3.58(a). In Fig. 3.58(b), the vertical axis indicates a slow scan in the range of 5 μs corresponding to a round trip of the ring, and the horizontal axis corresponds to the phase of the RF voltage for acceleration. Since the pulse widths are almost equal to the original electron bunch width in Fig. 3.58(b), it is found that the phase of the electron bunch is stable and strongly depends on the RF bucket number [Hara *et al.* (2001)]. We emphasize that the laser pulses are controlled to meet the SR pulses originating from a particular electron bunch (see Sec. 3.2.2.1). Thus, timing synchronization is enabled for any filling mode of the electron bunch, even when the phase shift due to beam loading is not negligible.

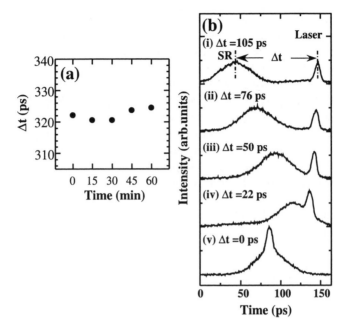

Fig. 3.59   (a) Long-term stability of the timing of the SR and laser pulses. Δt represents the interval between the two pulses. Data was taken about every 15 min. (b) Fine adjustment to achieve temporal overlap of the laser and SR pulse, as monitored with a streak camera.

(c)   Precision of synchronization between X-ray pulses and laser pulses

The precision of the laser–SR synchronization was estimated using a streak camera. Both long term stability of the synchronization and fine adjustment of the time interval between the laser and SR pulses are required to complete the timing control system.

Figure 3.59(a) shows the stability of the synchronization between the SR and laser pulses for an hour. Δt represents the interval between both pulses and the data was taken about every 15 min. The timing of both pulses are found to be stable for hours within a distribution of ± 2 ps, which is much smaller than the SR pulse width. The long term stability obtained here is especially useful for experiments in which integration must be carried out because of small signal intensity.

Figure 3.59(b) demonstrates control of the time interval between the

SR and laser pulses, as monitored with a streak camera. As an example, tuning of the laser pulse timing to overlap perfectly with the SR pulse is shown here. The tuning can be performed by a phase shifter, which varies the timing of the reference to the lasers. Δt represents the observed time interval between the SR and laser pulse. Starting from $\Delta t = 105$ ps [Fig. 3.59(b)(i)], we made Δt gradually smaller [Fig. 3.59(b)(iv)]. It should be noted that the position of the laser pulses on the screen changes slightly for Fig. 3.59(b)(i)–(iv), because of an abrupt drift of the trigger circuit in the streak camera. The trigger timing to the sweep of the streak camera was changed between Fig. 3.59(b)(iv) and (v) to get both pulses in the center of the screen. Perfect overlap is obtained in Fig. 3.59(b)(v) with a precision of ≈ 1 ps.

Temporal overlap of both pulses as well as a fixed time interval can, consequently, be achieved and both are stable with a precision of ± 2 ps for some hours. The synchronization system can be used for time-resolved measurements with a resolution of ± 2 ps, when deconvolution analysis is carried out for obtained data.

3.2.2.3 *Synchronous mechanical chopper*

Although a mechanical chopper [Wulff *et al.* (1997)] and a rotating mirror [Kosciesza *et al.* (1999)] are useful tools for reducing the repetition rate of SR, the shutter speed of such mechanical choppers is not high enough to extract single pulses from the SR pulse train with a full-filling electron bunch mode, in which the interval is a few ns. For this a special bunch mode of the storage ring, in which the intervals between the electron bunches are longer than the shutter-opening duration, is employed so as to make use of a mechanical chopper.

A fast mechanical chopper used at ID09 of ESRF for time-resolved white Laue diffraction [Bourgeois *et al.* (1996); Wulff *et al.* (1997)], is briefly described here. The chopper wheel consists of a triangular titanium disk that rotates about a horizontal axis perpendicular to the X-ray beam, as illustrated in Fig. 3.60. The chopper rotates synchronously with the RF master oscillator with a jitter of 10 ns rms at a frequency of 896.6 Hz, which is the 396th sub-harmonic of the ring frequency. A femtosecond laser is phase locked to the master oscillator with the method described in Sec. 3.2.2.1, and has a repetition rate of 896.6 Hz. The chopper has a 165-mm long tunnel with a trapezoidal cross section on the edge of the side of the triangle. By displaying the chopper laterally, the opening time

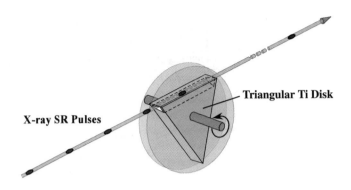

Fig. 3.60 Sketch of a mechanical chopper with a triangular wheel. The wheel rotates with a jitter of ∼ 10 ns at ∼ 1 kHz, so as to pick up single pulses from the SR pulse train with a single/hybrid bunch mode.

can be varied from 0.105 μs to 1.892 μs, according to the height of the tunnel (0.050–0.9 mm). The opening time of the chopper is adjustable for the single bunch mode or hybrid mode, in which the pulses are at least 0.053 μs apart. A shorter opening time of the chopper requires a smaller height of the tunnel. Thus, a toroidal mirror [Susini *et al.* (1995)] is used to focus the X-ray beams at the chopper, so that the vertical size is well below the height, to ensure reasonable transmission.

As the repetition rates of the SR and the femtosecond laser pulses are the same, an X-ray detector does not need a high speed. The time-resolved Laue pattern is then obtained by changing the time delay of the laser flash on the sample using an electric phase shifter in the RF signal to synchronize the laser oscillator.

3.2.2.4 *Time-resolved measurements using an X-ray streak camera*

An ultrafast X-ray streak camera, synchronized with the laser, makes studies with a time resolution in the 1-ps range possible. Here we describe the experimental scheme using the streak camera that was performed at ID09 of ESRF for investigation on the ultrafast lattice dynamics of InSb [Larsson *et al.* (2002)].

Figure 3.61 schematically shows the experimental setup. With the same synchronization techniques as described in Sec. 3.2.2.1, a Ti:Al$_2$O$_3$-based, 100-fs, 900-Hz repetition rate laser was synchronized to the SR pulses of the

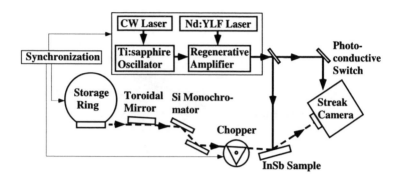

Fig. 3.61 Schematic of the time-resolved X-ray diffraction experiment using a fast X-ray streak camera at the SR facility.

ESRF storage ring with a jitter of the order of 10 ps. The undulator beam-line provides tunable X-rays using a double-crystal monochromator and the mean X-ray energy was set at 8.453 keV. The X-rays were focused using a toroidal mirror yielding an X-ray spot size of approximately 0.1 mm^2 on the sample. A mechanical chopper, again synchronized with the 100-ps X-ray SR pulses was used to reduce the pulse repetition rate of the X-rays to about 900 Hz in order to match the laser repetition rate (see Sec. 3.2.2.3). In order to obtain a time resolution better than that dictated by the X-ray pulse duration (or the jitter between the laser and the SR pulses), an X-ray streak camera with a photoconductive switch was used. The streak camera was also operated at a repetition rate of 900 Hz, with a time resolution of 500 fs. In the averaging mode, the effective time resolution of the streak camera system was found to be a few ps rms. A beam splitter placed in the laser beam produced two pulses, one of which was used to initiate the event of interest in the sample, and the other to irradiate the GaAs photoconduc-tive switch triggering the sweep voltage to the deflection plates of the streak camera (see Sec. 3.1.7). It is important to emphasize that the overall time resolution is determined by the jitter between the streak camera and the laser, and not that between the laser and the SR pulses. In collecting the data (often averaged over tens of thousands of shots), the delay between the laser and X-rays was slowly varied during the data collection process, by using an electronic phase shifter. The slow variation of the time recording of the X-ray pulse, as multiple pulses were collected, typically resulted in an averaged X-ray pulse with an approximately top-hat temporal profile and an effective duration several times that of the individual X-ray pulses. In this manner, it was possible to record the reflectivity of the perturbed crystal with a time resolution of a few ps over a time period longer than that of the individual X-ray pulses.

3.2.2.5 *Prospects for femtosecond timing control*

The techniques described above mainly have picosecond time scales. For shorter X-ray pulses, there are two methods to approach the synchronization with laser pulses in a femtosecond time scale: the use of slicing techniques in producing femtosecond X-ray pulses, as shown in Sec. 2.6, and achieving a more precise synchronization with the same techniques described in Sec. 3.2.2.1.

The former ensures synchronization between X-ray and laser pulses when the laser for slicing is also used for the irradiation of a sample. The latter may have practical difficulties in SR facilities, although subpicosesond control is achievable under ideal conditions with stable temperature and a noiseless environment. However, this technique is attractive in being directly and widely available for accelerator-based subpicosecond X-ray sources such as X-FELs (free electron lasers) and X-ERLs (energy recovery linacs), developments of which have recently been planned and started, as described in Secs. 2.7 and 2.8.

For the investigation of femtosecond phenomena using accelerator-based femtosecond X-ray sources in combination with short pulsed lasers, a time dispersive scheme for single-shot measurements may also help overcome the technical difficulties in femtosecond synchronization.

Bibliography

Akre, R., Bentson, L., Emma, P. and Krejcik, P. (2002). Bunch length measurements using a transverse rf deflecting structure in the SLAC linac, *Proc. of EPAC 2002*, Paris, France, pp. 1882–1884.

Baker, R. J. (1991). High voltage pulse generation using current mode second breakdown in a bipolar junction transistor, *Rev. Sci. Instrum.* **62**, p. 1031.

Barry, W. (1991) An autocorrelation technique for measuring picosecond bunch length using coherent transition radiation. *Proc. of the Workshop on Advanced Beam Instrumentation*, Tsukuba, Japan.

Bell, R. J. (1972) Intruductory Fourier Transform Spectroscopy, Academic, London, Chap. 9.

Bolotovskii, B. M. and Voskresenskii, G. V. (1966) Diffraction radiation, *Sov. Phys. Usp.* **9**.

Bourgeois, D., Ursby, T., Wulff, M., Pradervand, C., Legrand, A., Schildkamp, W., Labouré, S., Srajer, S., Teng, T. Y., Roth, M. and Moffat, K. (1996). Feasibility and realization of single-pulse Laue diffraction on macromolecular crystals at ESRF, *J. Synchrotron Rad.* **3**, pp. 65–74.

Catravas, P. *et al.* (1999). Measurement of electron beam bunch length and emittance using shot-noise-driven fluctuations in incoherent radiation, *Phys. Rev. Lett.* **82**, p. 5261.

Chang Z. (2002). Ultrafast X-ray source and detector, *2002 Research Meeting of the BES AMOS*.

Chang, Z. *et al.* (1997). Demonstration of a 0.54 picosecond X-ray streak camera, *SPIE Proc.* **2869**, pp. 971–976

Chantry, G. W. (1971) Submillimetre Spectroscopy, Academic, London, Appendix A.

Dewa, H. *et al.* (1998). Experiments of high energy gain laser wakefield acceleration, *Nucl. Instr. Meth. A* **410**, pp. 357–363.

Dnestrikovskii, Y. N. and Kostomorov, D. P. (1959). A study of ultrarelativistic charges passing through a circular aperture in a screen, *Sov. Phys. Dokl.* **4**, pp. 132 and 158.

Akre, R., Bentson, L., Emma, P. and Krejcik, P. (2001). A transverse rf deflecting structure for bunch length and phase space diagnostics, *Proc. of the 2001*

Particle Accelerator Conf., Chicago, pp. 2353–2355.

Finch, A., Liu, Y., Niu, H., Sibbett, W., Sleat, W. E., Walker, D. R., Yang, Q. L. and Zhang, H. (1988). Development and evaluation of a new femtosecond streak camera, Ultrafast Phenomena VI, Yajima, T., Yoshihara, Harris, C. B. and Shinonoya, S. eds., Springer Series Chemical Physics 48 (Springer, New York), p. 159.

Fitch, M. J., Melissinos, A. C., Colestock, P. J., Carneiro, J. -P., Edwards, H. T. and Hartung, W.H. (2001). Electro-optic measurement of the wake fields of a relativistic electron beam, Phys. Rev. Lett. 87, 034801.

Gatzke, J., Bellmann, R., Hertel, I., Wedowski, M., Godehusen, K., Zimmermann, P., Dohrmann, T., Borne, A. V. d. and Sonntag, B. (1995). Photoionization of Ca atoms in a time resolved pump-probe experiment with synchronized undulator- and Ti:Saphir laser pulses, Nucl. Instrum. Methods Phys. Res. A 365, pp. 603–606.

Happek, U., Blum, E. B. and Sievers, A. J. (1991). Observation of coherent transition radiation, Nucl. Instr. Meth. 60, p. 568.

Hara, T., Tanaka, Y., Kitamura, H. and Ishikawa, T. (2000). Performance of a CsI photocathode in a hard x-ray streak camera, Rev. Sci. Instrum. 71, pp. 3624–3626.

Hara, T., Tanaka, Y., Kitamura, H. and Ishikawa, (2001). Observation of hard X-ray pulses with a highly sensitive streak camera, Nucl. Instrum. Meth. A, 467-468, pp. 1125–1128.

Hecht, E. and Zajac, A. (1974). Optics, Addison-Wesley, Reading, Massachusetts, Sec. 9.7.

Henke, B. L., Knauer, J. P. and Premaratne, K. (1981). The characterization of x-ray photocathodes in the 0.1–10-keV photon energy region, J. Appl. Phys. 52, pp. 1509–1520

Jackson, J. D. (1975). Classical Electrodynamics, John Wiley and Sons.

Kim, K. J. (2000). AIP Conf. Proc. 413, p. 3.

Kinosihta, K., Ito, M. and Suzuki, Y. (1987). Femtosecond streak tube, Rev. Sci. Instrum. 58, p. 932.

Kinosihta, K., Suyama, M., Inagaki, Y., Ishihara Y. and Ito, M. (1990). Femtosecond streak tube, Proc. of 19th Int. Congress on High-Speed Photography and Photonics (ICHSPP), Cambridge, p. 490.

Knippels, G. H. M. et al. (1998). Two-color facility based on a broadly tunable infrared free-electron laser and a subpicosecond-synchronized 10-fs-Tisapphire laser, Opt. Lett. 23, 22, pp. 1754–1756.

Kobayashi, K. et al. (1999). Sub-100-fs timing jitter of mode-locked laser, SPIE Proc. 3616, pp. 156–164.

Kondo, Y. et al. (1993) Annual Report, Tohoku University 28(2), pp. 323 (in Japanese).

Korobkin, V. V., Maljutin, A. A. and Schelev, M. Ya. (1969). Time resolution of an image converter camera in streak operation, J. Photogr. Sci. 17, p. 179.

Kosciesza, D. and Bartunik, Hans D. (1999). Extraction of single bunches of synchrotron radiation from strorage ring with an X-ray chopper based on a rotating mirror, J. Synchrotron Rad. 6, pp. 947–952.

Krejcik, P., Decker, F. J., Emma, P., Hacker, K., Hendricson, L., O'Connell, C. L., Schlarb, H., Smith, H. and Stanek, M. (2003). Commisioning of the SPPS linac bunch compressor, **SLAC-PUB-9858**.

Krzywinski, J., Saldin, E. J., Schneidmiller, E. A. and Yurkov, M. V. (1997). A new method for ultrashort electron pulse-shape measurement using synchrotron radiation from a bending magnet, *DESY Rep.* **DESY-TESLA-FEL-97-03**.

Kung, P. H., Lihn, H. C., Bocek, D. and Wiedmann, H. (1994). Generation and measurement of 50-fs (rms) electron pulses, *Phys. Rev. Lett.* **73**, p. 967.

Lacoursiere, J., Meyer, M., Nahon, L., Morin, P. and Larzilliere, M. (1994). Time-resolved pump-probe photoelectron spectroscopy of helium using a mode-locked laser sycnchronized with synchrotron radiation pulses, *Nucl. Instrum. Meth. A* **351**, (1994) pp. 545–553.

Lai, R., Happek, U. and Sievers, A. J. (1994). Measurement of the longitudinal asymmetry of a charged particle bunch from the coherent synchrotron or transition radiation spectrum, *Phys. Rev. E* **50**, p. R4294.

Lai, R. and Sievers, A. J. (1994). Determination of a charged-particle-bunch shape from the coherent far infrared spectrum, *Phys. Rev. E* **50**, p. R3342.

Lai, R. and Sievers, A. J. (1995a). Phase problem associated with the determination of the longitudinal shape of a charged particle bunch from its coherent far-ir spectrum, *Phys. Rev. E* **52**, p. 4576.

Lai, R. and Sievers, A. J. (1995b). Determination of bunch asymmetry from coherent radiation in the frequency domain, *Micro Bunches Workshop*, AIP Conf. Proc. 367 (NY), p. 312.

Lai, R. and Sievers, A. J. (1997). On using the coherent far IR radiation produced by a charged-particle bunch to determine its shape: I Analysis, *Nucl. Instrum. Meth. A* **397**, p. 221.

Larsson, J., Allen, A., Bucksbaum, P. H., Falcone, R. W., Lindenberg, A., Naylor, G., Missalla, T., Reis, D. A., Scheidt, K., Sjögren, A., Sondhauss, P., Wulff, M. and Wark, J. S. (2002). Picosecond X-ray diffraction studies of laser-excited acoustic phonons in InSb, *Appl. Phys. A* **75**, pp. 467–478.

Lichtenberg, A. J. and Sesnic, S. (1967). *J. Opt. Soc. Am.* **56**, p. 75.

Lihn, H. C. (1996). Stimulated transition radiation, PhD Thesis, Applied Physics Department, Stanford University.

Lihn, H. C., Kung, P. H., Settakorn, C. and Wiedemann, H. (1996). Measurement of subpicosecond electron pulses, *Phys. Rev. E* **53**, p. 6413.

von der Linde, D. (1986). Characterization of the noise in continuously operating mode-locked lasers, *Appl. Phys. B* **39**, pp. 201.

Loudon,R. (1983). The Quantum Theory of Light, Clarendon Press, Oxford.

Lumpkin A. H. I. and Bingxin, Y. (1998) First multi-GeV particle-beam measurements using a synchroscan and dual-sweep X-ray streak camera, *Beam Instrumentation Workshop '98*.

Maksimchuk, A. *et al.* (1996). Signal averaging x-ray streak camera with picosecond jitter, *Rev. Sci. Instr.* **67**, pp. 697–699.

Meddens, B. J. H. *et al.* (1993). *Rev. Sci. Instrum.* **64**, pp. 2306.

Mizutani, M., Tokeshi, M., Hiraya, A. and Mitsuke, K. (1997). Development of a tunable UV laser system synchronized precisely with synchrotron radiation pulses from UVSOR, *J. Synchrotron. Rad.* **4**, pp. 6–13.

Mourou, G. *et al.* (1996). Jitter-free accumulating streak camera with 100 femtosecond time resolution, *Instrumentation reports ESRF-JULY 1996*, pp. 32–34.

Nakajima, K. *et al.* (1995). Observation of ultrahigh gradient electron acceleration by a self-modulated intense short laser pulse, *Phys. Rev. Lett.* **74**, 22, pp. 4428–4431.

Nakamura, K. *et al.* (2004). Measurement and numerical analysis of ultrashort electron bunch using fluctuaion in incoherent Cherenkov radiation, *J. of Nucl. Sci. Tech.*, in press.

Nakazato, T., Oyamada, M., Miimura, N., Urasawa, S., Konno, O., Kagaya, A., Kato, R., Kamiyama, T., Torzuka, Y., Nanda, T., Kondo, Y., Shibata, Y., Ishi, K., Ohsaka, T. and Ikezawa, M. (1989). Observation of coherent synchrotron radiation, *Phys. Rev. Lett.* **63**, p. 1245.

Niu, H., Degtyareva, V. P., Platonov, V. N., Prokhorov, A. M. and Schelev, M. Ya. (1998). A specially designed femtosecond streak image tube with temporal resolution of 50 fs, *SPIE Proc.* **1032** High Speed Photography and Photonics p. 79.

Novdick, J. S. and Saxon. D. S. (1954). Suppression of coherent radiation by electrons in a synchrotron, *Phys. Rev.* **96**, p. 180.

O'Shea, D. C. (1985). Elements of Modern Optical Design, Wiley-Interscience, p. 324.

Rowe, E. M. (1979). Topics in Current Physics, Synchrotron Radiation, Kunz, C. ed., Springer-Verlag, Berlin, p. 30.

Sajaev, V. (2000). Determination of longitudinal bunch profile using spectral fluctuations of incoherent radiation, *Proc. of EPAC 2000*, p. 1806.

Saldin, E. L., Schneidmiller, E. A. and Yurkov, M. V. The Physics of Free Electron Lasers, Chap. 6.

Sands, M. (1970). The physics of electron storage rings. An introduction, *SLAC Rep.* **SLAC-121**.

Scheidt, K. and Naylor, G. (1999). 500 fs streak camera for UV-hard X rays in 1 Khz accumulating mode with optical-jitter free-synchronization, *4. European Workshop on Beam Diagnostics and Instrumentation for Particle Accelerators*, DIPAC 99.

Settakorn, C. (2001). Generation and use of coherent transition radiation from short electron bunches, PhD Thesis, Applied Physics Department, Stanford University.

Shibata, Y., Hasebe, S., Ishi, K., Takahashi, T., Ohsaka, T., Ikezawa, M., Makazato, T., Oyamada, M., Urasawa, S. and Yamakawa, T. (1995). Observation of coherent diffraction radiation from bunched electrons passing through a circular aperture in the millimeter and submillimeter-wavelength regions, *Phys. Rev. E* **52**, p. 6787.

Shintake, T. (1996). Beam-profile monitors for very small transverse and longitudinal dimensions using laser interferometer and heterodyne technique, *KEK-PREPRINT-96-81*,, 20p. Invited talk at 7th Beam Instrumentation

Workshop (BIW 96), Argonne, IL.

Srinivasan-Rao, T., Amin, M., Castillo, V., Lazarus, D. M., Nikas, D., Ozben, C., Semertzidis, Y. K., Stillman, A., Tsang, T. and Kowalski, L. (2002). Novel single shot scheme to measure submillimeter electron bunch lengths using electro-optic technique, *Phys. Rev. ST-AB* **5**, 042801.

Stern, F. (1963). *Solid State Physics*, Seitz. F. and Turnbull, D., eds., Academic Press, New York, **15**, p. 300.

Susini, J., Baker, R., Krumrey, M., Schwegle, W. and Kvick Å. (1995). Adaptive X-ray mirror prototype: First results, *Rev. Sci. Instrum.* **66**, pp. 2048–2052.

Takahashi A. *et al.* (1994). New femtosecond streak camera with temporal resolution of 180 fs, *SPIE Proc.* **2116**, pp. 275–284

Takasago, K. and Kobayashi, K., FESTA, private communication.

Takahashi, A., Nishizawa, M., Inagaki, Y., Koishi M. and Kinoshita, K. (1993). A new femtosecond streak camera, *SPIE Proc.* **2002**, Ultrahigh- and High-Speed Photography, Videography, and Photonics p. 93.

Takahashi, A., Nishizawa, M., Inagaki, Y., Koishi, M. and Kinoshita, K. (1994). New femtosecond streak camera with temporal resolution of 180 fs, *SPIE Proc.* **2116**.

Takeshita, A. *et al.* (1999). Study of the 100 fs 10 kA X-band linac, *Nucl. Instr. Meth. A* **421**, pp. 43.

Tanaka, Y. (2001). Laser-SR synchronization, *SPring-8 Research Frontiers 2000/2001*, pp. 88–90.

Tanaka, Y., Hara, T., Kitamura, H. and Ishikawa, T. (2000). Timing control of an intense picosecond pulse laser to the SPring-8 synchrotron radiation pulses, *Rev. Sci. Instrum.* **71**, pp. 1268–1274.

Tanaka, Y., Hara, T., Kitamura, H. and Ishikawa, T. (2001). Synchronization of picosecond laser pulses to the target X-ray pulses at SPring-8, *Nucl. Instrum. Meth. A* **467-468**, pp. 1451–4154.

Tanaka, Y., Hara, T., Kitamura, H. and Ishikawa, T. (2002). Synchronization of short pulse laser with the SPring-8 synchrotron radiation pulses and its application to time-resolved measurements, *Rev. Laser Engineering* **30**, No. 9, pp. 525–530 (in Japanese).

Thompson D. J. and Dykes, D. M. (1994) *Synchrotron Radiation Sources – A Primer*, Winick, H. ed., World Scientific, Singapore, p. 91.

Toll J. S. (1956). Causality and the dispersion relation: Logical foundations, *Phys. Rev.* **104**, p. 1760.

Tsuchida, H. (1998). Stabilization of the output power of a 20-MW klystron to within 0.1-percent, *Opt. Lett.* **23**, pp. 286.

Uesaka, M. *et al.* (1997). Production and utilization of synchronized femtosecond electron and laser single pulses, *J. Nucl. Mat.* **248**, pp. 380–385.

Uesaka, M. *et al.* (2000). Experimental verification of laser photocathode RF gun as an injector for a laser plasma accelerator, *Trans. Plasma Sci.* **248**, 4, pp. 1133–1141.

Uesaka, M. *et al.* (2001). Hundreds- and tens-femtosecond time-resolved pump-and-probe analysis system, *Rad. Phy. Chem.* **60**, pp. 303–306.

Uesaka, M. and Knippels, G. H. M., private communication.

Watanabe, T., Sugahara, J., Yoshii, K., Ueda, T., Uesaka, M., Kondo, Y., Yoshimatsu, T., Shibata, Y., Ishii, K., Sasaki, S. and Sugiyama, Y. (2002). Overall comparison of subpicosecond electron beam diagnostics by the polychromator, the interferometer, and the femtosecond streak camera, *Nucl. Instrm. Meth. A* **480**, p. 63.

Wiedemann, H., Bocek, D., Hernandez, M. and Settakorn, C. (1997). Femtosecond electron pulses from a linear accelerator, *J. Nucl. Mat.* **248**, p. 374.

Williamson, D. E. (1952) *J. Opt. Soc. Am.* **42**, p. 712.

Wolf, E. (1962) *Proc. Phys. Soc.* **80**, p. 1269.

Wooten, F. (1972). Optical Properties of Solids, Academic Press, New York.

Wulff, M., Schotte, F., Naylor, G., Bourgeois, D., Moffat, K. and Mourou, G. (1997). Time-resolved structures of macromolecules at the ESRF: Single-pulse Laue diffraction, stroboscopic data collection and femtosecond flash photolysis *Nucl. Instrum. Meth. Phys. Res. A* **398**, pp. 69–84.

Yan, X., Macleod, A. M., Gillespie, W. A., Knippels, G. M. H., Oepts, D. and van der Meer, A. F. G. (2000). Subpicosecond electro-optic measurement of relaticistic electron pulses, *Phys. Rev. Lett.* **85**, pp. 34043407.

Yariv, A. (1989). Quantum Electronics, Wiley, New York.

Zolotorev, M. S. and Stupakov, G. V. (1996). Fluctuational interferometry for measurement of short pulses of incoherent radiation, *SLAC Rep.* **SLAC-PUB-7132**.

Zolotorev, M.S. and Stupakov, G.V. (1998). Spectral fluctuations of incoherent radiation and measurement of longitudinal bunch profile, *Proc. of the 1997 Particle Accelerator Conf.*, Vancouver, B. C., Canada, **2**, p. 2180.

Chapter 4

Applications

4.1 Radiation Chemistry

4.1.1 *Subpicosecond pulse radiolysis*

S. TAGAWA,

T. KOZAWA,

Nanoscience and Nanotechnology Center,

Institute of Scientific and Industrial Research (ISIR),

Osaka University,

8-1 Mihogaoka, Ibaraki,

OSAKA 567-0047, JAPAN

The economic scale of radiation utilization for industrial, agricultural and medical fields is now very large [Tagawa *et al.* (2002); Kume *et al.* (2002); Inoue *et al.* (2002); Yanagisawa *et al.* (2002)]. It is very important to make clear the detailed mechanisms of radiation-induced reactions for the development of advanced industrial and medical fields such as next generation lithography, nanotechnology and ion beam cancer therapy. Pulse radiolysis has been the most powerful method to make clear the mechanisms of radiation-induced reactions; transient species with short lifetimes in reactions can be detected directly by pulse radiolysis.

4.1.1.1 *History of picosecond and subpicosecond pulse radiolysis*

The history of picosecond and subpicosecond pulse radiolysis is shown in Fig. 4.1.

The first picosecond pulse radiolysis experiment was carried out in the late 1960's by the so-called stroboscopic method (essentially a pump-and-

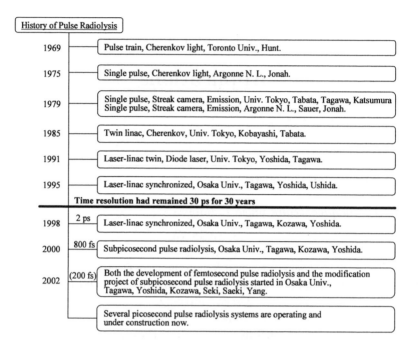

Fig. 4.1   History of picosecond and subpicosecond pulse radiolysis.

probe method) at the University of Toronto [Bronskill *et al.* (1970)]. Picosecond electron pulse trains generated with an S-band linear accelerator
(linac) were used as an irradiation source, and Cherenkov light induced by
the electron pulse trains was used as an analyzing light. The system was
very sophisticated and cheap, but the S/N ratio was not high. A similar
system was constructed at Hokkaido University more than 10 years later
[Sumiyoshi *et al.* (1982)]. In this system, the combination of picosecond
electron pulse trains with high-time-resolved detection systems was applied
to the picosecond pulse radiolysis with optical [Beck and Thomas (1972)]
and conductivity [Beck (1979)] spectroscopy. The measurable time domain
in both cases was limited by the time resolution of the system to 350 or
770 ps, determined by the pulse interval in the electron pulse trains of
S-band or L-band linacs respectively.

    The combination of picosecond single electron bunches with streak
cameras, developed independently in 1979 at Argonne National Laboratory [Jonah and Sauer (1979)] and at University of Tokyo by the authors

[Tagawa *et al.* (1979a)], enabled very high time resolution for emission spectroscopy. This extended the available research fields to organic materials such as liquid scintillators [Jonah and Sauer (1979); Tagawa *et al.* (1979a); Tagawa *et al.* (1982a)], polymer systems [Tagawa *et al.* (1979b)] and pure organic solvents [Katsumura *et al.* (1982)].

A subnanosecond pulse radiolysis system with a very fast response photodiode and transient digitizer was developed in 1982 [Tagawa *et al.* (1982b)] and could measure the reactive intermediates on a subnanosecond time scale using only one picosecond electron single bunch. It was a very powerful system for research on radiation chemistry [Tagawa *et al.* (1989)], because it was not limited to long time scale measurements and the radiation damage of the samples was much less than other picosecond pulse radiolysis systems.

A combination system of a picosecond pulse radiolysis system composed of a very fast response photodiode and a transient digitizer [Tagawa *et al.* (1982b)] with a specially designed picosecond pulse radiolysis system [Kobayashi *et al.* (1984)] composed of a streak camera with a time resolution of 50 ps was developed in 1984.

A twin linac system, where one linac produces picosecond electron pulses as a radiation source and the other linac produces Cherenkov light as a monitor light, was constructed in 1987 [Kobayashi *et al.* (1987)]. The strong Cherenkov light in the UV region enabled the measurement of the very rapid formation of alkyl radicals in irradiated liquid alkanes [Tagawa *et al.* (1989)]. Although the twin linac system was very interesting, operation of the system to obtain data of radiation chemistry took considerable effort. Radiation damage of the samples was also very large.

The LL twin picosecond pulse radiolysis system, which used a picosecond semiconductor laser as monitor light instead of the Cherenkov light of the twin linac pulse radiolysis, was developed by the authors in 1991 [Yoshida *et al.* (1991)]. It was the first successful picosecond pulse radiolysis system with laser monitor light, although several trials of picosecond pulse radiolysis with laser monitor light had been attempted. The operation of the LL twin picosecond pulse system was much easier than the twin linac pulse radiolysis and the use of a laser diode as monitor light enabled absorption spectroscopy from the visible to the infrared region in a picosecond time range for the first time [Yoshida *et al.* (1991)]. The kinetics of the geminate recombination of both electrons and cation radicals were measured for the first time on a picosecond time scale in the same sample and were found to agree with each other from 50 ps after a pulse [Yoshida *et al.* (1991)]. However, it was very difficult to obtain complete transient absorption spectra with this system because the laser diodes were not tunable light sources.

We solved the lack of tunability of the LL twin pulse radiolysis system by the development of a laser synchronized picosecond pulse radiolysis system in 1995 at Osaka University [Yoshida and Tagawa (1995); Yoshida *et al.* (2001)]. A femtosecond monitoring laser pulse was synchronized with both the electron pulses from the electron gun and accelerating electron pulses in the accelerating tubes in the laser synchronized picosecond pulse radiolysis system; in the LL twin linac system the picosecond laser pulse was synchronized with only the electron pulse from the electron gun. Although the time resolution of this system was 30 ps in 1995, it made further development of femtosecond pulse radiolysis possible.

The time resolution of pulse radiolysis remained at about 30 ps for 30 years since the late 1960's, shown in Fig. 4.1, although a lot of modifications were made for picosecond pulse radiolysis techniques during those 30 years. A time resolution of 2 ps [Kozawa *et al.* (2000)] and about 800 fs for 2 and 0.5-mm optical path sample cells respectively was obtained by using laser synchronized subpicosecond pulse radiolysis and a jitter compensation system with a 200-fs streak camera [Kozawa *et al.* (2000)]. The S/N ratio of laser synchronized subpicosecond pulse radiolysis was drastically improved by a double pulse method [Kozawa *et al.* (2002)].

Both the generation and application of femtosecond electron pulses have been carried out in many places, as described in Chapters 2 and 4. The time resolution of pulse radiolysis is now subpicosecond and several pulse radiolysis systems with about 10–30-ps time resolution are operating, in the Argonne National Laboratory [Bartels *et al.* (2000)], Brookhaven National Laboratory (Sec. 4.1.2.1) and University of Tokyo [Muroya *et al.* (2002a)], and are under construction in the University of Paris South [Belloni *et al.* (1998)] and Waseda University [Washio *et al.* (2002)], although the electron pulse bunch in the University of Tokyo is of femtosecond length. Recently both an improvement of a subpicosecond pulse radiolysis system and the construction of a femtosecond pulse radiolysis system for advanced material technology such as nanotechnology, was started at Osaka University.

4.1.1.2 *Time resolution of pulse radiolysis*

As the time resolution of a pulse radiolysis system is physically limited by the pulse width of the irradiation source, the generation of ultrashort electron pulses is a key technology for the observation of ultrafast radiation-induced reactions. There were two further difficulties in realizing higher time resolution pulse radiolysis. One is the problem of finding probe light that is precisely synchronized to the electron pulse. The other is the degradation of the time resolution caused by the velocity difference between an electron pulse and the probe light. In the previously reported picosecond pulse radiolysis systems, Cherenkov radiation was often used as a jitterless probe light. However, adopting this method with a compressed pulse required compressing an electron pulse twice, at the sample and at the Cherenkov radiation generator. From the viewpoint of beam optics, it is technically difficult to design optics which have two longitudinal focal points. On the other hand, the use of probe light other than radiation emitted from an electron pulse requires synchronization of the probe light with the linac and causes timing jitter and timing drift problems. In order to reduce jitter and drift, it is necessary not only to increase the stability of the equipment, but also to reduce environmental factors such as mechanical vibrations and fluctuations of the coolant temperature, room temperature and power supply. However, it does not seem feasible to reduce such factors for the whole accelerator system because the accelerator consists of many pieces of equipment and occupies a large space. To deal with the second problem, it is necessary to use a thin sample, which is discussed in the following section. Because the use of a thinner sample leads to a degradation of the S/N ratio, a high S/N ratio detection system is required for a high time resolution system. However, the construction of such a system has been difficult because pulse radiolysis requires measurement under radiation. It was because of these problems that the highest time resolution remained at about 30 ps for three decades. In this section, methods for observing radiation-induced reactions within 30 ps are described.

To detect the optical absorption of short-lived intermediates in a picosecond time region, a so-called *stroboscopic technique* is used. The short-lived intermediates produced in a sample by an electron pulse are detected by measuring the optical absorption of very short probe light such as Cherenkov radiation, or a femtosecond laser as in our system. The time profile of the optical absorption can be obtained by changing the delay between the electron pulse and the probe light.

The time resolution of stroboscopic pulse radiolysis does not depend on the time resolution of the detection system, such as a photodiode or an oscilloscope, and hence a high time resolution can be achieved. The time resolution is principally determined by the length of the electron pulse and of the probe light and also the timing accuracy or jitter between them (see Fig. 4.2). It is assumed that the pulse shapes of both the electron pulse

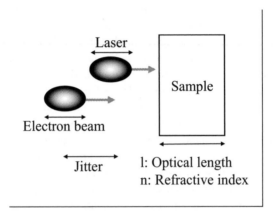

Fig. 4.2 Factors that limit the time resolution of stroboscopic pulse radiolysis.

and the probe light and the timing accuracy are Gaussian. The response function can then be expressed as

$$g_1(t) = \frac{1}{\sqrt{2\pi}\sigma} e^{-\frac{t^2}{2\sigma^2}}, \tag{4.1}$$

$$\sigma^2 = \sigma_i^2 + \sigma_b^2 + \sigma_j^2, \tag{4.2}$$

where σ_b is the length of the electron pulse, σ_i is the length of the probe light and σ_j is the timing accuracy.

However, the time resolution is degraded by the velocity difference between the light and the electron pulse in a sample. The transition time of an electron pulse through a sample is given by $L/(\beta c)$, where L is the optical length of a sample and β is the ratio of the velocity of the electron to that of light in vacuum, and c is the velocity of light in vacuum. In the case of light, the transition time is Ln/c, where n is the refractive index of a sample. Therefore, the time resolution is degraded by the thickness of a sample. The response function of the effect of the velocity difference can be deduced from the definition of a response function [Kozawa *et al.* (2002)]:

$$g_2(t) = \frac{c}{\sqrt{\pi}L\left(n - \frac{1}{\beta}\right)} \int_{\frac{2ct-L\left(n-\frac{1}{\beta}\right)}{2\sqrt{2}\sigma c}}^{\frac{2ct+L\left(n-\frac{1}{\beta}\right)}{2\sqrt{2}\sigma c}} e^{-x^2} dx . \tag{4.3}$$

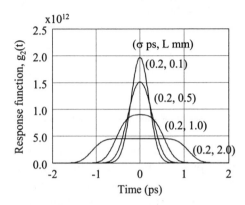

Fig. 4.3 Response functions calculated from Eq. (4.3) for $n = 1.33$.

Figure 4.3 shows response functions calculated from Eq. (4.3). When the optical length of the sample is much shorter than other factors, such as the length of the electron pulse, the response function takes on a Gaussian shape. As the optical length increases, the shape of the response function becomes trapezoidal. The effect of the velocity difference can be precisely deconvoluted because the velocity of a relativistic electron and the refractive index and optical length of a sample can be measured precisely. A thick sample causes an elongation of an electron pulse by electron scattering, which leads to a degradation of the time resolution. Therefore, the thickness of samples should be as small as possible.

4.1.1.3 *Subpicosecond pulse radiolysis system*

Figure 4.4 shows the outline of the newly developed subpicosecond pulse radiolysis system. The system consists of three key systems, a femtosecond linac system as an irradiation source, a femtosecond laser system as a probe light and a jitter compensation system. The femtosecond linac system

344 Femtosecond Beam Science

uses the L-band linac at the Institute of Scientific and Industrial Research
(ISIR), Osaka University, with a magnetic pulse compressor. The shortest
achievable pulse length is in subpicoseconds. The pulse width was roughly

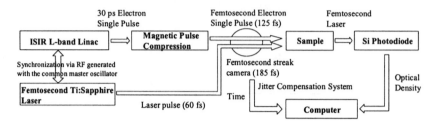

Fig. 4.4 Diagram of subpicosecond pulse radiolysis system.

estimated from an analysis of a spectrum of coherent transition radiation,
assuming a Gaussian pulse shape. The femtosecond laser system is made
up of a mode-locked ultrafast Ti:Sapphire laser (Tsunami, Spectra-Physics
Lasers, Inc.), an Ar ion laser (BeamLok 2080, Spectra-Physics Lasers, Inc.)
used to pump the Ti:Sapphire laser, a pulse selector (Model 3980, Spectra-
Physics Lasers, Inc.) and other electrical equipment. The Ti:Sapphire laser
is synchronized to the ISIR L-band linac using a phase-lock loop [Yoshida
et al. (2001)]. A RF pulse of 81 MHz, which is a multiple of the frequency
of the master oscillator, 27 MHz, is fed to the phase lock loop. The shortest
pulse length of the laser is 60 fs FWHM. A sample is irradiated by a femtosecond electron single pulse produced by the femtosecond linac system.
The time-resolved optical absorption is detected with a femtosecond laser
synchronized with the electron pulse. The intensity of the laser pulse is
measured with a Si photodiode. The rough timing between the electron
pulse and the laser pulse is controlled by a RF system. A jitter compensation system was designed to achieve a higher time resolution [Kozawa et al.
(2000)]. Obtaining the time profile of an intermediate requires two kinds of
information; time and optical density. Information about the time density
is provided by the jitter compensation system with an accuracy of 185 fs.
The optical density is calculated from the laser intensities. The time profile
of the optical absorption can be obtained by changing the phase of the RF
with an electrical phase shifter. All equipment described above is controlled
by a personal computer. The acquisition time is about 1 s/shot. A time
profile is made up of several thousands of measurement points.

4.1.1.4 *Jitter compensation system for highly time-resolved measurements*

In order to reduce the effect caused by timing jitter and drift between a laser pulse and an electron pulse, a jitter compensation system was designed, as shown in Fig. 4.5. The measurement cycle to obtain one point of the

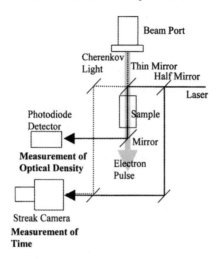

Fig. 4.5 Schematic drawing of the jitter compensation system.

time profile starts with generating a trigger signal. The first trigger is generated by a pulse generator connected to a personal computer via a GPIB interface. After passing through a synchronization circuit with an RF pulse, the trigger is used to produce an electron pulse and to pick out one single laser pulse or several laser pulses (when the double pulse method is applied) from the continuous pulses emitted from the Ti:Sapphire laser. After being accelerated and compressed, the electron pulse is applied to a sample and induces reactions in it. Before entering the sample, the electron pulse emits Cherenkov light in the air between the exit of the beam line and the sample. The Cherenkov light is reflected by a thin mirror placed in front of the sample and transported to a femtosecond streak camera (FESCA-200) by an optical system arranged on optical stages. The laser pulse is divided into two pulses by a half mirror. One is used to measure the optical density of intermediates induced in the sample. Another is transported to the streak camera to determine the time interval between the Cherenkov light from the electron pulse and the laser pulse. The time interval is obtained by a computational analysis of the image provided by the streak camera. The analysis of the streak image is carried out within several hundreds of milliseconds. This system measures the pulse's time of arrival to 185-fs precision. The precise measurement of the time interval between both pulses cancels the effect of jitter. By repeating this procedure several thousands of times, a time profile is obtained.

4.1.1.5 Early processes of radiation chemistry

Using the subpicosecond pulse radiolysis system, new research into condensed matter has started [Kozawa et al. (2002); Saeki et al. (2001); Saeki et al. (2002)]. On the exposure of materials to high energy electron beams, energy is deposited in materials mainly through ionization processes. An electron ejected from a parent molecule looses its energy via interactions with surrounding molecules and is eventually thermalized. The subsequent reactions of the thermalized electron depend on the material. The details of these reactions are unknown.

For nonpolar liquids, the time dependent behavior of a geminate ion pair in n-dodecane has been reported [Saeki et al. (2001); Saeki et al. (2002)]. Figure 4.6 shows the time dependent behavior of cation radicals of n-dodecane monitored at 790 nm. The time dependent behavior of cation radicals of n-dodecane has been investigated in detail over a time range longer than several tens of picoseconds [Yoshida et al. (1993)]. It was reported that the behavior can be well reproduced by the Smoluchowski equation:

$$\frac{\partial w}{\partial t} = D\nabla\left(\nabla w + w\frac{1}{kT}\nabla V\right), \qquad (4.4)$$

where w, D, k, T and V represent the survival probability, the sum of the diffusion constants, the Boltzmann constant, the absolute temperature and the Coulomb potential, respectively. The solid line in Fig. 4.6 represents a kinetic trace calculated with previously reported parameters, convoluted with the response function, superimposed in Fig. 4.6. In the simulation, the parameters of the electron diffusion coefficient $(De^-) = 6.4 \times 10^{-4}$ cm^2/s, the cation radical diffusion coefficient $(D+) = 6.0 \times 10^{-6}$ cm^2/s, the relative dielectric constant $\varepsilon = 2.01$, the reaction radius $R = 0.5$ nm and an exponential function with $r_0 = 6.6$ nm were used. The response function is mainly determined by the velocity difference in the sample and the time

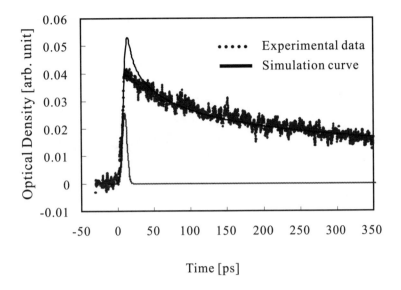

Fig. 4.6   The time dependent behavior of cation radicals of *n*-dodecane obtained by pulse radiolysis of neat dodecane, monitored at 790 nm. The dots and the solid line represent the experimental data and a simulation curve, respectively. The response function is superimposed.

resolution of the streak camera, in the 500-ps range. The kinetic trace agreed with the experimental data beyond 50 ps. However, a clear discrepancy is observed within 50 ps. Although the reason for the discrepancy is still unknown, several causes, such as the relaxation of the cation radical and the failure of the Smoluchowski equation in the early time range, have been suggested [Saeki *et al.* (2001)]. Further research is under way.

For polar liquids, the time dependent behavior of solvated electrons in methanol is being studied. It has been reported that presolvated electrons and solvated electrons are formed in alcohol on exposure by an electron beam [Kenney-Wallace and Jonah (1982)]. However, the details of the formation process of solvated electrons in methanol have not been made clear, as it occurs within 30 ps after irradiation. Recently, the details of the time dependent behavior of presolvated electrons and solvated electrons have been investigated using laser photolysis [Wiesenfeld and Ippen (1980); Hirata *et al.* (1989); Hirata and Mataga (1990); Shi *et al.* (1995); Pepin *et al.* (1994)]. Continuous spectral shifts of solvated electrons and presolvated electrons have also been reported [Pepin *et al.* (1994)]. The time dependent

Femtosecond Beam Science

behavior of intermediates obtained in the subpicosecond pulse radiolysis of
methanol is shown in Fig. 4.7. The fast and slow components of the decay

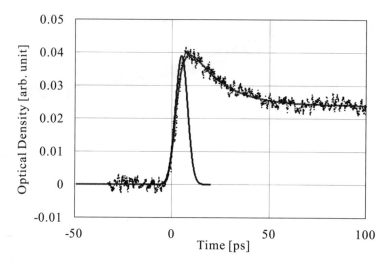

Fig. 4.7 Time dependent behavior of intermediates obtained in the subpicosecond pulse
radiolysis of neat methanol, monitored at a wavelength of 790 nm. The dots represent
experimental data and the solid line represents a kinetic trace. The response function is
superimposed.

curve are due to presolvated electrons and solvated electrons, respectively.
Although an estimation of the decay time of a presolvated electron requires
an analysis at all wavelengths where the presolvated electron is observed
because of the continuous spectral shift, the kinetic trace shown in Fig. 4.7
was fitted to the experimental data according to a strictly stepwise model
independent from other wavelengths with Eqs. (4.3), (4.5) and (4.6) as
follows:

$$f_1(t) = A_1 \frac{k_1}{k_1 - k_2} \left(e^{-k_2 t} - e^{-k_1 t} \right) , \qquad (4.5)$$

$$f_2(t) = A_2 \frac{k_1 k_2}{k_1 - k_2} \left(\frac{e^{-k_3 t} - e^{-k_2 t}}{k_2 - k_3} - \frac{e^{-k_3 t} - e^{-k_1 t}}{k_1 - k_3} \right) , \qquad (4.6)$$

where $f_1(t)$ is the time dependent behavior of a presolvated electron, $f_2(t)$
is the time dependent behavior of a solvated electron, k_1 and k_2 are the rate
constants of the formation of presolvated electrons and solvated electrons,

k_3 is the rate constant of the decay of solvated electrons and A_0 and A_1 are constants. 3.0 ps was tentatively used as the rise time of a presolvated electron, as reported in a femtosecond photolysis experiment [Pepin et al. (1994)]. The decay time of a presolvated electron monitored at 790 nm was estimated to be 16.4 ± 1.0 ps.

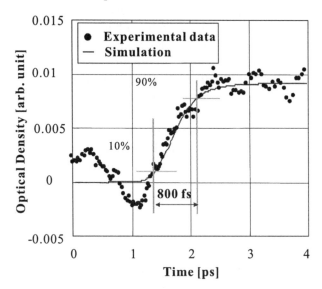

Fig. 4.8 Time dependent behavior of intermediates obtained in the subpicosecond pulse radiolysis of neat water. The optical length of the cell is 0.5 mm.

Hydrated electrons are also interesting. Figure 4.8 shows the time dependent behavior of hydrated electrons. The experimental data were tentatively fitted with Eq. (4.3). The 10–90% rise-time was 800 fs. This is slightly larger than that expected from the values shown in Fig. 4.4. This may reflect the formation time of hydrated electrons or its precursor.

Radiation chemistry in the subpicosecond time region is now in the nascent stage. Future research will make reveal more about the early processes in radiation chemistry.

4.1.1.6 *Application to materials for nanotechnology*

Recently, nanotechnology has attracted much attention and has been intensively investigated. Nanotechnology is expected to realize the greatest industrial revolution since the 1760s. There are two approaches to fabricating structures with nanoscale dimensions. One is called *bottom-up technology* and the other is *top-down technology*. The fusion of these two technologies will transform the now nascent technology into a fully fledged nanoindustry. Top-down technology especially will play a key role in connecting the nanoscale world to the real world. Nanolithography is the most promising top-down technology. With a decrease in fabrication dimension, the exposure sources of lithography have changed toward shorter wavelengths from the g-line of an Hg lamp, at 436 nm, to the i-line of an Hg lamp, at 365 nm, to KrF excimer lasers, at 248 nm. ArF excimer lasers, at 193 nm, are being deployed as an exposure tool for mass production. However, the resolving power of light sources will fall short of the anticipated market demand in the near future. Radiation sources such as electron beams (EB) and X-rays are expected to take the place of light sources. The research and development of each element of the radiation sources, such as resist, optics and masks for EB and X-ray lithography, have been pursued. The resist especially is a key technology for mass production and has been investigated enthusiastically.

Pulse radiolysis is the most effective method to investigate radiation-induced reactions and has been used to make clear the reaction mechanisms of resist materials. For the development of high performance resists, it is important to understand their reaction mechanisms. Recently, it has been pointed out that as the dimension of resist patterns gets closer to the spur size, the fast processes of radiation chemistry in the subpico- and pico-second time region become even more important. A new type of resist called the *chemically amplified resist* is being developed as a next-generation resist for EB lithography. On exposure of a chemically amplified resist to an electron beam, energy is deposited mostly on a base polymer via an ionization process. An electron generated, by the ionization, reacts with an acid generator after thermalization and a counter anion of acid is released via dissociative electron attachment. A proton is also generated from the radical cation of the base polymer. Thus, a proton and a counter anion are produced at different places. Therefore, in a chemically amplified resist, how far the electron migrates from its parent radical cation is a critical issue, especially from the viewpoint of nanofabrication. These reactions take place in the subpico- and pico-second time region and therefore high resolution pulse radiolysis can give much information essential to resist development. Figure 4.9 shows the time dependent distribution function of electrons in *n*-dodecane. This distribution can be obtained from an analysis of the time dependent behavior of geminate ion pairs. To effectively develop a new resist for nanolithography, it is important to properly understand reactions

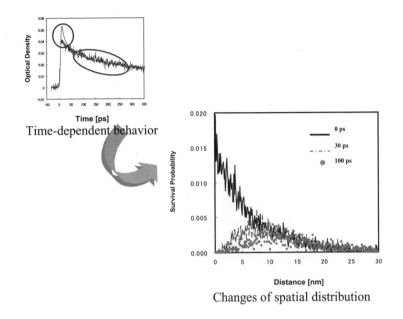

Time-dependent behavior

Changes of spatial distribution

Fig. 4.9 The changes of a thermalized electron distribution can be obtained from the time dependent behavior observed by subpicosecond pulse radiolysis. The solid line, dashed line and dots represent the distribution of cation radical electrons at distances at 0 ps, 30 ps and 100 ps after irradiation, respectively.

in nanospace that occur immediately after exposure.

4.1.2 *Radiolysis by RF gun*

4.1.2.1 *Supercritical xenon chemistry*

J. F. WISHART
Collider-Accelerator Department Building 817,
Brookhaven National Laboratory, Upton,
NY 11973, USA

(a) LEAF accelerator and experimental detection systems

The Laser–Electron Accelerator Facility (LEAF) accelerator at the Brookhaven National Laboratory (BNL) [Wishart (1998); Wishart (2001)] consists of a 3.5-cell RF photocathode electron gun, as shown in Fig. 4.10. The accelerator is driven by 15 MW of peak power from a 2.856-GHz S-band klystron to produce accelerating gradients of up to 100 MV/m. At the opti-

mal point in the RF cycle a 266-nm laser pulse from the frequency-stabilized
femtosecond laser system excites photoelectrons from the magnesium cathode in the back plate of the gun. The photoelectron bunch is accelerated
to 9 MeV within the 30-cm length of the gun, after which it is transported
to the target areas for chemical kinetics measurements.

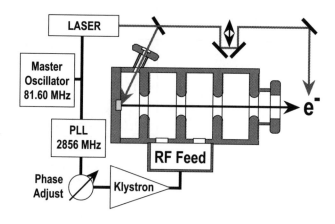

Fig. 4.10   Representation of the LEAF electron gun and synchronization system.

Photocathode electron gun accelerators have become quite numerous in
recent years for many applications such as free electron lasers and accelerator science. LEAF was the first electron-gun-based facility to be designed
specifically for pulse radiolysis. It differs from previous designs in that it
was designed to work with higher charges per-pulse (10–20 nC in 5–30 ps)
in order to provide large signals for kinetics experiments. Reliability and
ease of operation were important considerations for maximum productivity
of the physical chemists using the facility. The LEAF accelerator is unique
among picosecond radiolysis facilities because it is composed of a single
combined cathode and acceleration section, without additional booster or
linear accelerator (linac) sections. One advantage of this feature is the
simplicity of the required control system. There are only two accelerator
parameters to adjust during routine operation: the microwave power and
the relative phase of the microwaves and the laser pulse used to excite the
photocathode. The computerized beam transport system is controlled with

graphical LabVIEW software (National Instruments) that provides the operator with an intuitive display of the entire system. The 9-MeV beam energy was selected to provide adequate sample penetration depth without inducing radiological activation of common materials, many of which have photonuclear thresholds just above 10 MeV.

The initial design of LEAF includes two target areas: a short electron pulse width ($\sim$ 5 ps, 10 nC) station suitable for stroboscopic electron-pulse–laser-probe transient absorption measurements and a higher charge, longer pulse station (20 nC, 30 ps) for digitizer-based transient absorption measurements with time resolution down to 150 ps. The dual-beam pulse–probe detection scheme is shown in Fig. 4.11.

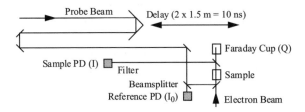

Fig. 4.11 Pulse–probe detection system at LEAF. All unlabeled components are mirrors. The electron beam passes through the sample and two mirrors before being stopped in the Faraday cup. This allows the radiolytic dose to be measured for each shot.

The probe laser beam (typically 50–100-fs long) is introduced with a mirror to be collinear with the electron beam. The measured charge in each electron bunch, Q, is used to correct the measured absorbance $\log(I_0/I)$ for shot-to-shot fluctuations. As the delay stage is moved, a time profile of the transient absorbance is constructed for kinetic analysis. An optical parametric amplifier is used to generate any color of probe light between 200 and 2200 nm, although experiments to date have only been done in the range accessible with silicon photodiodes, $\leq$ 1000 nm. Typical samples for pulse–probe radiolytic experiments are liquids or supercritical fluids, either in their pure form or with compounds to be studied dissolved in them. Often, the sample solutions are circulated between a reservoir and a flow cell placed in the beam, to dilute the effects of cumulative radiation damage on the chemical reactions that are being studied. A typical pulse–probe run may consist of 220 absorbance data points (baseline-corrected and spaced unevenly in time over 9 ns), each of which is the average of 20 accelerator pulses, and takes 30–40 minutes to complete.

(b)   Recent results

An example of the valuable information that can be obtained through pulse–probe radiolysis is the observation of Xe_2^* excimer formation upon irradiation of supercritical xenon [Holroyd *et al.* (2003)]:

$$e^- + Xe_2^+ \rightarrow Xe^* + Xe \qquad (4.7)$$

$$Xe^* + Xe \rightarrow Xe_2^* . \qquad (4.8)$$

A transient species that absorbs broadly throughout the region of visible light is assigned to the excimer species Xe_2^* on the basis of lifetime and kinetic data. The formation of excimers by electron–ion recombination was time-resolved by pulse–probe measurements at the LEAF accelerator using an 800-nm probe beam. Absorption grew over a period of 100 ps at 41 bar, and faster at higher pressures, as shown in Fig. 4.12. A trace showing the prompt absorbance rise of hydrated electrons in water is included in the figure as an indication of the overall time response of the experiment.

The observed rates of excimer formation correlate well with the expected rates for electron–ion recombination from Eq. (4.7). The rate of excimer growth slows as the density is reduced because the concentrations of e^- and Xe_2^+ ions are lower, since there are fewer ionization events per unit volume as the density decreases.

LEAF's pulse–probe detection system has been used to study a number of short timescale processes in radiation chemistry, including recombination of geminate ion pairs in hydrocarbons and ionic liquids, dissociative electron attachment to halogenated aromatic hydrocarbons and excited states of radical ions.

(c)   Future directions

The time resolution of LEAF's pulse–probe detection system is presently limited in part by the electron pulse width of ~ 5 ps. The electron pulse width may be further reduced in the future, but a fundamental limitation remains in the fact that high energy electrons travel faster through the sample than does the analyzing light pulse, due to the refractive index of the material. The thicker the sample, the further the light falls behind, roughly one picosecond for every millimeter of path length. In order to improve temporal resolution, sample depth and signal must be reduced. For sub-picosecond pulse–probe radiolysis, only chemical systems with large molar absorption coefficients are practical, and large radiolytic doses and signal averaging are needed to improve the S/N ratio.

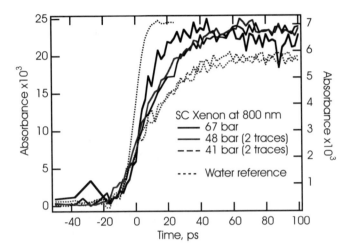

Fig. 4.12   Growth of absorbance in supercritical xenon due to excimer formation. The water and 67 bar Xe data are displayed vs. the left-hand scale, the 48 and 41 bar data vs. the right-hand scale.

There is another way to use accelerator facilities such as LEAF to study ultrafast reactions of radiolytically generated transient species. The pulse–pump–probe method uses the accelerator's electron pulse to prepare a free-radical precursor species. After a suitable delay (up to a few nanoseconds) an intense *pump* laser pulse is used to photolyze the precursor species and initiate the chemical reaction of interest. A tunable *probe* pulse is then used to follow the kinetics. The pump and probe phases of the experiment take full advantage of the femtosecond laser pulse widths. The experimental applications of pulse–pump–probe include excited state photophysics and photochemistry of short-lived radical species, probing ion recombination processes and ultrafast electron transfer reactions [Miller (1998)].

The outlook for growth of ultrafast (pico- and sub-picosecond) pulse-radiolysis facilities is very promising. LEAF has been joined by several new accelerator installations worldwide based on RF electron gun technology, at the Université de Paris-Sud (Orsay, France), University of Tokyo (Tokai-mura, Japan), Osaka University (Osaka, Japan), and Waseda University (Tokyo, Japan). Even more facilities are under consideration in several other countries. It is hoped that this cluster of state-of-the-art instruments will sustain and invigorate the valuable field of radiation chemistry for many years to come.

Acknowledgments
The author wishes to thank Dr. Richard Holroyd, Dr. Andrew Cook, Dr. Alison Funston and Dr. John Miller for helpful discussions.

4.1.2.2 *Ultrafast water chemistry*

Y. MUROYA,

Y. KATSUMURA,

Nuclear Engineering Research Laboratory, University of Tokyo,

2-22 Shirane-shirakata, Tokai, Naka,

IBARAKI 319-1188, JAPAN

(a)   Pulse radiolysis by S-band 1.5-cell photoinjector

During the last decade, there has been remarkable progress in miniaturizing femtosecond lasers and shortening electron pulses. Besides the development of intense ultrashort lasers, electron guns with lower emittance have also been developed, such as the laser photocathode RF gun. In addition, a magnetic pulse-compressor has been introduced to the traditional linear accelerator (linac), which has enabled us to generate electron pulses less than 1 ps, especially in the case of lower emittance [Kobayashi *et al.* (2002)]. At the Nuclear Engineering Research Laboratory (NERL), construction has started on a higher time-resolved pulse radiolysis system composed of a femtosecond laser, which takes the place of Cherenkov light as analyzing light, and a photocathode RF gun [Muroya *et al.* (2002a)]. Outlines of similar systems have been reported elsewhere [Wishart and Nocera (1998); Aoki *et al.* (2000); Belloni *et al.* (1998)].

Time resolution in pump-and-probe pulse radiolysis is largely controlled by the following factors: the pulse widths of the electron beam, the pump, and the analyzing light, the probe, the precision of their synchronization and the difference in the flight time between the laser and electron pulse in the sample. Although relativistic electrons travel with the velocity of light in a medium, the speed of light in a medium is c/n, where c is the speed of light in a vacuum and n is the refractive index of the medium. Therefore, all these factors must be addressed to attain higher time resolution. Stabilization of the RF signal and passive mode-locked laser oscillator are key issues for precise synchronization, details of which are discussed in Sec. 3.2.1.

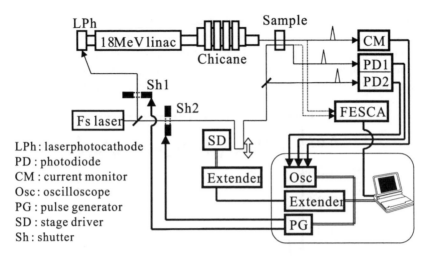

Fig. 4.13   NERL pulse radiolysis measurement system.

(b)   Measurement system

The measurement system is shown in Fig. 4.13. To measure the temporal behavior of radiation-induced species, an absorption spectroscopy method has been adopted. Before reaching the sample cell, the analyzing laser pulse is split into two by an aluminum mirror. One beam passes through the sample as analyzing light in the same flight path as the electron beam. The other is used as a reference. Absorption can be calculated as the intensity ratio I/I_0 of the light passing through the sample I to the reference light I_0. The signal intensity is detected by two pin photodiodes, PD1 for I_0 and PD2 for I, with an oscilloscope. For adjustment of the time difference between the electron beam and the analyzing light, an optical delay stage is used. In addition, to eliminate high voltage noise and Cherenkov light, two shutters are employed. Reliable absorbance is derived by using four different values of I/I_0 under the shutter conditions for pump/probe of on/on, off/on, on/off and off/off. A time profile can be produced by repeating the following processes on a computer: operating the shutters, acquiring intensity data from the oscilloscope, calculating the absorption and adding a timing delay to the optical stage. Normally, a femtosecond streak camera (Hamamatsu, FESCA200) is used for beam adjustment and monitoring the time difference between the electron beam and the laser before starting the pump-and-probe experiment.

(c)   Performance of the pump-and-probe pulse radiolysis system
A pulse radiolysis experiment using the above system has been done with
an electron pulse of 2 ps FWHM with $\sim$ 1 nC/pulse and a full size length of
3 mm. Millipore water is used as a sample, flowing through a high quality
quartz cell with an optical pass length of 5 mm. Figure 4.14 shows the time

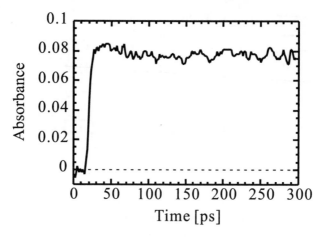

Fig. 4.14   Time profile of hydrated electrons in a picosecond time scale.

profile of hydrated electrons in water measured at the fundamental wavelength of the laser, 795 nm. The rise time of 9 ps obtained is equivalent
to the time resolution of the system. When the 5-mm cell is replaced by
a 1.8-cm cell, the time resolution decreases to 20 ps, instead of increasing
absorption. In general, the time resolution of the system δt_{total} can be ex-
pressed as $\delta t_{\text{total}} = \delta t_{\text{cell}} + (\delta t_E^2 + \delta t_L^2 + \delta t_{\text{sync}}^2)^{1/2}$, where δt_{cell} is the flight
time difference between the electron and laser pulse through the sample
cell, δt_E and δt_L are the pulse durations of the electron and the laser pulse,
respectively, and δt_{sync} is the precision of synchronization. The experimental results clearly indicate that the dominant factor is δt_{cell} under the
present conditions. The value of 10 ps is equivalent to the time difference in
a 1.0-cm cell. Hence, it is important to use thinner sample cells for higher
time resolution, but it should be noted that absorbance decreases with a
decrease of the optical pass length.

(d)   Pump-and-probe pulse radiolysis experiment; G-value of hy-
      drated electron

Water is one of the most popular and important chemical compounds in a
great variety of fields, and radiation research is not an exception. Up to
now, much study on radiation-induced reactions in water has been done
and it is well known that water molecules which have been decomposed
by ionizing radiation give several transient species, such as e_{aq}^-, H_3O^+, H,
OH, H_2 and H_2O_2, through radiation-induced processes. With regards to
the primary yield of hydrated electrons at 0.1–1 μs, a G-value of 2.7 has
been clearly established. The G-value is defined as the number of species
produced or disappearing per 100 eV energy absorbed. On the other hand,
in the picosecond time range, values from 4.0 to 4.8 have been reported.
There are several reasons for this discrepancy, for example; difficulty of
dose evaluation in the picosecond time range. Here, to determine the initial yield of hydrated electrons, an improved method based on the combination of pump-and-probe and conventional kinetic methods is introduced.
The time-based behavior up to the microsecond time scale is measured as
follows; in the picosecond time scale, the pump-and-probe method at the
fundamental wavelength of the femtosecond laser, 795 nm, was used, then,
under exactly the same conditions, the time-based behavior in the sub-
micro and microsecond time range was measured by the kinetic method by
substituting a He–Ne laser, at 633-nm wavelength, for the analyzing light.
Figure 4.15 shows the time profile of hydrated electrons. The ratio of the
absorbance gives 4.1 as the G-value of hydrated electrons at 20 ps, based
on a primary yield of 2.7 [Draganic and Draganic (1971)]. Here, one should
note that the two analyzing lights should be adjusted so as to pass along
the same axis through the sample. The beam profile, evaluated by using
Radcolor^{TM} film, gave a sharp spatial distribution of the dose. Therefore,
it is important to analyze within a 1-mm diameter; the size of each of the
analyzing lights was adjusted by several pinholes.

A Monte Carlo numerical simulation was employed to elucidate the experimentally obtained results. In a previous report, illustrated as a dashed
line in Fig. 4.15, G-values of 5.2 and 4.8 at 1 and 20 ps, respectively, were derived [Frongillo *et al.* (1998)]. Therefore, reexamination of the simulation
code was done in order to reconcile the reported results. Some parameters,
such as the mean free path, the cross section of geminate ion recombination
and the branching ratios of dissociative and non-dissociative reactions in
physicochemical processes, were adjusted and the final evaluation is shown
as a solid line in Fig. 4.15, together with the experimental result. This

360   *Femtosecond Beam Science*

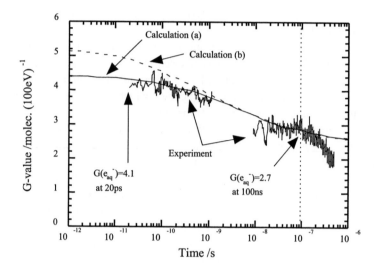

Fig. 4.15   Time dependent yield of hydrated electrons from the picosecond to microsecond ranges. The solid line [Calculation (a)] and the dashed line [Calculation (b)] show recent and previous Monte Carlo simulation results, respectively.

Monte Carlo calculation gave a value of 4.4 as the initial yield at 1 ps [Muroya *et al.* (2002b)].

(e) Future work

Some improvements are still necessary for the present pulse radiolysis system. For example, a more stable synchronization, shorter than 1 ps, and the introduction of a white light continuum converted from the fundamental light. Improvements are being made slowly but surely. The system is promising in that it enables us to reveal not only time dependent fast phenomena, but also the chemical properties of ultrashort-lived species.

Acknowledgments
We thank Prof. M. Uesaka, Dr. T. Watanabe and Mr. T. Ueda for their valuable help during the development of the system.

4.1.3 *Supercritical water*

Y. KATSUMURA
M. LIN
Nuclear Engineering Research Laboratory, University of Tokyo
2-22 Shirane-shirakata, Tokai, Naka,
IBARAKI 319-1188, JAPAN

4.1.3.1 *Supercritical water and its importance*

For light water, above 374°C and 22.1 MPa, the liquid and gas phases are merged into a single phase, called supercritical water (SCW), as shown in Fig. 4.16.

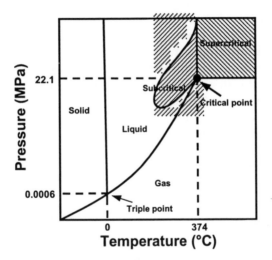

Fig. 4.16 Schematic phase diagram of light water, H_2O.

It is well known that SCW possesses some peculiar properties compared with water at room temperature. Its density is dependent on pressure, for example, 0.166 g/cm^3 under 25 MPa and 0.523 g/cm^3 at 40 MPa. The pKw of water is also strongly dependent on temperature and pressure in SCW. The dielectric constant is less than 10, which is much lower than that of water at room temperature, for which $\varepsilon = 79$, and it behaves like a non-polar solvent. The hydrogen network is broken in supercritical water and the water molecules are clustered.

Since SCW has so many unusual properties, some potential applications are expected. SCW oxidation (SCWO) is an effective and powerful technology that can be used for the destruction of hazardous organic wastes, e.g. PCBs (polychlorobiphenyl), dioxins, organic materials in sludge and other environmentally unfriendly materials; the products of the destruction mainly consist of atmospheric gases O_2 and CO_2. This method could be also used to destroy chemical weapons. Here we focus only on the pulse radiolysis of SCW, in order to accumulate basic data for SCW-cooled nuclear reactors, which is one of the most promising next generation reactors.

4.1.3.2 *Pulse radiolysis experimental setup for supercritical water*

Principally, pulse radiolysis techniques at ambient temperature can be used for the study of SCW. However, special attention should be paid to the structure of the cell so that it can stand up to at least 400°C and 40 MPa. Figure 4.17 shows an experimental setup for the pulse radiolysis study of SCW combined with a spectroscopic detection technique. A sample solution, loaded by an HPLC pump, passes through a preheater then is fed into the high temperature cell made of Hastelloy and sapphire windows. The pressure is adjusted by a backpressure regulator.

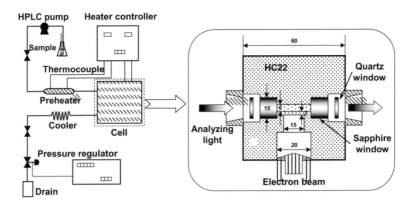

Fig. 4.17   Schematic arrangement of pulse radiolysis experiment and high temperature cell.

4.1.3.3 *Examples of pulse radiolysis studies on supercritical water*

Pulse radiolysis of SCW is still in its infancy; the first work was reported by Ferry and Fox in 1998 and 1999 [Ferry and Fox (1998); Ferry and Fox (1999)]. To date, reported data is still limited.

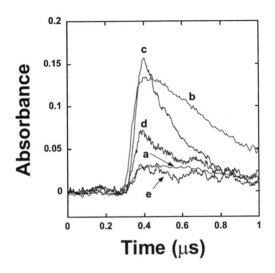

Fig. 4.18   Time profiles of e^-_{aq} at 1100 nm in D_2O. a 46 Gy/pulse; b 25°C, 0.1 MPa; c 100°C, 10 MPa; d 250°C, 20 MPa; e 350°C, 25 MPa; f 400°C, 35 MPa.

The hydrated electron e^-_{aq} is one of the most important water decomposition products. Figure 4.18 shows time profiles of e^-_{aq} in D_2O at 1100 nm in the range 25–400°C. e^-_{aq} is formed under all the applied conditions, though the absorption intensity and decay rate of e^-_{aq} varies with conditions [Wu *et al.* (2000)]. The absorption spectrum of e^-_{aq} is very sensitive to the solvent environment. The absorption peak of e^-_{aq} in D_2O at room temperature is at 700 nm. However, at 400°C/35 MPa, the spectrum broadens and its peak position shifts to around 1200 nm. It has also been shown that spectra of hydrated electrons not only have temperature dependence but also pressure dependence [Wu *et al.* (2000); Cline *et al.* (2000)].

The reaction of hydrated electrons with Ag^+ to form silver atoms Ag^0 and charged dimers Ag_2^+ has been studied up to 380°C [Mostafavi *et al.* (2002)]. Measurements of transient absorption spectra show that the wavelength maxima of the absorption spectra of Ag^0 have a red shift of 15 nm with increasing temperature from 25 to 200°C, while those of Ag_2^+ exhibit a red shift of about 60 nm at 380°C, as shown in Fig. 4.19. The shift in the absorption spectrum of Ag_2^+ is much more pronounced than that of Ag^0 because the solvation energy of a neutral silver atom is much less than that of the charged dimer. The reaction of hydrated electrons with benzophenone to form anion radicals and ketyl radicals, as well as their temperature dependent absorption spectra, have been observed up to 400°C [Wu et al. (2002a)]. The absorption spectrum of benzophenone anion radical shows a red shift, but that of ketyl radical shows a blue shift.

The reactions of OH with SCN^- and CO_3^{2-} to produce $(SCN)_2^{\bullet-}$ and $CO_3^{\bullet-}$ have been observed from room temperature to 400°C [Wu et al. (2001); Wu et al. (2002b)]. The $(SCN)_2^{\bullet-}$ radical also shows a red shift in its spectrum with increasing temperature, while $CO_3^{\bullet-}$ is almost temperature independent. The formation and decay behaviors of the $CO_3^{\bullet-}$ radical were measured precisely and a dimer model for $CO_3^{\bullet-}$ has been proposed.

Using a methyl viologen solution in the presence of *tert*-butanol, $G(e_{aq}^-)$ has been estimated up to 400°C [Lin et al. (2002)]. Figure 4.20 shows that the radiolytic yield of hydrated electrons increases up to 300°C, which agrees well with reported results [Elliot (1994); Elliot et al. (1996)]. Above 300°C, $G(e_{aq}^-)$ decreases with increasing temperature, and in supercritical conditions (374°C/22.1 MPa), a significant pressure or density effect on the G value has been observed.

4.1.3.4   *Future subjects*

The study of the radiation chemistry of SCW is just at its beginning. Many problems are waiting to be resolved experimentally and theoretically. A major unanswered issue is the change that occurs in the radiation-induced chemistry of water as the temperature and pressure are raised beyond the critical point. Other issues are:

- How the spatial distribution of the pairs and excited molecules changes when ionizing radiation interacts with this unusual medium.
- The solvation or energy relaxation of the excited molecules.
- Whether the ionizing radiation affects the yield and the rate constant of reactions.
- The nature of pressure or density effects.

A comprehensive understanding of the influence of this special medium on the early events of radiolysis is strongly expected. In this respect, pulse radiolysis systems with a higher time resolution, i.e. picosecond or even femtosecond, would be very useful.

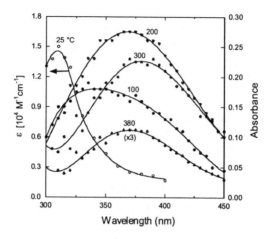

Fig. 4.19 Absorption spectra of Ag_2^+ at different temperatures.

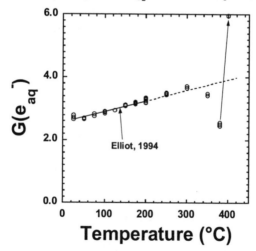

Fig. 4.20 G value of hydrated electrons as a function of temperature at 25 MPa.

Acknowledgments

The authors are grateful to Dr. G. Wu, Mr. Y. Muroya, Mr. T. Morioka, Dr. H. He and Dr. H. Kudo for their collaboration in the experiments and fruitful discussions. This study is partly supported by the Innovative Basic Research Program in the field of high temperature engineering using High Temperature Engineering Test Research conducted by the Japan Atomic Energy Research Institute.

4.2   Time-Resolved X-ray Diffraction

4.2.1   *Phonon dynamics in semiconductors*

K. KINOSHITA[1],
M. UESAKA,
Nuclear Engineering Research Laboratory,cUniversity of Tokyo,
2-22 Shirane-shirakata, Tokai, Naka,
IBARAKI 319-1188, JAPAN
[1]Division of Accelerator Physics and Engineering,
National Institute of Radiological Sciences,
4-9-1, Anagawa, Inage-ku, Chiba-chi,
CHIBA 263-8555, JAPAN

4.2.1.1   *Ultrafast microscopic dynamics*

Remarkable new possibilities have become available in time-resolved material analysis by the development of laboratory scale sources of ultrashort X-ray pulses, which have been made accessible by progress in ultra-intense and ultrashort laser technologies such as table-top terawatt lasers. Nowadays, laser intensities of more than 10^{18} W/cm^2 are achievable by focusing a high power laser to a small spot. Such intense laser pulses can easily generate hot plasmas by interacting with solid targets, making energetic electrons that produce X-rays by hitting the solid. These X-rays can have short pulse durations because of the short interaction time between the laser and the plasma, and the rapid cooling of the plasma. Therefore we can utilize these X-rays as probes to investigate ultrafast phenomena in materials with the time-resolved X-ray diffraction method. This method has been used in combination with synchrotron radiation (SR), with a time resolution longer than picoseconds [Larson *et al.* (1986)]. Recently pico- to subpico-second X-ray pulses have been produced by Thomson scattering of ultrashort laser pulses synchronized to relativistic electron beams [Leemans *et al.* (1997)]. It was shown that laser-produced plasma X-rays also generate pulses as short as several picoseconds [Yoshida *et al.* (1998)] or subpicoseconds in appropriate conditions [Reich *et al.* (2000)]. Several experiments have been also carried out on Thomson-scattering X-ray generation by colliding subpicosecond electron pulses with terawatt laser pulses [Uesaka *et al.* (2000)], and on characteristic X-ray generation by hitting solid targets with energetic electron beams [Harano *et al.* (2000)] in order to generate ultrashort X-ray pulses for time-resolved analysis. These X-rays, having been applied to the dynamic analysis of materials [Rischel *et al.* (1997); Rose-Petruck *et al.* (1999); Kinoshita *et al.* (2001)], are expected to provide a promising tool for the investigation of ultrafast unknown phenomena and experimental verification of solid state physics. Figure 4.21 shows fast

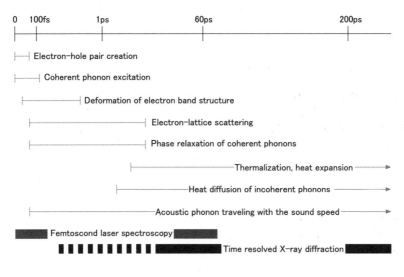

Fig. 4.21 Ultrafast phenomena in laser–solid interactions.

phenomena at each timescale in laser–solid interactions. Time-resolved X-ray diffraction provides structural information of samples in a timescale of several picoseconds to hundreds of picoseconds, while femtosecond laser spectroscopy covers phenomena from several tens of femtoseconds to several tens of picoseconds. Time-resolved X-ray diffraction has been studied at many institutes around the world. Table 4.1 shows several of those institutes, together with information such as the X-ray sources and samples used at the institute.

Table 4.1   Research institutes studying time-resolved X-ray diffraction.

Research Institute	X-ray source	Time Resolution	Crystal	Observed phenomena
UC SanDiego	Laser plasma	10 ps	GaAs	Thermal expansion, shock wave
LBNL	SR	10 ps	GaAs	Acoustic phonon
U. Essen	Laser plasma	10 ps	Ge+Si	Melting, acoustic phonon
U. Oxford/ESRF	SR	1 ps	InSb	Acoustic phonon
LOA	Laser plasma	<1 ps	LB crystal, InSb	Destruction, melting
Tokyo Inst. Tech.	Laser plasma	100 ps	Si	Shock wave
NERL, U. Tokyo	Laser plasma	10 ps	GaAs	Thermal expansion, shock wave

4.2.1.2   *Strain wave in crystals*

Laser-induced strain derived by Thomsen [Thomsen *et al.* (1986)] is assumed in calculating the temporal evolution of X-ray diffraction patterns. The laser energy absorbed in a crystal causes a temperature increase, distributed according to the absorption length ζ [Blakemore (1982)]. Thus, the temperature increase ΔT can be described as

$$\Delta T(z) = \Delta T_s \, e^{-z/\zeta} \,, \qquad (4.9)$$

where z is the distance from the surface into the crystal and ΔT_s is the temperature increase at the surface just after the laser irradiation. This temperature distribution gives rise to a thermal stress given by

$$-3B\beta\Delta T(z) \,, \qquad (4.10)$$

where B is the bulk modulus and β is the linear expansion coefficient. Then, the thermal elastic equations are given by

$$\sigma_{33} = 3 \,\frac{1-\nu}{1+\nu}\, B\eta_{33} - 3B\beta\Delta T(z) \,, \qquad (4.11)$$

$$\rho \,\frac{\partial^2 u_3}{\partial t^2} = \frac{\partial \sigma_{33}}{\partial z} \,, \qquad (4.12)$$

$$\eta_{33} = \frac{\partial u_3}{\partial z}, \tag{4.13}$$

where η_{33} is the elastic strain tensor, ν is Poisson's ratio, u_3 is the displacement in the z direction and ρ is the density. The initial condition is $\sigma(z) = 0$ everywhere. These equations can be solved analytically. The solution is

$$\eta_{33}(z,t) = \Delta T_s \beta \frac{1+\nu}{1-\nu} \left[e^{-z/\zeta} \left(1 - \frac{1}{2} e^{-vt/\zeta} \right) \right.$$
$$\left. - \frac{1}{2} e^{-|z-vt|/\zeta} \operatorname{sgn}(z - vt) \right], \tag{4.14}$$

where v is the sound velocity. This solution gives two components, a strain wave propagating inward with the speed of sound and a thermal expansion of the lattice at the surface. The dashed lines in Fig. 4.22 show X-ray diffraction patterns calculated based on kinetic diffraction theory with the corresponding crystal strains. An additional peak arises at a small diffraction angle due to the X-ray diffraction of $K_{\alpha 1}$ from the expanded lattice; it approaches the original angle as the strain wave goes deeper into the sample, as shown in Fig. 4.22.

4.2.1.3 *Experiments*

Figure 4.23 shows an experimental setup for time-resolved X-ray diffraction. A laser pulse entering the vacuum chamber is divided into a main pulse and a pump pulse in a ratio of nine to one by the beam splitter. The main pulse is focused onto the copper target with the off-axis parabolic mirror (focal length 162 mm) to generate the laser-produced plasma X-rays that are used for the X-ray diffraction from the single crystal as the probe pulse. Focused by the spherical lens, the pump pulse induces transient phenomena inside the crystal after passing through the delay path, with which the time interval between the pump and probe pulses can be controlled. For easy alignment, large wafers or disks of single crystals are employed as samples for X-ray diffraction and mounted on the automatic positioner. Two-dimensional diffraction images are taken by X-ray imaging plates with a spatial resolution of 50 μm. The samples are surrounded by lead plates for shielding of the background X-rays scattered from the circumference. The X-rays may reach the sample surface only through the 3-mm wide slit. Polyvinylidene chloride film is put in front of the parabolic mirror as protection against ablated debris from the copper target. Diffracted X-rays pass through the beryllium window and are measured by the imaging plates

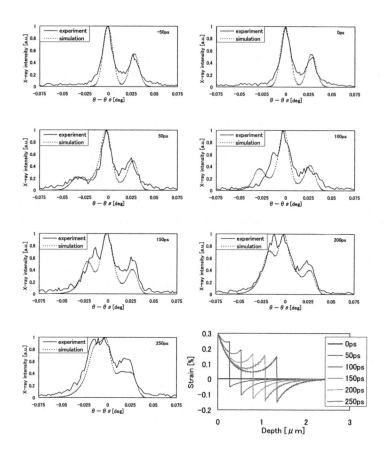

Fig. 4.22   Picosecond changes in diffracted X-ray profiles from a GaAs(111) single crystal.

outside of the chamber. Another 1 cm-wide slit is put at the exit of the beryllium window to reduce the background X-rays.

A femtosecond streak camera (FESCA200) is used to determine the zero point of the delay time where the pump pulse and the probe pulse reach the crystal surface at the same time. A mirror with a hole at the center is put in place of a crystal with an angle to guide the light from the focal point of the parabolic mirror reflected and scattered by the copper target toward the streak camera. This light from the copper target substitutes for

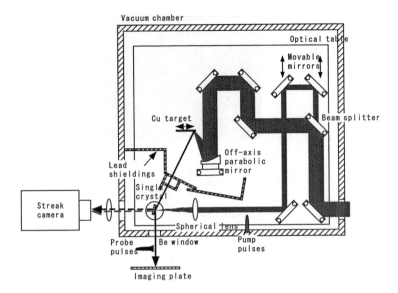

Fig. 4.23   Experimental setup for time-resolved X-ray diffraction with laser plasma X-rays.

the X-ray pulse while the pump pulse can directly reach the streak camera after passing through the hole of the mirror (the dashed rays in Fig. 4.23). Figure 4.24 shows an image of the streak camera and its temporal profile. The pump pulse arrives 7.2 ps before the light from the copper in Fig. 4.24. We can adjust the distance between the pump pulse and the probe pulse to zero by changing the delay path. The pulse width is elongated due to the resolution of 500 fs at that range and other optical errors.

Figure 4.25 shows a diffraction image taken by an imaging plate and its horizontal profile of X-ray intensity from a GaAs wafer of (111) orientation with pump pulses of 1.3 mJ/mm^2 with 150 ps delay. The accumulation time of this image was about 30 s at 10 Hz repetition. The two thick lines in Fig. 4.25 are characteristic X-rays from copper, i.e. CuK$_{\alpha1}$ (1.5407 Å) and CuK$_{\alpha2}$ (1.5439 Å). The left-hand side of the image corresponds to smaller angles of diffraction and the right-hand side to larger angles. The lattice distance of GaAs(111) is 3.254 Å, which makes the diffraction angles for CuK$_{\alpha1}$ and CuK$_{\alpha2}$ 13.65° and 13.68°, respectively. We can see an extra line on the left of the original lines in the image at the laser irradiated region. This line is caused by deformation of the lattice of the GaAs crystal induced

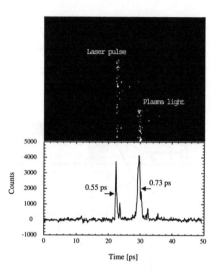

Fig. 4.24   Timing adjustment with a femtosecond streak camera.

by the laser pulses, resulting in a subpeak in the profile.

The solid lines in Fig. 4.22 show the temporal evolution of the diffraction profiles. The negative delay in Fig. 4.22 means that the probe pulse arrives prior to the pump pulse. Naturally, we can see only two peaks, $K_{\alpha 1}$ and $K_{\alpha 2}$, at -50 ps from the undisturbed crystal and there is no change at 0 ps. A small subpeak can be seen on the left of $K_{\alpha 1}$ at 50 ps. At 100 ps the subpeak is small and narrow while $K_{\alpha 2}$ is blurred. The position of the subpeak approaches the peak of $K_{\alpha 1}$ as the delay time increases to 150 ps and 200 ps, and it merges into $K_{\alpha 1}$ at 250 ps.

These changes of the diffraction profiles in Fig. 4.22 are directly brought about from the ultrafast evolution of the lattice in the crystal in the picosecond timescale. The laser entering the crystal excites electrons in the valence band to the conduction band, and thus is absorbed in making electron–hole pairs. Therefore, the deposited energy is distributed according as the absorption length. An energy transfer from carriers to the lattice occurs by scattering. This brings a rapid temperature increase in the crystal, resulting in thermal stress in the lattice. This thermal stress triggers a strain wave propagating into the crystal with the speed of sound. Time-resolved X-ray diffraction can pick out snapshots of the crystal undergoing the above process. Figure 4.26 shows an image of such a phenomenon. The lattice

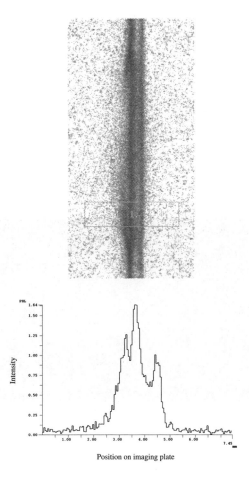

Fig. 4.25   Diffracted X-ray image from a laser-irradiated crystal taken with an image plate.

distance near the surface is expanded by the temperature increase and the strain wave propagates inward as acoustic phonons.

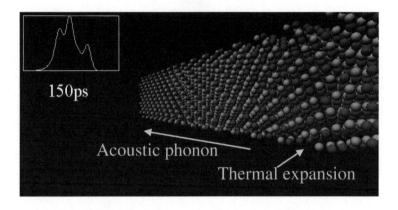

Fig. 4.26   Snapshot of the atom distribution in a crystal irradiated with an ultrashort laser pulse.

4.2.2 Shock wave propagation in semiconductors

K. G. NAKAMURA,
Y. HIRONAKA,
K. KONDO,
Materials and Structures Laboratory, Tokyo Institute of Technology,
4259 Nagatsuta, Yokohama,
KANAGAWA 226-8503, JAPAN

4.2.2.1 Shock compression science

Shock compression produces very high pressures, above Mbar, and is studied to define high pressure equations of states on hundreds of materials. Also, shock compression data are still the basis of the pressure scale. Shock compression science provides essential contributions to solid mechanics, geophysics, astrophysics, materials science, solid state physics and solid state chemistry [Asay and Shahinpoor (1993)]. Physical and chemical phenomena such as phase transitions, shock demagnetization, shock induced opacity, fragmentation, material failure and shock induced electrical conduction are found under high pressures. Technological applications to materials science such as metal cutting and shaping with explosives, explosive welding, powder consolidation and synthesis of diamond have also been developed. Since shock compression, which is often produced by explosion and high velocity impact, is a transient and irreversible event, high speed measurement techniques are required to study physical and chemical properties of shock-compressed materials. Shock and particle velocities are often measured by conventional shock wave measurements, such as pin techniques and laser interferometers. Elastic-plastic deformation and phase transition in solids are determined from changes in the Hugoniot curves (see Fig. 4.27). However, such methods cannot provide direct information about microscopic mechanisms governing shock-induced structural changes. X-ray diffraction is the most commonly used technique for determining crystal structures and is widely used in both physical and material sciences. Although X-ray diffraction techniques are routinely applied in the analysis of recovered samples from shock compression, very little work has been performed in the field of diffraction during transient compression itself. Ultrafast time resolutions are necessary for the dynamics of shock-induced phase transitions and propagation of shock waves, because the events move at high speed.

Recent developments of intense femtosecond lasers enable the generation of ultrashort pulsed X-rays. Combined with laser-induced shock compres-

Femtosecond Beam Science

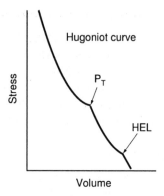

Fig. 4.27 When solids undergo a shock-induced phase transition, a volume change at the transition causes a significant change in the Hugoniot curve. P_T and HEL are the pressures of the phase transition and Hugoniot elastic limit, respectively.

sion and a pump-and-probe technique, ultrafast time-resolved X-ray diffraction measurements have recently been used to investigate structural dynamics under transient shock compression. This section summarizes the field of time-resolved X-ray diffraction studies of crystals under shock compression. A brief history of transient X-ray diffraction studies of shock-compressed materials is first described. Next, the basic aspects of experiments, including laser shock compression, laser plasma X-ray sources and time-resolved X-ray diffraction using the pump-and-probe technique, are described. Picosecond time-resolved X-ray diffraction studies on laser shocked Si(111) crystal are described.

4.2.2.2 *X-ray diffraction of shocked solids*

Pioneering work in the application of X-ray diffraction to shock compression of solids was performed by Johnson *et al.* in the early 1970s [Johnson *et al.* (1970)]. They produced flash X-rays with a pulse width of 20 ns by using a high current and high voltage transmission line to a cathode–anode gap and measured X-ray diffraction of the transient shock-compressed state of solids. Subsequent work was done by Kondo *et al.* [Kondo *et al.* (1979)] in 1979. The generated shock waves using a gas gun and flash X-rays were synchronized to the gun. They studied shock-compressed LiF at around 20 GPa using flash X-ray diffraction and found that the single-crystal transforms to an imperfect one, like a mosaic crystal with disorder in the range of several degrees. Several different studies have been reported in the literature. Gupta *et al.* [Gupta *et al.* (1998)] studied elastic-plastic deformation in shocked LiF by flash X-rays generated by a 300-kV flash X-ray tube, and plate impact experiments at around 5 GPa. Although the transient structure in shocked crystals was investigated using flash X-ray diffraction studies and a gas gun, only a snapshot was taken at arbitrary timing.

Wark *et al.* [Wark *et al.* (1989)] generated both shock waves and X-rays using a high power pulsed laser and studied the structural change of shocked Si crystals by time-resolved X-ray diffraction. They used a giant laser system, which was made for laser fusion research. Using an X-ray streak camera with a resolution of 20–50 ps, time-resolved X-ray diffraction measurements were performed on the laser-shocked Si crystals. Although laser shock compression was used, the experiment was a single-shot experiment because the high energy lasers could emit only a few times a day.

Recent developments of intense femtosecond lasers enable the generation of ultrashort X-rays pulses using table-top size laser systems at 10 Hz or much higher repetition rates and can realize ultrafast time-resolved X-ray diffraction measurements [Kmetec *et al.* (1992); Guo *et al.* (1997); Yoshida *et al.* (1998); Fujimoto *et al.* (1999); Rischel *et al.* (1997); Rose-Petruck *et al.* (1999); Siders *et al.* (1999)]. High quality data can be obtained by accumulation with repetitive experiments. Combining laser induced shock compression and ultrafast time-resolved X-ray diffraction measurements can be performed in a typical laboratory.

4.2.2.3 *Laser shock*

Laser shock is a technique to generate shock waves by pulsed laser irradiation of solids. Several different techniques have been developed using an ablator, a hohlraum and a flyer [Wark *et al.* (1989); Cauble *et al.* (1998); Kadono *et al.* (2000); Matsuda *et al.* (2002); Wakabyashi *et al.* (2000)]. The most simple technique is the direct irradiation method. Shock waves are generated inside the solid by momentum conservation, since particles are ejected from the surface with some momentum by laser ablation. Several empirical equations for estimating the pressure induced by laser irradiation have been reported. The formula reported by Fabbro *et al.* [Fabbro *et al.* (1990)] is $P = 0.393\, I^{0.7}\lambda^{-0.3}t^{-0.15}$, where P [GPa] is the pressure, λ [mm] the laser wavelength, I [GW/cm^2] the laser power density and t [ns] the pulse duration. Very high pressure can be generated by laser shock using a giant pulsed laser, e.g. 1.7 TPa is generated in tantalum by irradiation with a 2-ns laser with 460 J on a spot size of 500 μm in diameter [Wakabyashi *et al.* (2000)]. The shock duration is almost the same as the laser pulse width.

4.2.2.4 *Laser plasma hard X-ray pulses*

Ultrashort hard X-ray pulses are generated by the interaction between femtosecond intense laser fields and solids. Intense laser fields above 10^{16} W/cm^2 can be generated by focusing femtosecond laser beams from a table-top terawatt T^3 laser system [Yoshida *et al.* (1998); Fujimoto *et al.* (1999)]. The T^3 laser system consists of a Ti:Sapphire laser oscillator and a chirp pulse amplification (CPA) system. 25-fs laser pulses are generated at 76 MHz by the Ti:Sapphire laser oscillator, pulse stretched to 300 ps, regeneratively amplified and amplified again by two stages of multipath amplifiers using nanosecond pulsed YAG lasers (10 Hz). The amplified 300 ps pulse, which has a maximum energy of 400 mJ/pulse, is compressed to 50 fs in a vacuum chamber using a pair of gratings. The femtosecond pulse is focused on a metal target within a spot size of approximately 40 μm in a vacuum chamber (10^{-5} Torr) using a parabolic mirror of focal length 150 mm at an incident angle of 60° to the surface normal. The obtained laser intensity can be controlled between 10^{16} and 10^{18} W/cm^2. The target is a disk 70 mm in diameter and 5 mm thick and is rotated and translated in order to expose a fresh surface for each laser shot. Energy spectra are measured using a direct detection X-ray CCD camera with an energy resolution of 120 eV and the temporal profiles are measured using an X-ray streak camera with a time resolution of 2 ps.

Figure 4.28 shows X-ray spectra from medium-Z targets (Ti(22), Fe(26), Ni(28), Cu(29) and Zn(30)) at a laser intensity of 2×10^{16} W/cm^2. The spectra consist of strong characteristic line emissions K$_\alpha$ and K$_\beta$ and a very weak continuum at around 3 keV. The intensity ratio of the K-shell emissions to the continuum decreases as Z increases. The spatial distribution of the K-shell emission is almost isotropic. In the case of a Cu target, the generated photon number of the Cu-K$_\alpha$ line emission (8.04 keV) was measured to be approximately 10^{11} photons/4πsr/pulse. Figure 4.29 shows the temporal profile of X-rays from the Cu target measured by the X-ray streak camera. The pulse width of total X-rays (both K-shell emission and continuum) was measured to be 6.4 ps FWHM, which is approximately 100 times longer than the incident laser pulse.

The mechanism of X-ray generation by an intense femtosecond laser is

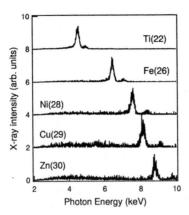

Fig. 4.28 Energy spectra of X-rays generated by focusing a 50-fs laser on a metal target at 10^{16} W/cm^2.

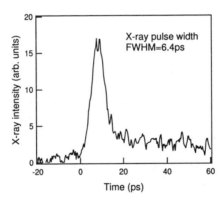

Fig. 4.29 Temporal profile of X-rays from a Cu target measured by an X-ray streak camera with a temporal resolution of 2 ps.

mainly explained as follows [Guo *et al.* (2001)]. Electrons generated by a pre-pulse or a pedestal are accelerated to high energy by the strong ponderomotive potential of the main pulse. The high energy electrons collide with the metal and induce K-shell emission and bremsstrahlung, as in a conventional X-ray tube. The high energy electrons are measured using a magnet type spectrometer. Figure 4.30 shows electron energy spectra for 50-fs laser irradiation of a thin-film Cu target at 2.7×10^{18} W/cm^2 [Oishi

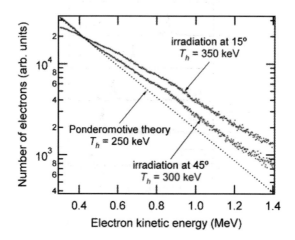

Fig. 4.30   Energy spectra of high energy electrons generated by focusing a 50-fs laser on a Cu film at 2.8×10^{18} W/cm^2. Two irradiation angles are examined: 15° and 45° normal to the surface. The dashed line is a ponderomotive theory calculation.

et al. (2001)]. The electrons have energies up to 1 MeV and their temperature is approximately 300 keV. The pulse width of the X-rays may be affected by a plasma, which depends on the laser pulse shape and the target, because the X-rays are generated by the collision of high energy electrons. Femtosecond pulsed X-rays can be generated by using a steep-raised pulse without pre-pulse and a thin target.

4.2.2.5   *Ultrafast time-resolved X-ray diffraction of shock compressed silicon*

(a)   Rocking curve measurements
Picosecond time-resolved X-ray diffraction experiments have been performed on laser-shocked silicon crystals using a pump-and-probe technique [Hironaka *et al.* (2000)]. Figure 4.31 shows a schematic drawing of the experimental setup. The 300-ps output from the laser is divided into two beams by a beam splitter. One is directed to the pulse compressor and used for X-ray generation, and the other is irradiated on the sample for shock generation after passing through the optical delay line. The X-rays are generated by focusing a 50-fs laser beam on the iron target at 10^{17} W/cm^2. The sample used is an *n*-type Si(111) wafer with a thickness of 860 μm. The 300-ps laser is focused on the sample at an intensity of approximately

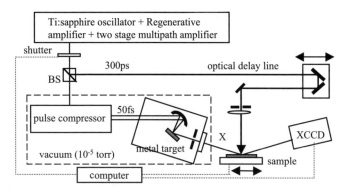

Fig. 4.31 Schematic drawing of experimental setup for time-resolved X-ray diffraction.
BS: beam splitter, X: X-rays.

4 GW/cm^2. To maintain the shock-compressed state for a long time (sub-nanosecond order), a 300-ps pulse is used instead of a 50-fs pulse. The generated X-ray pulse is irradiated at the center of the laser focal spot on the Si(111) and the diffracted X-rays are detected with an X-ray linear sensor or X-ray CCD and accumulated for 600 shots. The sample is irradiated 200 times, since the sample is moved at a slow speed. The sensor is protected from visible and infrared light by a 350-μm-thick filter of Be.

Figure 4.32 shows the measured diffraction profiles as a function of angle (rocking curves) at various delay times between 0 and 1380 ps with a time interval of 60 ps. Zero delay is defined by the initial deviation of the diffracted signals. The diffracted $K_{\alpha 1}$ (1.9360 Å) and $K_{\alpha 2}$ (1.9399 Å) lines from the pristine sample are well resolved. When the pump laser beam irradiated the sample, the intensities of these lines decreased and signals at higher angles appeared. After the end of pump beam irradiation (after 300 ps), the shift and intensity of the higher angle diffraction gradually decreased. The shift of the diffracted signals towards higher angles corresponds to the decrease of the crystal lattice spacing and demonstrates that the silicon lattice is compressed by irradiation of the pump laser beam.

(b) Shock wave propagation
X-ray diffraction from the laser-irradiated silicon crystal was analyzed by a code based on the dynamical diffraction theory applied to strained and laser-shocked crystals [Hironaka *et al.* (2000)]. The Takagi-Taupin equation was solved using a one-dimensional layer approximation and the method of Wie *et al.* [Wie *et al.* (1986)]. In the calculation, the depth profile of

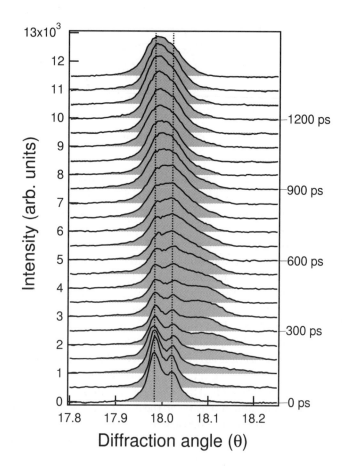

Fig. 4.32 Time-resolved X-ray diffraction of 300-ps-laser-irradiated Si(111) at 4 GW/cm^2 at every 60 ps. The X-rays used are Fe K$_{\alpha 1}$ and K$_{\alpha 2}$ emissions.

the strained crystal was divided into 100 layers with a total thickness of 10 μm. The depth profile was approximated by a combination of three Gaussian functions. The strain profiles were determined by assuming that the calculated rocking curve represents the measured curve. Figure 4.33

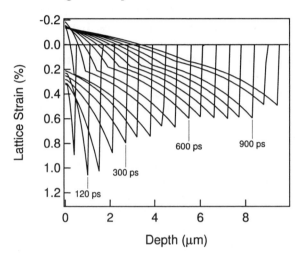

Fig. 4.33   Strain profiles at various delay times for 300-ps-laser-irradiated Si(111) at 4 GW/cm^2.

shows strain profiles as a function of depth at various delay times. The calculated rocking curves using the strain profiles are in good agreement with the experimentally obtained ones. The velocity of the propagating shock wave was estimated to be 9.4 km/s from the changes in the strain profiles. Since the lattice compression at the Hugoniot elastic limit is reported to be 2.6% at about 5.4 GPa [Gust and Ryce(1971)], the maximum lattice strain of −1.05% (compression) corresponds to a maximum pressure of 2.18 GPa. The pressure estimated from the empirical formula given in Ref. [Fabbro *et al.* (1990)] is 1.52 Gpa. The rise of the pressure to the peak value takes about 180 ps, which is comparable to that of the pump laser beam. For delay times longer than 420 ps, a small expansion of about 0.15%, due to both rarefaction and thermal expansion effects, is observed near the sample surface.

Using picosecond pulsed X-rays, lattice deformation accompanied with shock wave propagation inside the 300-ps-laser-irradiated crystal at 4 GW/cm^2 can be directly measured without any disturbance. Dynamics of elastic-plastic transformation and phase transitions at much higher pressure can be studied with more intense laser irradiation. In the present experiment, each X-ray diffraction has a spatial resolution of about 0.06 μm in depth, because the X-ray pulse width is about 7 ps and the shock wave propagates at 9.4 km/s. The structural changes in the shock front can be obtained using femtosecond probe X-ray pulses. 100 ps are required for homogeneous excitation of a 1-μm-thick sample to a shock-compressed state if the excitation pulse is a femtosecond pulse, because the response of atomic displacement to pulse excitation propagates with a shock velocity. Excitation by shock compression is different from that by a light pulse because instantaneous excitation is possible for excitation by a femtosecond pulse.

4.2.2.6 *Summary*

Shock wave dynamics in solids is studied by ultrafast time-resolved X-ray diffraction. Pulsed X-rays have been generated by focusing 43-fs laser beams onto metal targets above 10^{16} W/cm^2. The pulse width of the generated X-rays from a thick Cu disk target was measured to be 6.4 ps. Picosecond time-resolved X-ray diffraction has been performed on 300-ps laser shocked Si(111) 60 ps steps. Lattice deformation, accompanied with shock wave propagation, inside the crystal at a shock velocity of 9.4 km/s was directly measured without any disturbance.

4.2.3   *Fast X-ray shutter using laser-induced lattice expansion at SR source*

Y. TANAKA,

T. ISIKAWA,

SPring-8/RIKEN Harima Institute,

1-1-1 Kouto, Mikazuki-cho, Sayo-gun,

HYOGO 679-5198, Japan

A fast shutter of X-rays has the potential to produce very short X-ray pulses, depending on its speed. Neither mechanical choppers (see Sec. 3.2.2.3) nor surface acoustic wave (SAW) devices, have a shutter speed with a response time of less than nanoseconds. X-ray shutters with speeds faster than mechanical and SAW devices can use short pulsed lasers as a trigger to switch X-ray diffraction. A theoretical proposal for switching using Bragg reflection on phonon induced crystal lattice vibrations by laser irradiation [Buckbaum and Merlin (1999)], and the demonstration of laser triggered phonon switching in the Laue configuration [DeCamp *et al.* (2001)] have been reported.

This section describes a fast shutter for X-ray synchrotron radiation (SR) using laser-induced crystal lattice expansion [Tanaka *et al.* (2002b); Tanaka *et al.* (2002a)]. After a brief explanation of the pump-and-probe scheme using a laser–SR synchronization system to observe the time response of laser-induced crystal lattice expansion, we describe the application of lattice expansion to X-ray switching and the demonstration of the isolation of a single pulse from a SR pulse train in a double-crystal configuration.

4.2.3.1   *Optical switching of X-rays using transient expansion of crystal lattice*

Figure 4.34 shows the pump-and-probe experimental scheme to observe laser-induced lattice expansion using the laser–SR synchronization system at SPring-8 described in Sec. 3.2.2. In Fig. 4.34(a), a GaAs crystal is irradiated by a pulsed laser synchronized with SR pulses. The laser has a wavelength of 800 nm and a pulse duration of 130 fs. The laser beam size on the crystal has a diameter of about 3 mm, which is larger than the X-ray beam size. The incident photon energy of the SR is 21.3 keV, which is monochromatized with a Si(111) double-crystal monochromator.

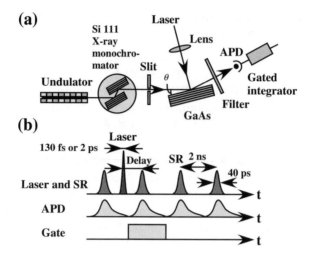

Fig. 4.34 (a) Experimental setup for time-resolved X-ray diffraction from a GaAs crystal. (b) Timing chart of laser pulses, SR pulses, signal from APD detector, and gate of boxcar integrator.

The X-rays diffracted by the 004 diffraction plane are detected with an avalanche photodiode (APD). Figure 4.34(b) shows the timing chart of the measurement system in which the target probe SR pulse is isolated from the SR pulse train by adjusting the gate of the boxcar integrator. The dependence of the diffraction intensity on the laser pulse delay is obtained by changing the timing.

Figure 4.35(a) shows the time-resolved rocking curves for the 004 reflection from laser-irradiated GaAs. The laser pulse hits the crystal surface at $t = 0$ in Fig. 4.35(a). Laser irradiation shifts the Bragg peak by -2.5 arc sec. without any observable change in the profile. This implies a lattice expansion of $\Delta d/d = 4 \times 10^{-5}$, which is derived by the differentiation of Bragg's law:

$$\Delta\theta = -(\Delta d/d)\tan\theta_B, \qquad (4.15)$$

where $\Delta\theta$ is a deviation from the Bragg angle θ_B for a certain fixed X-ray wavelength. The Bragg peak shift induced by the laser pulse is reproducible when the laser power is moderate enough to cause no damage to the GaAs surface in air. At the sample surface the laser power density $P_L = 10$ mJ/cm^{-2}. Fixing the crystal at an offset angle of -7 arc sec.,

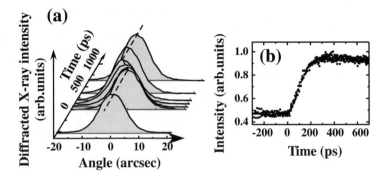

Fig. 4.35 (a) Time-resolved rocking curves for 004 reflection from laser-irradiated GaAs.
(b) Time dependence of diffracted X-ray intensity at an offset angle of −7 arc sec.

the diffracted X-ray intensity is drastically changed according to the Bragg
peak shift by the lattice expansion with a response time of a few hundred
picoseconds, as shown in Fig. 4.35(b). This change in the diffracted X-ray
intensity is used for *switching on* the X-rays.

4.2.3.2 *X-ray shutter using optical switch*

When the deviation of the Bragg angle exceeds the relevant rocking curve
width according to Eq. (4.15), the laser pulse causes the X-rays to be reflected (*switching on*) or to cease being reflected (*switching off*). A higher
laser power density gives a larger deviation than the Bragg angle, although
the response time is not changed so much. Hence, a higher laser power density is preferable when performing switching, as long as it remains below
the surface damage threshold. For the 004 diffraction of GaAs, the Bragg
peak shift necessary for switching cannot be observed practically. Hence,
we have adopted a double-crystal configuration for optical-switching, because the arrangement of the two crystals in a $(+, -)$ geometry makes the
reflectance of X-rays highly sensitive to the crystal angles.

Figure 4.36 shows a schematic diagram of the experimental setup to
isolate a single pulse from the SR pulse train. The crystal angles are ad-
justed to have deviations $\Delta\theta_1$ and $\Delta\theta_2$ from θ_B. The X-ray beam emitted
from the undulator has a photon energy of 16.4 keV. A 800-nm laser with
a pulse duration of 2 ps is guided through a beam splitter and lenses onto
the crystals. The power density P_L at the crystals was about 70 mJ/cm^2,

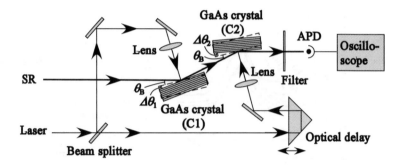

Fig. 4.36   Experimental setup for optical-switching of X-ray SR pulses. Two crystals C1 and C2 are located in a $(+,-)$ geometry. Picosecond laser beams are guided onto the crystals. The throughput SR pulses are detected by an APD through an aluminum foil filter. With a high precision diffractometer, the crystal angles are finely adjusted to have deviations $\Delta\theta_1$ and $\Delta\theta_2$ from the Bragg angle θ_B. The positions of C1 and C2 that satisfy the Bragg condition without lattice expansion are drawn as dashed lines.

chosen to be just below the surface damage threshold. The timing of laser shots onto the second crystal C2 was controlled with an optical delay apparatus. The X-rays diffracted by the crystals were observed with a 1-GHz oscilloscope through an APD detector.

Figure 4.37(a) shows the result of *switching on* the SR pulse train using a laser shot at the first crystal C1. The crystal was set at an angle corresponding to the Bragg angle for the expanded crystal lattice by laser irradiation. The second crystal C2 was placed to satisfy the on-Bragg condition for the X-rays reflected by C1. Here, the offset angles from the Bragg angle are $\Delta\theta_1 = -12$ arc sec. and $\Delta\theta_2 = +35$ arc sec. In Fig. 4.37(a), an SR pulse train with an interval of about 24 ns appears after the laser irradiation at $t = 0$ s. Then, *switching off* is shown in Fig. 4.37(b). Both GaAs crystals were set to satisfy the Bragg condition without lattice expansion: $\Delta\theta_1 = \Delta\theta_2 = 0$. A laser pulse onto C1 at $t = 25$ ns reduced the throughput of the SR pulse train drastically. A combination of these two techniques enables us to isolate a single SR pulse, as shown in Fig. 4.37(c). For *switching off* the X-ray beam, C1 was set at the Bragg angle without laser irradiation. C2 was adjusted to satisfy the Bragg condition for the expanded lattice, *switching on* the X-ray beam using a laser shot. Thus C2 should have a minus offset angle. Here, the offsets for C1 and C2 were tuned to give a high contrast, and were set at $\Delta\theta_1 = +7$ arc sec. and $\Delta\theta_2 = -20$ arc sec. The timing of the laser shots used to irradiate C1 and C2 was controlled

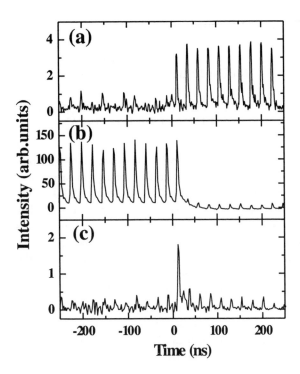

Fig. 4.37 Optical-switching of X-ray SR pulses. Laser irradiation of the first crystal determines the (a) on or (b) off conditions of the switch. (c) A combination of both states of the optical switch selects a single pulse from the SR pulse train.

to occur, respectively, just after and just before the moment the target SR pulse arrived at the crystals. Then, a single pulse was extracted from the SR pulse train with a throughput efficiency of about 10^{-4}.

It is to be noted that more rapid changes in the reflectivity of X-rays were observed for an asymmetric geometry. Further development of a faster X-ray switch will allow for an ultrashort femtosecond X-ray pulses to be shaped from a single SR pulse.

4.3   Protein Dynamics

M. UESAKA,
Nuclear Engineering Research Laboratory,
University of Tokyo,
22-2 Shirane-shirakata, Tokai, Naka,
IBARAKI 319-1188, JAPAN.

Many fundamental biological events take place over picosecond or femtosecond time scales, so that ultrafast biology in principle offers vast fields of study [Srajer *et al.* (1996); Techert *et al.* (2001); Oka *et al.* (2000); Perman *et al.* (1998)]. However, it is experimentally possible to study these processes only under conditions that can be initiated by a flash of light, which is quite a restrictive requirement. Ultrafast structural transitions in biology are studied by pump-and-probe analysis with optical pumping.

In this section we introduce two examples of time-resolved X-ray diffraction for photoinduced protein dynamics. In these methods pulsed X-rays and X-ray analysis are used as a pump-and-probe analysis. There are several methods of X-ray analysis, such as radiography, Bragg-, Laue-, Debye-Scherer-diffractions and others. Femtosecond lasers are used to supply an optical pump pulse. Synchrotron radiation (SR) and laser plasma X-rays have been used to supply an X-ray probe pulse. For SR, a femtosecond laser and a synchrotron must be synchronized, as discussed in Sec. 3.2.2. Laser pulses from a TW laser are split into an optical pump pulse and an intense laser pulse. The latter is used to generate laser plasma X-ray pulses (see Sec. 2.4.3). Recently, new pulsed X-ray sources such as X-ray FEL (see Sec. 2.7.2) and Compton scattering X-ray sources (see Sec. 2.5) are being adopted as X-ray sources. Typical pulse durations and intensities of several pulsed X-ray sources are summarized in Table 4.2. The X-ray sources can be selected according to the pulse duration and phenomenon of interest (see Fig. 4.21 in Sec. 4.2.1). Another important factor is the X-ray intensity for generating the X-ray image for X-ray analysis. The threshold intensities and possible ranges available for typical sources are schematically depicted in Fig. 4.38. The threshold intensities are evaluated assuming an updated imaging plate and CCD camera as detector. Although the intensity at the point source of the laser plasma X-rays is as high as 10^8–10^{10} photons per shot, the X-rays are spread over a wide solid angle. Then, the number of X-rays per shot at a sample of mm scale would be about 10^4 or less. If

Table 4.2   Typical pulsed X-ray sources and characteristics.

	Pulse duration (FWHM)	Intensity per pulse	Peak intensity	Average intensity per s	Frequency	Divergence
XFEL (LCLS)	~ 230 fs	10^{12} photons	8.6×10^{24} photons/s	2.4×10^{14} photons/s / 310 mW	120 Hz	1 μrad
Inverse Compton Scattering X-ray (BNL)	500 fs to 20 ps	2.8×10^{6} photons	8×10^{18} photons/s	2.8×10^{7} photons/s / 30 nW	10 Hz	8 mrad
Laser Plasma X-ray (Tokyo Inst. of Tech.)	6 ps to 33 ps	10^{4} photons /cm^2	10^{15} photons /s/cm^2	10^{5} photons/s/cm^2 / 0.13 nW/cm^2	10 Hz	2π sr
X-ray Laser (JAERI)	~ 7 ps	10^{12} photons	10^{23} photons/s		1 shot /20 min	7 mrad
Synchrotron Radiation (SPring8)	>50 ps	10^{9} photons	10^{19} photons/s	10^{14} photons/s / 160 mW	>10^5 Hz	50 μ rad
(KEK PF)		10^{6} photons		10^{12} photons/s / 2.8 mW	>10^6 Hz	

the intensity of the source is higher than the limit, then pump-and-probe analysis by a single shot is possible. Hence, snap shots can be obtained at different times during destructive phenomena such as damage formation, irreversible phenomena such as ablation and chaotic phenomena such as turbulent flow in a blood capillary. In the low intensity case, stacking of the X-rays images is needed. Therefore only reversible phenomena such as linear phonon dynamics and reversible phase transition are available (see Sec. 4.2).

We introduce the molecular mechanism of light driven proton pump by bacteriorhodopsin (bR). The structures of photoreaction intermediates are analyzed with SR X-ray diffraction at SPring8 [Oka *et al.* (2000); Oka *et al.* (2002)]. Some proteins absorb light and make use of the energy to induce reactions in photoinactive proteins, causing photoinduced isomerization. Absorption of light by some part of a protein molecule creates an excited state that relaxes by the rotation of one or more atoms around a

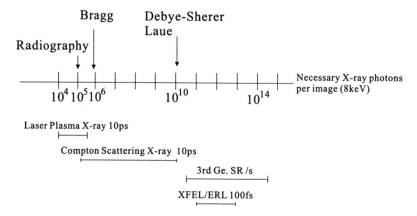

Fig. 4.38 X-ray intensity required to obtain image data for several methods of analysis and available ranges of pulsed X-ray sources.

chemical bond, switching between *cis* and *trans* isomers. Determination of the detailed structure of the excited state prior to isomerization as well as of the isomerized state, would provide crucial information on the mechanism.

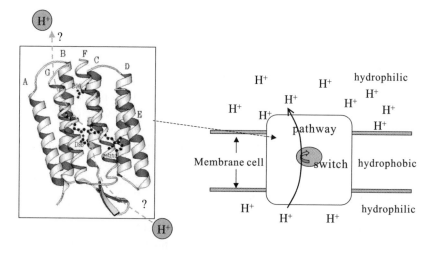

Fig. 4.39 Schematic view of proton pump of bR drawn by Molscinpt [Kranlis (1991)].

Bacteriorhodopsin (bR) is a light driven proton pump that transports protons from a cytoplasmic to a extracellular medium by using absorbed photon energy, as shown in Fig. 4.39. This induces a pH change by which the light stimulation is transported. The photocycle is characterized in the bR, K, L, M, N and O states by their spectra in the visible range [Oka *et al.* (2002)] (see Fig. 4.40).

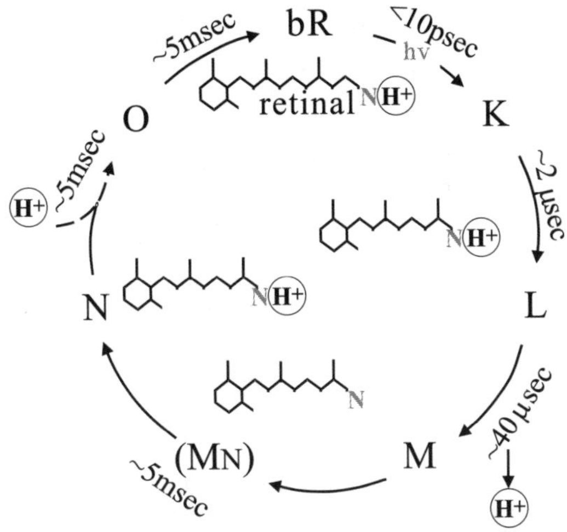

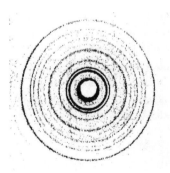

Fig. 4.40   Photocycle of bacteriorhodopsin. The isomer state of the retinal and protonation state of the Shiff base are shown with the time constant of each reaction.

Fig. 4.41   Typical X-ray diffraction pattern of purple membrane (bR).

bR includes a retinal as a light acceptor. When bR is irradiated by light, the retinal isomerizes within pico- to subpico-seconds. The isomer-

Femtosecond Beam Science

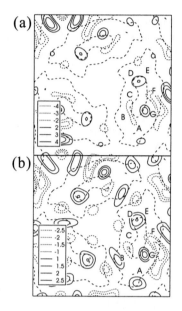

Fig. 4.42 Difference Fourier maps of slow and fast components of wild-type bR (pH 9, 10°C). (a) Map of slow components (B1). (b) Map of fast component (B2). The peak heights of the F and G helices are 5.0 and 3.6 in (a) and 1.9 and 2.3 in (b). The estimated errors of these maps are 0.75 and 0.47, respectively.

ization of a retinal in bR can be observed, with a time resolution of 5 fs, by pulse laser photolysis [Kobayashi *et al.* (2001)]. This isomerization induces a total structural change within micro- to milli-seconds. The total structural change of bR can be observed, with a time resolution of 122 ms, by time-resolved X-ray diffraction analysis at SPring8 [Oka *et al.* (2000); Oka *et al.* (2002)]. Time-resolved diffraction experiments promise to reveal the structural change during the photoreaction cycle. The light-induced structural change is triggered by deprotonation of the Schiff base. Based on these results, a model to elucidate the vectorial proton transport has been proposed.

Fig. 4.41 shows a typical X-ray diffraction pattern of purple membrane (bR) and the calculated difference Fourier map is shown in Fig. 4.42.

We now describe the time-resolved Laue analysis of myoglobin with CO down to 50 ps resolution at the European Synchrotron Radiation Facility (ESRF) [Wulff *et al.* (1997); Schotte *et al.* (2000)]. In-vacuum low-K undulators are used and can produce a spectral flux of 4.0×10^8 photons/0.1%

bandwidth per pulse at 15 keV, a factor of 5 higher than standard in-air undulators. Studies shows that high level photolysis is possible with laser pulses as short as 100 fs and that radiation damage from repeated X-ray and laser exposure is small.

The first studied biological reaction was the dissociation and rebinding of CO in myoglobin (Mb). The study used single-bunch Laue diffraction and the photolysis was done using a 10-ns Nd:YAG laser [Srajer *et al.* (1996); Eaton *et al.* (1996)]. The biological function of myoglobin is to store oxygen in muscle cells, just as hemoglobin transports oxygen in blood cells. Muscle cells need myoglobin as peak load buffer when the blood circulation cannot supply oxygen fast enough, for example during muscle contraction.

Myoglobin as not normally seen as a photoactive protein, but one of its lesser known properties is that when exposed to an intense flash of light, it temporarily releases its oxygen. The structure of myoglobin is very similar to that of one of the four sub-units of hemoglobin, i.e. hemoglobin is a tetramer of four myoglobin units. Myoglobin is a compact globular molecule with a molecular weight of 17800 Dalton (atomic mass unit, approximately the mass of a hydrogen atom, $1\,\text{Da} = 1\,\text{u} = 1.66 \times 10^{-27}$ kg) and consists of a single polypeptide chain of 153 amino acids. The oxygen is bound by a Fe^{2+} atom in the center of a porphyrin ring embedded within the hydrophobic interior of the protein. The porphyrin ring and its iron atom constitute the *heme* group, which is anchored to the protein through a covalent link to a proximal histidine. As Fig. 4.43 shows, the binding site is buried in the protein with no obvious opening to the outside. The motivation for doing time-resolved structural studies on myoglobin is to learn more about the pathway of the oxygen entering and escaping than can be inferred based on the static structure, and to identify rapid structural changes associated with ligand release and rebinding. The distal histidine is supposed to act as a *doorstop* with two stable positions, opening and closing a channel through which the ligand can enter or exit.

Myoglobin binds CO 30 times more strongly than oxygen. This relative affinity should be compared to the relative affinities of pure heme in solution without the protein environment, where CO binds to the free heme 1500 times more strongly than oxygen.

The X-ray diffraction data was collected over a range of 180°. For each setting of the crystal, 64 single-bunch shots were accumulated on the detector at a rate of 0.5 Hz, limited to this value due to the laser heat load on the sample. The detector was an image-intensified CCD camera, cooled by liquid nitrogen to eliminate dark current during the accumulation of data.

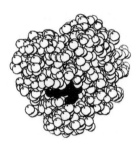

Fig. 4.43   Myoglobin molecule and its binding site. The drawing on the left shows a
ribbon representation of the protein backbone with the CO molecule on top of the heme
plane. A realistic space-filling is shown on the right where each atom is shown with its
Van-der-Waals radius. The molecule is rendered with the program Molscript and the
data taken from Kurian,Karplus & Petsko 1986-87, PDB 1mbc.

To maximize the photon flux, two insertion devices were used in tandem:
the W70 (20.6 mm gap) and the U46 (16.3 mm gap). These produce a
broad spectrum between 8 to 38 keV. To obtain a smoother spectrum, the
gaps of both insertion devices were tapered by 1 mm (the difference in gap
from input to output). A Laue pattern based on the accumulation of 10
single-bunch shots is shown in Fig. 4.44.

New 100 ps data is being scrutinized at the time of writing, and hence we
will limit our discussion to the nanosecond maps shown in Fig. 4.45. These
difference maps were calculated from 50 000 structure factors analyzed to
1.8 Å resolution. The map shows the difference in electron density of the
MbCO molecule at five time points between 4 ns and 350 μs. The difference
map is displayed relative to the protein backbone shown by frames. Inside
the black regions, the number of electrons decreases and inside the gray
region it increases. The contour level of the maps was chosen to be $3.5\,\sigma$
of the peak electron density in the 4 ns map. Note that the density is gone
in the 350 μs map, which shows that CO takes 350 μs to recombine. The
positive peak below the heme plane is the heme doming: a movement of the
iron atom out of the plane caused by a change in co-ordination from six to
five when the Fe–C bond is broken. Above the heme plane, one finds three
pockets of positive density. The first is positioned next to the CO hole and
the second is placed between the CO hole and the 107 residue. The third
feature is near the His 64 residue and could be assigned to a relaxation of
the His 64 *doorstop*, although the corresponding negative density is missing.
The first pocket is potentially a docking site. The second pocket coincides

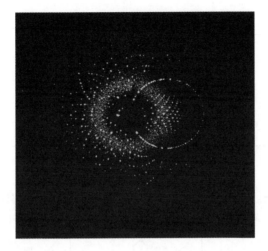

Fig. 4.44   Laue diffraction pattern of a myoglobin crystal. 64 single X-ray pulses with 500-ps time delay to the laser pulse were accumulated by the detector at a repetition rate of 0.5 Hz. The detector was a liquid nitrogen cooled CCD with X-ray image intensifier, the distance was 125 mm and the X-ray source was a U46 and W70.

with that found in the cryostructures. Its occupancy is small, however. Note that if an atom moves from one position to another, the difference map will show a dipole distribution provided that the map is of sufficient resolution.

Time-resolved Laue diffraction is likely to become a powerful and useful method for structural investigations of biological molecules, although the two examples described here were done in a time region slower than femtoseconds. The success of experiments indicates that the time resolution of macro-molecular crystallography may reach 200 fs with future free electron lasers, such as XFEL.

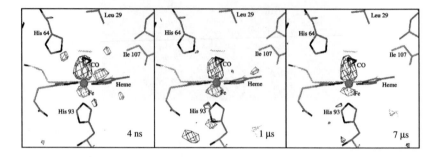

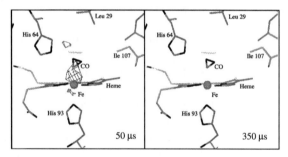

Fig. 4.45 Photolysis and rebinding of the CO ligand in myoglobin. This series of difference maps is based on pulsed Laue data collected at the ESRF on the ID9 beamline [Srajer *et al.* (1996)]. In the 4 ns map, the positive gray features around the black hole left by the CO could be indications of docking sites. The electron density maps are contoured at $+3.5\,\sigma$ and $-3.5\,\sigma$. Picture prepared by Thomas Ursby, ESRF.

4.4 Molecular Dynamics Simulation

T. KUNUGI
Department of Nuclear Engineering, Kyoto University,
Yoshida, Sakyo,
KYOTO 606-8501, Japan

M. SHIBAHARA
Department of Mechanical Engineering, Osaka University,
Yamadaoka, Suita,
OSAKA 565-0871, Japan

4.4.1 *Ultrafast phenomena and numerical modeling*

This section's main focus is numerical modeling of ultrafast phenomena, i.e. femtosecond order phenomena. The progresses of both laser technology and nanotechnology have made necessary a greater understanding of light interaction processes at the molecular or nanometer scales within the order of femtoseconds. Molecular dynamics (MD) simulation of laser interactions during very short time periods is important for applications of technologies such as femtosecond laser development, laser micro-fabrication and laser annealing processes for ultra-shallow junctions.

Some MD simulations of laser irradiation and thermal radiation have been developed to deal with phase change processes. Makino and Wakabayashi [Makino and Wakabayashi (1994)] characterized thermal radiation by its wavelength, phase and energy based on electromagnetic wave theory. They calculated the electronic field of radiation in the liquid and solid phase regions and added the energy of the radiation to the kinetic energy of the atoms. Kotake and Kuroki [Kotake and Kuroki (1990); Kotake and Kuroki (1993)] expressed the excitation of an atomic energy state due to light irradiation as a change in the potential energy acting between the lattice atoms. They used some parameters of the excitation strength, i.e. the laser energy of irradiation and the excitation life, and solved the Hamiltonian equations of the atoms with an MD method. Chokappa *et al.* [Chokappa *et al.* (1989)] assumed *energy carrier* particles to have mass to supply a given energy over a given pulse duration to the atomic system. Atoms within the first few layers of the surface absorbed the laser energy from the energy carriers. These methods cannot investigate the influence of changes of the atomic configuration on the evaporating and melting processes. In the first part of this section, MD simulations are introduced that allow the study of the thermal behavior of multiple-layered thin films irradiated by high power ultrashort (HPUS) pulsed lasers. A Monte Carlo (MC) technique is applied to simulate the laser irradiation of atomic systems [Ezato *et al.* (1996)].

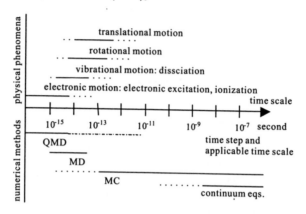

Fig. 4.46 Relationship between physical phenomena and numerical methods along a timescale.

With the progress of computational hardware, the processes of light interaction with matter have been recently studied by solving the Schrödinger equation directly [Heatherand and Metiu (1998); Haug and Metiu (1991); Bandrauk (1988); Shibahara and Kotake (1997); Shibahara and Kotake (1998)]. Since light absorption is related to changes in the rotational, vibrational or electronic states of molecules, it is possible to apply the MD method to the thermal process of light absorption. The potential energy used in the MD method should be changed by light absorption at visible or ultraviolet energy levels during the process. This makes it difficult to analyze such phenomena with MD methods, as shown in Fig. 4.46. A quantum molecular dynamics (QMD) simulation for light interaction problems with matter [Shibahara and Kotake (1997); Shibahara and Kotake (1998); Shibahara *et al.* (2000)], applied to simulate heat absorption of a thin film under light irradiation, is introduced in the latter half of this section. Light is treated as an electric field and its effects are included in the potential terms in the Schrödinger equation and Newton's equations. For the numerical calculation of the time-dependent Schrödinger equation, the split operator method [Feit *et al.* (1982)] is employed and the wave function of the valence electrons is evaluated at grid points with a time step of 10^{-17} s. By using the QMD method, the time history of the thermal energy in the system during a few hundred femtoseconds can be calculated numerically by changing the light characteristic parameters.

4.4.2 Molecular dynamics simulation including light interactions

Assuming that a classical system consists of N atoms and that the interatomic force $\mathbf{F}_i$ acting on the i-th atom is expressed by using an N-body potential function E_c, the motion of the i-th atom is expressed as

$$m\frac{d\mathbf{r}_i(t)}{dt} = \mathbf{F}_i(t) = -\sum_i \operatorname{grad} E_c^i, \tag{4.16}$$

$$\mathbf{v}_{i(t)} = \frac{d\mathbf{r}_i}{dt}, \qquad i = 1, 2, \cdots, N, \tag{4.17}$$

where m, $\mathbf{r}_i$ and $\mathbf{v}_i$ are the mass, coordinates and velocity vectors of the i-th atom, respectively. In MD simulations, the potential energy function between atoms and molecules should be selected carefully in order to get useful and meaningful calculation results. For example, we can apply the N body potential function proposed by Rosato *et al.* [Rosato *et al.* (1989)], which can describe the bulk and defect properties of a transition metal, as

$$E_c^i = E_b^i + E_r^i, \tag{4.18}$$

$$E_b^i = -\left\{\sum_j \xi^2 \exp\left[-2q\left(r_{ij}/r_0 - 1\right)\right]\right\}^{1/2}, \tag{4.19}$$

$$E_r^i = -\sum_j A \exp\left[-2p\left(r_{ij}/r_0 - 1\right)\right], \tag{4.20}$$

where E_c^i, E_b^i and E_r^i are the cohesive, binding and repulsive energy of the i-th atom, respectively. r_{ij} $(= |r_{ij}| = |\mathbf{r}_i - \mathbf{r}_j|)$ is the distance between the i-th and j-th atoms and r_0 is the equilibrium distance. For instance, the potential parameters of Cu are $\xi = 1.35$ eV, $A = 0.39$ eV, $p = 10.08$ and $q = 2.56$, and the atom diameter of Cu is $\sigma = 2.52$ Å. Those of iridium (Ir) are $\xi = 1.38$ eV, $A = 0.40$ eV, $p = 14.53$ and $q = 2.90$ and $\sigma = 2.71$ Å. In calculating the interaction between Ir and Cu, the harmonic mean is used for the potential parameters with energy units (ξ and A) and the arithmetic mean for the others (p, q and r_0).

We assume the incident laser energy is transferred to the atoms by many *energy carriers* with no mass but having a given energy [Ezato *et al.* (1996)]. The energy carriers can be absorbed or scattered by the atoms. However, the optical characteristics of the laser beam and the excitation of an atom cannot be dealt with because the interatomic potential used here does not express the electron's excited state due to absorption of the laser. We trace the trajectories of the energy carriers with the *test particle* MC method, as follows: The energy carriers, $N_p = 300\,000$ for instance, are distributed uniformly in a laser-irradiated plane at the top boundary of the simulation cell. One energy carrier is then emitted into the atomic system along the z-direction as shown in Fig. 4.48, and is allowed to search for an atom with which to collide. After it is absorbed, scattered out of the laser-irradiated plane or has penetrated the solid region, the next energy carrier is emitted. If an energy carrier with a direction cosine $\mathbf{c}_p$ at $\mathbf{r}_p$ defined by Eq. (4.25) collides with an atom at $\mathbf{r}_j$, they satisfy the following equality, as shown in Fig. 4.47,

$$|\mathbf{r}_{pj} + \mathbf{t}_{pj}\mathbf{c}_p| = \frac{1}{2}\sigma, \tag{4.21}$$

where $\mathbf{r}_{pj} = \mathbf{r}_p - \mathbf{r}_j$ and $\mathbf{t}_{pj}\mathbf{c}_p$ is the distance between the energy carrier and the collision point at the fictitious surface of the atom. The parameter t_{pj} is given by solving Eq. (4.21), giving

$$t_{pj} = \frac{1}{c_p^2}\left[-b_{pj} - \left\{b_{pj}^2 - c_p^2\left(\mathbf{r}_{pj} - \frac{1}{4}\sigma^2\right)\right\}^{1/2}\right], \tag{4.22}$$

where $b_{pj} = |\mathbf{r}_{pj}\cdot\mathbf{c}_p|$. The collision between the atom and the energy carrier occurs when $b_{ij} < 0$ and $b_{pj}^2 - c_p^2(\mathbf{r}_{pj} - 1/4\sigma^2) > 0$.

When the energy carrier collides with an atom, whether it is absorbed or scattered is decided using a uniform random number R_ε in the MC method. We assume an absorption rate $\varepsilon = 0.2$ of the energy carrier at the fictitious surface of an atom. The energy carrier is absorbed by the atom if $R_\varepsilon < \varepsilon$ or scattered and finds the next atom for collision if $R_\varepsilon > \varepsilon$. The scattered direction (η, θ), as shown in Fig. 4.47, can be determined by using two uniform random numbers R_η and R_θ based on the isotropic scattering assumption:

$$\eta = \cos^{-1}(1 - 2R_\eta), \tag{4.23}$$

$$\theta = 2\pi R_\theta. \tag{4.24}$$

The direction cosine of the scattered energy carrier is then expressed by

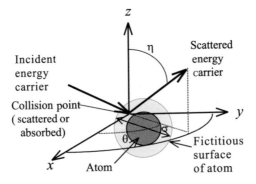

Fig. 4.47   Scattered direction of energy carrier after collision with an atom.

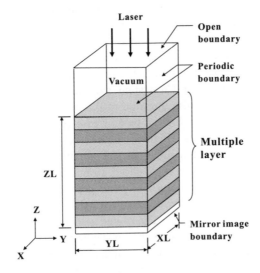

Fig. 4.48   Simulation system and boundary conditions.

$$\mathbf{c}_p = (C_x, C_y, C_z) = (\cos\theta\sin\eta, \sin\eta, \cos\eta). \qquad (4.25)$$

This tracing of the trajectory of the energy carrier with the MC method is repeated for all of the energy carriers.

We assume that an energy carrier can transfer its energy to atomic kinetic energy instantaneously if an atom absorbs it. The energy of the individual energy carrier I_p can be expressed by using the irradiated laser energy in the atomic system I_0 and the x-y cross section of the simulation cell S_{xy}:

$$I_p = I_0/(S_{xy}\Delta t N_p).\qquad(4.26)$$

The velocity of the i-th atom after absorbing energy from an energy carriers is expressed by

$$\mathbf{v}_i^{\text{new}} = \mathbf{v}_i^{\text{old}} \left\{ \left(\sum I_p/K_i \right) + 1 \right\}^{1/2},\qquad(4.27)$$

where $\sum I_p$ is the total laser energy absorbed by the i-th atom and K_i is the kinetic energy of the i-th atom.

An ordinary multiple thin film consists of alternating heavy and light atomic layers. We chose Ir as the heavy atom and Cu as the light atom. The simulation systems and boundary conditions are shown in Fig. 4.48. An MD/MC simulation with HPUS pulsed laser irradiation was performed for different simulation systems to investigate the effects of the configuration of the atomic layers and the irradiation intensity on the thermal behavior of the multiple thin film. The configurations and intensities considered were case (a) an Ir–Cu system with an irradiation intensity of $I_0 = 10^{15}$ W/m^2 and a pulse duration of 0.1 ps, case (b) an Ir–Cu system with $I_0 = 10^{16}$ W/m^2 for 0.1 ps and case (c) a Cu–Ir system with $I_0 = 10^{15}$ W/m^2 for 0.1 ps. Figures. 4.49(a)–(c) show snapshots of the atomic configuration corresponding to the simulation cases (a), (b) and (c). These snapshots show vertical sliced atomic layers containing atoms located in a cross section of depth 4σ normal to the paper. For case (a) in Fig. 4.49(a), some void regions are observed between the first Ir layer and the first Cu layer at 6 ps. These void regions expand from 6 ps to 8 ps and eventually delamination between the first Ir layer and the first Cu layer occurs. In case (b), which is the same atomic system, but with a 10 times larger intensity of the laser irradiation than case (a), there is intense evaporation at the surface of the first Ir layer and an expansion in the first Cu layer at 3 ps, which is compressed at 0.5 ps, as shown in Fig. 4.49(b). In case (c), despite the irradiation intensity being the same as case (a), intense evaporation at the surface of the first Cu layer occurs, seen from an comparison of the number of atoms in the vacuum region at 8 ps for case (c) with that of case (a), as shown in Figs. 4.49(c) at 8 ps and (a) at 4 ps.

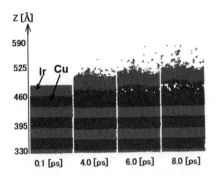

(a) Ir–Cu system with $I_0 = 10^{15}$ W/m^2 irradiated for 0.1 ps.

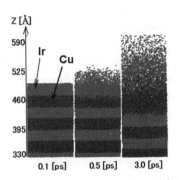

(b) Ir–Cu system with $I_0 = 10^{16}$ W/m^2
irradiated for 0.1 ps.

(c) Cu–Ir system with $I_0 = 10^{15}$ W/m^2
irradiated for 0.1 ps.

Fig. 4.49 Snapshots of atomic configurations of (a) Ir–Cu system with irradiation of 10^{15} W/m^2 for 0.1 ps, (b) Ir–Cu system with irradiation of 10^{16} W/m^2 for 0.1 ps and (c) Cu–Ir system with the irradiation of 10^{15} W/m^2 for 0.1 ps.

4.4.3 Quantum molecular dynamics simulation including light interactions

For an atomic system composed of atom nuclei and electrons, the time dependent Schrödinger equation under light irradiation can be expressed as

$$i\hbar\frac{\partial}{\partial t}\Psi = H\Psi = (K + U + U_{\text{light}})\Psi, \qquad (4.28)$$

where E_0, H, K, U, U_{light} and Ψ are the magnitude of the light electric field [V/m], the Hamiltonian of the system, the kinetic energy operator, the potential energy of the system, the potential energy under light irradiation and the wave function of the system, respectively. The magnetic field of light can be ignored in comparison with the electric field. The light electric field $\mathbf{E}$ is expressed as

$$\mathbf{E} = E_0 \cdot \cos\theta \cdot \cos(\omega t + \varphi), \qquad (4.29)$$

where θ, ω and φ are the direction angle, angular frequency and the phase angle of the light, respectively. By considering the motion of nuclei and electrons, the wave function can be assumed to be separable into a wave function of nuclei and one of electrons. Free electrons can be considered as uncoupled due to a weak coupling potential energy to the nucleus, so that the bound electrons and the nucleus can be considered to form an ion for a metallic atom. For a neutral atom, all the electrons can be considered to be bound electrons. The wave function can then be assumed to be

$$\Phi = \Psi_i \cdot \Psi_{fe} \cdot \Psi_a, \qquad (4.30)$$

where Ψ_i, Ψ_{fe} and Ψ_a are the wave functions of the ions, free electrons and neutral atoms, respectively. The Hamiltonian can be written as follows by considering interactions between particles,

$$H = -K_{fe} - K_i - K_a$$
$$+U_{i-i} + U_{i-fe} + U_{fe-fe} + U_{i-a} + U_{fe-a} + U_{a-a} + U_{light}. \quad (4.31)$$

The subscripts 'fe', 'i' and 'a' represent the free electrons, ions and neutral atoms, respectively. U_{A-B} is the potential energy between particle A and B.

Newton's equation can be obtained for ions and neutral atoms by ignoring the distribution of Ψ_i and Ψ_a and assuming $\hbar \to 0$. For a system composed of an ion, an electron and molecules (one metallic atom and some neutral atoms), when the light is highly convergent and irradiates only the metallic atom (see Fig. 4.50), the equations for an ion and other atoms can be written as

$$M_i \frac{\partial^2 \mathbf{R}_i}{\partial t^2} = -\frac{\partial}{\partial \mathbf{R}_i} \left\{ U_{i-fe} + U_{i-a} + \sum_{i,a} q_{i-a} \cdot \mathbf{E} \cdot (\mathbf{R}_{i-a} - \mathbf{R}_{i0-a0}) \right\},$$

$$(4.32)$$

$$M_a \frac{\partial^2 \mathbf{R}_a}{\partial t^2} = -\frac{\partial}{\partial \mathbf{R}_a} \left\{ U_{a-a} U_{a-fe} + U_{i-a} + \sum_{i,a} q_{i-a} \cdot \mathbf{E} \cdot (\mathbf{R}_{i-a} - \mathbf{R}_{i0-a0}) \right\},$$

$$(4.33)$$

where M_x, $\mathbf{R}_x$, q_{i-a}, $\mathbf{R}_{i-a}$ and R_{i0-a0} are the mass of particle x, the position of particle x, the effective dipole moment between the ion and neutral atom, the distance between the ion and neutral atom and the equilibrium distance between the ion and neutral atom, respectively. The Schrödinger equation for free electrons is finally described as

$$i\hbar\frac{\partial\Psi_{fe}}{\partial t} = \left\{ -K_{fe} + U_{fe-fe} + U_{i-fe} + U_{i-a} + e\cdot\mathbf{E}\cdot(\mathbf{R}_i - \mathbf{r}_{fe}) \right\}\Psi_{fe}, \quad (4.34)$$

where e, $\mathbf{R}_i$ and $\mathbf{r}_{fe}$ are the unit electric charge, the position of the ion and the coordinates of the free electron, respectively. The wave function of a free electron can be obtained by solving Eq. (4.34). With this Ψ_{fe}, Newton's equations, Eqs. (4.32) and (4.33), can be solved numerically to obtain the motion of the ions and atoms.

The electronic states of the bound electrons can change with the free electron field, which modifies the potential between ions and atoms. Light irradiation excites the electronic state of free electrons, say, from state 1 to 2. In state 1, free electrons are in ground states localized around ions, and in state 2 they are diffusely distributed and delocalized due to light energy absorption. Corresponding to the states 1 and 2, there are potential functions between ions and neutral atoms, denoted by $U_{local,i-a}$ and $U_{deloc,i-a}$, respectively. During excitation, the potential energy may take an intermediate between these functions and may be assumed to take the form of a linear summation:

$$U_{i-a} = \alpha\cdot U_{local,i-a} + (1-\alpha)\cdot U_{deloc,i-a}. \quad (4.35)$$

For example, $U_{local,i-a}$ and $U_{deloc,i-a}$ can be expressed simply by the Lennard–Jones potential having different parameters:

$$U_{local,i-a} = 4\varepsilon\left\{ \left(\frac{\sigma}{\mathbf{R}_{i-a}}\right)^{12} - \left(\frac{\sigma}{\mathbf{R}_{i-a}}\right)^6 \right\}, \quad (4.36)$$

$$U_{deloc,i-a} = 4\varepsilon'\left\{ \left(\frac{\sigma'}{\mathbf{R}_{i-a}}\right)^{12} - \left(\frac{\sigma'}{\mathbf{R}_{i-a}}\right)^6 \right\}. \quad (4.37)$$

The parameter α can be a function of the positions of the free electrons, which are related to the effective dipole moment of free electrons β:

$$\beta = \sum_{\text{all space}} e\cdot\rho(\mathbf{r}_{fe})\cdot|\mathbf{R}_i - \mathbf{r}_{fe}|, \quad (4.38)$$

where $\rho(\mathbf{r}_{\text{fe}})$ is the electronic density at the coordinates $\mathbf{r}_{\text{fe}}$. If the parameter α is assumed to be proportional to β, α is given by

$$\alpha = (\beta_{\text{deloc}} - \beta)/(\beta_{\text{deloc}} - \beta_{\text{local}}). \tag{4.39}$$

The potential energy between an ion and an atom is $U_{\text{local,i-a}}$ at $\beta = \beta_{\text{local}}$ and $U_{\text{deloc,i-a}}$ at $\beta = \beta_{\text{deloc}}$. For the potential energy among an electron, an ion and neutral atoms, the simplest and the most physical form of the Lennard-Jones and Coulomb-type potentials can be applied [Feit *et al.* (1982)].

The solution of Eq. (4.34) is a complex function of time t and position $\mathbf{r}$. Solving this equation involves a time integration and second-order space derivatives. For the time integration, the split operator method [Feit *et al.* (1982)] is often employed and is written as

$$\Phi_{\text{fe}}(t + \Delta t) = \exp\left(\frac{i\hbar\Delta t}{4m_e}\nabla^2\right) \exp\left(\frac{-i\Delta t U}{\hbar}\right) \exp\left(\frac{i\hbar\Delta t}{4m_e}\nabla^2\right) \Psi_{\text{fe}}(t), \tag{4.40}$$

where m_e and U are the mass of an electron and the potential energy of a free electron, respectively. In order to ensure sufficient precision of the second-order derivatives on fixed space grid points, it is necessary to prepare very fine grid points.

In most cases, a Fourier expansion [Feit *et al.* (1982)] or Chebyshev polynomials are employed. The velocity and position of ions and atoms can be calculated by the MD method using the central difference method or other methods.

The QMD method is applied to investigate the heat absorption of metallic atoms embedded in very thin films. The configuration of a metallic atom and other atoms in a thin film is shown in Fig. 4.50. The thin film consists of 5 atomic layers. The angular frequency ω of the light is expressed as the light energy $\hbar\omega$ in the present section. The total calculation time is 400 fs. The atoms studied here are not specially assigned but are generalized so as to allow a discussion of the general features of light energy absorption in thin films. The ionic and atomic energy, i.e. the potential energy and kinetic energy, and the free electron energy are defined as the energies of molecular dynamics (MD) and quantum dynamics (QD). Figure 4.51 shows the time history of the MD energy under light irradiation. The light energies are 0.03, 0.1 and 4.0 eV, with each of the three plots corresponding to a light energy value. The magnitudes of the light electric field are 2.5, 5.0, 7.5 and 10.0 GV/m. The frequency of the light electric field at 0.03 eV corresponds to the natural frequency of atoms under the initial conditions. The QMD simulation can be applied to make the light energy absorption model dependent on light characteristics in the molecular and nanometer scales for femto or nanosecond timescales. The MD simulation, including the light absorption model, can be applied to the prediction and estimation of phase changes and heating processes of molecular systems under femto and nanosecond laser irradiation as direct numerical simulations.

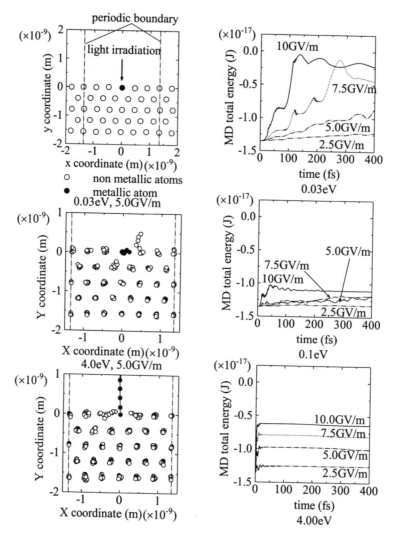

Fig. 4.50 Configuration of the calculation system and atomic trajectories under light irradiation for 400 fs. o: nonmetallic atom, •: metallic atom

Fig. 4.51 Time history of MD energy under light irradiation. The light electric field has the values 2.5, 5.0, 7.5 and 10.0 GV/m. The three plots have light energy values, from top to bottom, of 0.03, 0.1 and 4.0 eV.

Bibliography

Aoki, Y., Yang, J. F., Hirose, M., Sakai, F., Tsunemi, A., Yorozu, M., Okada, Y., Endo, A., Wang, X. J. and Ben-Zvi, I. (2000). *Nucl. Instr. Meth. A* **455**, p. 99.

Asay. J. R. and Shahinpoor, M. (1993). High Pressure Shock Compression of Solids, Springer, New York.

Bandrauk, A. D. (1988). Atomic and Molecular Processes with Short Intense Laser Pulses, Plenum Press, New York.

Bartels, D. M., Cook, A. R., Mudaliar M. and Jonah, C. D. (2000). Study on geminate ion recombination in liquid dodecane using pico- and subpicosecond pulse radiolysis, *J. Phys. Chem.* **104**, p. 1686.

Beck, G. (1979). Transient conductivity measurements with subnanosecond time resolution, *Rev. Sci. Instrum.* **50**, p. 1147.

Beck, G. and Thomas, J. K. (1972). Picosecond observations of some ionic and excited state processes in liquids, *J. Phys. Chem.* **76**, p. 3856.

Belloni, J., Marignier, J. L. and Gaillard, M. (1998). *Cahiers Radiobiol.* 8.

Blakemore, J. S. (1982). Semiconducting and other major properties of gallium arsenide, *J. Appl. Phys.* **53**, pp. R123–R181.

Bronskill, M. J., Taylor, W. B., Wolff, R. K. and Hunt, J. W. (1970). Design and performance of a pulse radiolysis system capable of picosecond time resolution, *Rev. Sci. Instrum.* **41**, p. 333.

Bucksbaum, P. H. and Merlin, R. (1999). The Bragg switch: A proposal to generate sub-picosecond X-ray pulses, *Solid State Commun.* **111**, pp. 535–539.

Cauble, R. *et al.* (1998). Absolute equation-of-state data in the 10–40 Mbar (1–4 TPa) regime, *Phys. Rev. Lett.* **80**, pp. 1248–125; Cauble, R. *et al.* (1993). Demonstration of 0.75 Gbar planar shocks in X-ray driven colliding foils, *Phys. Rev. Lett.* **70**, pp. 2102–2105.

Chokappa, D. K., Cook, S. J. and Clancy, P. (1989). Nonequilibrium simulation method for the study of directed thermal processing, *Phys. Rev. B* **39**, pp. 10075–10087.

Cline, J., Jonah, C. D., Bartels, D. M. and Takahashi, K. (2000). The solvated electron in supercritical water: Spectra, yields, and reactions, *Proc. of the 1st Int. Symposium on Supercritical Water-Cooled Reactors, Design and*

Technology, Univ. of Tokyo, Japan, p. 194.

DeCamp, M. F., Reis, D. A., Buckbaum, P. H., Adams, B., Caraher, J. M., Clarke, R., Conover, C. W. S., Dufresne, E. M., Merlin, R., Stoica, V. and Wahlstrand, J. K. (2001). Coherent control of pulsed X-ray beams, *Nature* **413**, pp. 825–827.

Draganic, I. G. and Draganic, Z. D. (1971). The Radiation Chemistry of Water, Academic Press, New York and London.

Eaton, W. A., Henry, E. R. and Hofrichter, J. (1996). Nanosecond crystallographic snapshot of protein structural changes *Science* **274**, pp. 1631–1632.

Elliot, A. J. (1994). Rate constants and G-values for the simulation of the radiolysis of light water over the range 0–300^{circ}C, AECL-11073, COG-94-167.

Elliot, A. J., Ouellette, D. C. and Stuart, C. R. (1996). The temperature dependence of the rate constants and yields for the simulation of the radiolysis of heavy water, AECL-11658, COG-96-390-1.

Ezato, K., Kunugi, T. and Shimizu, A. (1996). Monte Carlo / molecular dynamics simulation on melting and evaporation processed of material due to laser irradiation, *ASME HTD* **323**, pp. 171–178.

Fabbro et al. (1990). Physical study of laser-produced plasma in confined geometry, *J. Appl. Phys.* **68**, 775.

Feit, M. D., Fleck, Jr., J. A. and Steiger, Q. (1982). Solution of the Schrodinger equation by a spectral method, *J. Comput. Phys.* **47**, p. 412.

Ferry, J. L. and Fox, M. A. (1998). Effect of temperature on the reaction of HO$^\bullet$ with benzene and pentahalogenated phenolate anions in subcritical and supercritical water, *J. Phys. Chem. A* **102**, p. 3705.

Ferry, J. L. and Fox, M. A., (1999). Temperature effects on the kinetics of carbonate radical ractions in near-critical and supercritical water, *J. Phys. Chem. A* **103**, p. 3438.

Frongillo, Y., Goulet, T., Fraser, M.-J., Cobut, V., Patau, J.P. and Jay-Gerin, J.-P. (1998). *Radiat. Phys. Chem.* **51**, p. 245.

Fujimoto, Y. et al. (1999). Spectroscopy of hard X-rays (2–15 keV) generated by focusing femtosecond laser on metal targets, *Jpn. J. Appl. Phys.* **38**, p. 6754.

Guo, T. et al. (1997). Picosecond-milliangstrom resolution dynamics by ultrafast x-ray diffraction, *SPIE Proc.* **3157**, p. 84.

Guo, T., Spielmann, Ch., Walker, B. C. and Barty, C. P. J. (2001) Generation of hard x rays by ultrafast terawatt lasers, *Rev. Sci. Instrum.* **72**, p. 41.

Rigg, P. A. and Gupta, Y. (1998). Real-time x-ray diffraction to examine elastic plastic deformation in shocked lithium fluoride crystals, *Appl. Phys. Lett.* **73**, pp. 1655–1657; Gupta, Y. M. et al. (1989). Experimental developments to obtain real-time x-ray diffraction measurements in plate impact experiments, *Rev. Sci. Instr.* **70**, pp. 4008–4014.

Gust, W. H. and Ryce, E. B. (1971). Axial yield strengths and two successive phase transition stresses for crystalline silicon, *J. Appl. Phys.* **42**, p. 1897.

Harano, H., Kinoshita, K., Yoshii, K., Ueda, T., Okita, S. and Uesaka, M. (2000). Ultrashort X-ray pulse generation using subpicosecond electron linac, *J. Nucl. Mater.* **280**, pp. 255–263.

Haug, K. and Metiu, H. (1991). The absorption spectrum of a potassium atom in a Xe cluster, *J. of Chem. Phys.* **95**, pp. 5670–5680.

Heather, R. and Metiu, H. (1988). Multiphoton dissociation of a diatomic molecule: Laser intensity, frequency, and pulse shape dependence of the fragment momentum distribution, *J. Chem. Phys.* **88–89**, pp. 5496–5505.

Hirata Y. and Mataga, N. (1990). Solvation dynamics of electrons ejected by picosecond dye laser pulse excitation of p-phenylenediamine in several alcoholic solutions, *J. Phys. Chem.* **94**, p. 8503.

Hirata, Y., Murata, N., Tanioka Y. and Magata, N. (1989). Dynamic behavior of solvated electrons produced by photoionization of indole and tryptophan in several polar solvents, *J. Phys. Chem.* **93**, p. 4527.

Hironaka, Y., Tange, T., Inoue, T., Fujimoto, Y., Nakamura, K. G., Kondo, K. and Yoshida, M. (1999). Picosecond pulsed X-ray diffraction from a pulsed laser heated Si(111), *Jpn. J. Appl. Phys.* **38**, pp. 4950–4951.

Hironaka, Y. *et al.* (2000). Evolving shock-wave profiles measured in a silicon crystal by picosecond time-resolved x-ray diffraction, *Appl. Phys. Lett.* **77**, pp. 1967–1969.

Holroyd, R., Wishart, J. F., Nishikawa, M. and Itoh, K. (2003). Reactions of charged species in supercritical xenon as studied by pulse radiolysis, *J. Phys. Chem. B* **107**, p. 7281.

Inoue, T, Hayakawa, K, Shiotari, H, Takada, E, Torikoshi, M, Nagasawa, K, Hagiwara, K. and Yanagisawa, K. (2002). Economic scale of utilization of radiation (III): Medicine comparison between Japan and the U. S. A., *J. Nucl. Sci. Technol.* **39**, p. 1114.

Johnson, Q., Mitchell, A., Keeler, R. N. and Evans, L. (1970). X-ray diffraction during shock-wave compression, *Phys. Rev. Lett.* **25**, p. 1099; Johnson, Q., Mitchell, A. and Evans, L. (1971). X-ray diffraction evidence for crystalline order and isotropic compression during the shock-wave process, *Nature* **231**, p. 310; Johnson, Q. and Mitchell, A. (1972). First X-ray diffraction evidence for a phase transition during shock-wave compression, *Phys. Rev. Lett.* **29**, p. 1369.

Jonah, C. D. (1975). A wide-time range pulse radiolysis system of picosecond time resolution, *Rev. Sci. Instrum.* **46**, p. 62.

Jonah, C. D. and Sauer, M. C. (1979). Kinetics of solute excited state formation in the pulse radiolysis of 9,10-diphenyl anthracene and p-terphenyl in cyclohexane solutions, *Chem. Phys. Lett.* **63**, p. 535.

Kadono, T. *et al.* (2000). Flyer acceleration by a high-power KrF laser with a long pulse duration, *J. Appl. Phys.* **88**, pp. 2943–2947.

Katsumura, Y., Tagawa, S. and Tabata, Y. (1982). The formation process of the excited state of cycloalkane liquid using picosecond pulse radiolysis, *Radiat. Phys. Chem.* **19**, p. 267.

Kenney-Wallace, G. A. and Jonah, C. D. (1982). Picosecond spectroscopy and solvation clusters. The dynamics of localizing electrons in polar fluids, *J. Phys. Chem.* **86**, p. 2572.

Kinoshita, K., Harano, H., Yoshii, K., Ohkubo, T., Fukasawa, A., Nakamura, K., Uesaka, M. (2001). Time-resolved X-ray diffraction at NERL, *Laser Part.*

Beams **19**, pp. 125–131.

Kmetec, J. D. *et al.* (1992). MeV x-ray generation with a femtosecond laser, *Phys. Rev. Lett.* **68**, p. 1572.

Kobayashi, T., Saito, T. and Ohtani, H. (2001). Real-time spectroscopy of transition states in bacteriorhodopsin during retinal isomerization, *Nature* **414**, pp. 531–534.

Kobayashi, H., Tabata, Y., Ueda, T. and Kobayashi, T. (1987). A twin linac pulse radiolysis system (II), *Nucl. Instrum. Meth.* B **24/25**, p. 1073.

Kobayashi, T., Uesaka, M., Katsumura, Y., Muroya, Y., Watanabe, T., Ueda, T., Yoshii, K., Nakajima, K., Zhu, X. W. and Kando, M. (2002). *J. Nucl. Sci. Technol.* **39**, p. 6.

Kobayashi, H., Ueda, T., Kobayashi, T., Tagawa, S., Yoshida, Y. and Tabata, Y. (1984). Absorption spectroscopy system based on picosecond single electron beams and streak camera, *Radiat. Phys. Chem.* **23**, p. 393.

Kondo, K., Sawaoka, A. and Saito, S. (1979). Microscopic observation of the shock-compressed state of LiF by flash X-ray diffraction, *High-pressure Science and Technology,* Timmerhaus, K. D. and Barber, M. S. eds., AIP, New York, 905-910.

Kotake, S. and Kuroki, M. (1990). Molecular dynamics aspects of solid melting by laser beam radiation, *9th Int. Heat Transfer Conf.* **4**, pp. 277–282.

Kotake, S. and Kuroki, M. (1993). Molecular dynamics study of solid melting and vaporization by laser irradiation, *Int. J. Heat and Mass Transfer* **36**, pp. 2061–2067.

Kozawa, T., Mizutani, Y., Miki, M., Yamamoto, Y., Suemine, S., Yoshida, Y. and Tagawa, S. (2000). Development of subpicosecond pulse radiolysis system, *Nucl. Instrum. Meth.* A **440**, p. 251.

Kozawa, T., Yoshida, Y. and Tagawa, S. (2002). Study on radiation-induced reaction in microscopic region for basic understanding of electron beam patterning in lithographic process (I) – Development of Subpicosecond Pulse Radiolysis and Relation between Space Resolution and Radiation-Induced Reactions of Onium Salt – , *Jpn. J. Appl. Phys.* **41**, p. 4208.

Kume, T., Amano, E., Nakanishi, T. M. and Chino, M. (2002). Economic scale of utilization of radiation (II): Agriculture comparison between Japan and the U. S. A., *J. Nucl. Sci. Technol. 39*, p. 1106.

Larson, B. C., Tischler, J. Z. and Mills, D. (1986). Nanosecond resolution time-resolved X-ray study of silicon during pulsed-laser irradiation, *J. Mater. Res.* **1**, pp. 144–154.

Larsson, J., Allen, A., Bucksbaum, P. H., Falcone, R. W., Lindenberg, A., Naylor, G., Missalla, T., Reis, D. A., Scheidt, K., Sjögren, A., Sondhauss, P., Wulff, M. and Wark, J. S. (2002). Picosecond X-ray diffraction studies of laser-excited acoustic phonons in InSb, *Appl. Phys.* A **75**, pp. 467–478.

Leemans, W. P., Schoenlein, R. W., Volfbeyn, P., Chin, A. H., Glover, T. E., Balling, P., Zolotorev, M., Kim, K.-J., Chattopadhyay, S. and Shank, C. V. (1997). Interaction of relativistic electrons with ultrashort laser pulses: Generation of femtosecond X-rays and microprobing of electron beams, *IEEE J. Quantum Elect.* **33**, pp. 1925–1934.

Lin, M., Katsumura, Y., Wu, G., Muroya, Y., He, H., Morioka, T. and Kudo, H. (2002). A pulse radiolysis study to estimate the radiolytic yields of water decomposition products in high-temperature and supercritical water: Use of methyl viologen as a scavenger, *Proc. of Super Green 2002*, Suwon, Korea, p. 206.

Makino, T. and Wakabayashi, H. (1994). Numerical simulation on melting behaviour of an atomic layer irradiated by thermal radiation, *Thermal Sci. Eng.* **2-1**, pp. 158–165.

Matsuda, A., Nakamura, K. G. and Kondo, K. (2002). Time-resolved Raman spectroscopy of benzene and cyclohexane under laser-driven shock compression, *Phys. Rev. B* **65**, 174116.

Miller, J. R., Penfield, K., Johnson, M., Closs, G. and Green, N. (1998). Pulse radiolysis measurements of intramolecular electron transfer with comparisons to laser photoexcitation, *Photochemistry and Radiation Chemistry: Complementary Methods for the Study of Electron Transfer*, Wishart, J. F. and Nocera, D. G., eds., *Adv. Chem. Ser.* **254**, Ch. 11, American Chemical Society, Washington, DC.

Mostafavi, M., Lin, M., Wu, G., Katsumura, Y. and Muroya, Y. (2002). Pulse radiolysis study of absorption spectra of Ag^0 and Ag_2^+ in water from room temperature up to $380^{circ}C$, *J. Phys. Chem. A* **106**, p. 3123.

Muroya, Y., Lin, M., Watanabe, T., Kobayashi, T., Wu, G., Ueda, T., Yoshii, K., Uesaka, M. and Katsumura, Y. (2002a). *Nucl. Instr. Meth. A* **489**, pp. 554–562

Muroya, Y., Meesungnoen, J., Jay-Gerin, J.-P., Filali-Mouhim, A., Goulet, T., Katsumura, Y. and Mankhetkorn, S. (2002b). *Can. J. Chem.* **80**, p. 1367.

Oishi, Y. *et al.* (2001). Production of relativistic electrons by irradiation of 43-fs-laser pulses on copper film, *Appl. Phys. Lett.* **79**, pp. 1234–1236.

Oka, T. *et al.* (2000). *PANS* **97**, p. 14278.

Oka, T., Yagi, N., Tokunaga, F. and Kataoka, M. (2002). Time-resolved x-ray diffraction reveals movement of F helix of D96N bacteriorhodopsin during M-MN transition at neutral pH, *Biophysical Journal* **82**, pp. 2610–2616.

Pepin, C., Goulet, T., Houde, D. and Jay-Gerin, J.-P. (1994). Femtosecond kinetic measurements of excess electrons in methanol: Substantiation for a hybrid solvation mechanism, *J. Phys. Chem.* **98**, p. 7009.

Perman, B., Srajer, V., Ren, Z., Teng, T. Y., Pradervand, C., Ursby, T., Bourgeois, D., Schotte, F., Wulff, M., Kort, R., Hellingwerf, K. and Moffat, K. (1998). Energy transduction on the nanosecond time scale: Early structural events in a xanthopsin photocycle, *Science* **279**, p. 1946.

Reich, Ch., Gibbon, P., Uschmann, I. and Förster, E. (2000). Yield optimization and time structure of femtosecond laser plasma Kα sources, *Phys. Rev. Lett.* **84**, pp. 4846–4849.

Rischel, C., Rousse, A., Uschmann, I., Albouy, P.-A., Geindre, J.-P., Audebert, P., Gauthier, J.-C., Förster, E., Martin, J.-L. and Antonetti, A. (1997). Femtosecond time-resolved X-ray diffraction from laser-heated organic films, *Nature* **390**, 6659, pp. 490–492.

Rosato, V., Guillope M. and Legrand B. (1989). Thermo-dynamical and structural

416   *Femtosecond Beam Science*

properties of f.c.c. transition metal using simple tight-binding model, *Phil. Mag. A* **59**, pp. 321–336.

Rose-Petruck, C., Jimenez, R., Guo, T., Cavalleri, A., Siders, C. W., Ráksi, F., Squier, J. A., Walker, B. C., Wilson, K. R. and Barty, C. P. (1999). Picosecond-milliångström lattice dynamics measured by ultrafast X-ray diffraction, *Nature* **398**, 6725, pp. 310–312.

Saeki, A., Kozawa, T., Yoshida Y. and Tagawa, S. (2001). Study on geminate ion recombination in liquid dodecane using pico- and subpicosecond pulse radiolysis, *Radit. Phys. Chem.* **60**, p. 319.

Saeki, A., Kozawa, T., Yoshida, Y. and Tagawa, S. (2002). Study on radiation-induced reaction in microscopic region for basic understanding of electron beam patterning in lithographic process (II) – relation between resist space resolution and space distribution of ionic species –, *Jpn. J. Appl. Phys.* **41**, p. 4213.

Schotte, F., Techert, S., Anfinrud, P., Srajer, V., Moffat, K. and Wulff, M. (2000) Recent advances in the generation of pulsed synchrotron radiation suitable for picosecond time-resolved X-ray studies, *Handbook on Synchrotron Radiation* **IV**, Mills, D. ed.

Shi, X., Long, F. H., Lu. H and Eisenthal, K. B. (1995). Electron solvation in neat lcohols, *J. Phys. Chem.* **99**, p. 6917.

Shibahara, M., Katsuki, M. and Kotake, S. (2000). Quantum molecular dynamics study on heat absorption of materials by light irradiation, *Int. J. Heat and Technology* **18**, pp. 17–22.

Shibahara, M. and Kotake, S. (1997). Quantum molecular dynamics study on light-to-heat absorption mechanism: Two metallic atom system, *Int. J. Heat and Mass Transfer* **40**, pp. 3209–3222.

Shibahara, M. and Kotake, S. (1998). Quantum molecular dynamics study of light-to-heat absorption mechanism in atomic systems, *Int. J. Heat and Mass Transfer* **41**, pp. 839–849.

Siders, C. W. *et al.* (1999). Detection of nonthermal melting by ultrafast X-ray diffraction, *Science* **286**, p. 1360.

Srajer, V., Teng, T. Y., Ursby, T., Pradervand, C., Ren, Z., Adachi, S., Schildkamp, W., Bourgeois, D., Wulff, M. and Moffat, K. (1996). Photolysis of the carbon monoxide complex of myoglobin: Nanosecond time-resolved crystallography, *Science* **274**, p. 1726.

Sumiyoshi, T., Sawamura, S., Koshikawa, Y. and Katayama, M. (1982). Transient species in picosecond pulse radiolysis of liquid carbon tetrachloride, *Bull. Chem. Soc. Jpn.* **4**, p. 2341.

Tagawa, S., Hayashi, N., Yoshida, Y., Washio, M. and Tabata,Y. (1989). *Radiat. Phys. Chem.* **34**, p. 503.

Tagawa, S., Kashiwagi, M., Kamada, T., Sekiguchi, M., Hosobuchi, K., Tominaga, H., Ooka, N. and Makuuchi, K. (2002). Economic scale of utilization of radiation (I): Industry-comparison between Japan and U. S. A., *J. Nucl. Sci. Technol.* **39**, p. 1002.

Tagawa, S., Katsumura, Y. and Tabata, Y. (1979). The ultra-fast process of picosecond time-resolved energy transfer in liquid cyclohexane by picosecond single-pulse radiolysis, *Chem. Phys. Lett.* **64**, p. 258.

Tagawa, S., Katsumura, Y. and Tabata, Y. (1982a). Picosecond pulse radiolysis studies on the formation of excited singlet states of solute molecules in several saturated hydrocarbon and ethanol solvents, *Radiat. Phys. Chem.* **19**, p. 125.

Tagawa, S., Tabata, Y., Kobayashi, H. and Washio, M. (1982b). The formation of solute excited triplet states via geminate ion recombination in cyclohexane solutions of biphenyl and pyrene on subnanosecond and nanosecond timescales, *Radiat. Phys. Chem.* **19**, p. 193; Tagawa, S., Washio, M., Tabata, Y. and Kobayashi. (1982) Geminate recombination kinetics of solute radical ions.singlet excited state formation in cyclohexane solutions of biphenyl, *Radiat. Phys. Chem.* **19**, p. 277.

Tagawa, S., Washio, M. and Tabata, Y. (1979b). Picosecond time-resolved fluorescence studies of poly(N-vinylcarbazole) using a pulse-radiolysis technique, *Chem. Phys. Lett.* **68**, p. 276.

Tanaka, Y., Hara, T., Kitamura, H. and Ishikawa, T. (2002a). Synchronization of short pulse laser with the SPring-8 synchrotron radiation pulses and its application to time-resolved measurements, *Review of Laser Engineering* **30**, No. 9, pp. 525–530 (in Japanese).

Tanaka, Y., Hara, T., Yamazaki, H., Kitamura, H. and Ishikawa, T. (2002b). Optical-switching of x-rays using laser-induced lattice expansion, *J. Synchrotron Rad.* **9**, pp. 96–98.

Techert, S., Schotte, F. and Wulff, M. (2001). Picosecond x-ray diffraction probed transient structural changes in organic solids, *Phys, Rev. Lett.* **86**, 2030.

Thomsen, C., Grahn, H. T., Maris, H. J. and Tauc, J. (1986). Surface generation and detection of phonons by picosecond light pulses *Phys. Rev. B* **34**, pp. 4129–4138.

Uesaka, M., Kotaki, H., Nakajima, K., Harano, H., Kinoshita, K., Watanabe, T., Ueda, T., Yoshii, K., Kondo, M., Dewa, H., Kondo, S. and Sakai, F. (2000). Generation and application of femtosecond X-ray pulse, *Nucl. Instrum. Meth. A* **455**, pp. 90–98.

Von der Linde, D., Sokolowski-Tinten, K., Blome, C., Dietrich, C., Zhou, P., Tarasevitch, A., Cavalleri, A., Siders, C. W., Barty, C. P. J., Squier, J., Wilson, K. R., Uschmann, I. and Förster, E. (2001). Generation and application of ultrashort X-ray pulses, *Laser Part. Beams* **19**, pp. 15–22.

Wakabyashi, K. *et al.* (2000). Laser-induced shock compression of tantalum to 1.7 TPa, *Jpn. J. Appl. Phys.* **39**, pp. 1815–1816.

Wark, J. S. *et al.* (1989). Subnanosecond x-ray diffraction from laser-shocked crystals, *Phys. Rev. B* **40**, pp. 5705–5714; Whitlock, R. R. and Wark, J. S. (1995) Orthogonal strains and onset of plasticity in shocked LiF crystals, *Phys. Rev. B* **52**, pp. 8–11; Whitlock, R. R. and Wark, J. S. (1997). X-ray diffraction dynamics of shock-compressed crystals, *Time-resolved Diffraction* Helliwell, J. R. and Rentzepis, P. M. eds., Claredon Press, London, pp. 106–136.

Washio, M., Kashiwagi, S., Kawai, H., Hama, Y., Ishikawa, H., Kobayashi, M., Kuroda, R., Maeda, K., Mori, M., Nagasawa, F., Yada, A., Ben-Zvi, I.,

Wang, X. J., Hayano, H. and Urakawa, J. (2002). *Proc. EPAC2002*, Paris, p. 1774.

Wie, C. R., Tombrello, T. A. and Vreeland, T. Jr. (1986). Dynamical x-ray diffraction from nonuniform crystalline films: Application to x-ray rocking curve analysis, *J. Appl. Phys.* **59**, p. 3743.

Wiesenfeld, J. M. and Ippen, E. P. (1980). Dynamics of electron solvation in liquid water, *Chem. Phys. Lett.* **73**, p. 47.

Wishart, J. F. (1998). The Brookhaven Laser-Electron Accelerator Facility (LEAF), *Houshasenkagaku (Biannual J. of the Jap. Soc. of Rad. Chem.)* **66**, pp. 63–64.

Wishart, J. F. (2001). Accelerators for ultrafast phenomena, *Radiation Chemistry: Present Status and Future Trends*, Jonah, C. D. and Rao, B. S. M. eds., *Studies in Physical and Theoretical Chemistry* **87**, Ch. 2, Elsevier Science, pp. 21–35.

Wishart, J. F. and Nocera, D. G. (1998). *Adv. Chem. Ser.* **254**, p. 35.

Wu, G., Katsumura, Y., Muroya, Y., Li, X. and Terada, Y. (2000). Hydrated electron in subcritical and supercritical water: a pulse radiolysis study, *Chem. Phys. Lett.* **325**, p. 531.

Wu, G., Katsumura, Y., Muroya, Y., Lin, M. and Morioka, T. (2001). Temperature dependence of $(SCN)_2^{*-}$ in water at 25-400°C: Absorption spectrum, equilibrium constant and decay, *J. Phys. Chem. A* **105**, p. 4933.

Wu, G., Katsumura, Y., Lin, M., Morioka, T. and Muroya, Y. (2002a) Temperature dependence of ketyl radical in aqueous benzophenone solutions up to 400°C: A pulse radiolysis study, *Phys. Chem. Chem. Phys.* **4**, p. 3980.

Wu, G., Katsumura, Y., Muroya, Y., Lin, M. and Morioka, T. (2002b). Temperature dependence of carbonate radical in $NaHCO_3$ and Na_2CO_2 solutions: Is the radical a single anion?, *J. Phys. Chem. A* **106**, p. 2430.

Wulff, M., Schotte, F., Naylor, G., Bourgeois, D., Moffat, K. and Mourou, G. (1997). Time-resolved structures of macromolecules at the ESRF: "Single-pluse Laue diffraction, stroboscopic data collection and femtosecond flash photolysis" *Nucl. Instrum. Meth. Phys. Res. A* **398**, pp. 69–84.

Yanagisawa, K., Kume, T., Makuuchi, K., Tagawa, S., Chino, M., Inoue, T., Takehisa, M., Hagiwara, M. and Shimizu, M. (2002). An economic index regarding market creation of products obtained from utilization of radiation and nuclear energy (IV): Comparison between Japan and the U. S. A., *J. Nucl. Sci. Technol.* **39**, p. 1120.

Yoshida, M., Fujimoto, Y., Hironaka, Y., Nakamura, K. G., Kondo, K., Ohtani, M. and Tsunemi, H. (1998). Generation of picosecond hard x rays by tera watt laser focusing on a copper target, *Appl. Phys. Lett.* **73**, pp. 2393–2395.

Yoshida, Y., Mizutani, Y., Kozawa, T., Saeki, A., Seki, S., Tagawa, S. and Ushida, K. (2001). Development of laser-synchronized picosecond pulse radiolysis system, *Radiat. Phys. Chem.* **60**, p. 313.

Yoshida, Y. and Tagawa, S. (1995). Development of femtosecond pulse radiolysis system for material science, *Proc. Int. Workshop Femtosecond Tech.*, Tsukuba, p. 63.

Yoshida, Y., Tagawa, S., Washio, M., Kobayashi, H. and Tabata, Y. (1989).

Pulse radiolysis studies on liquid alkanes and related polymers, *Radiat. Phys. Chem. 34*, p. 493.

Yoshida, Y., Ueda, T., Kobayashi, T., Shibata, H. and Tagawa, S. (1993). Studies of geminate ion recombination and formation of excited states in liquid n-dodecane by means of a new picosecond pulse radiolysis system, *Nucl. Instr. Meth. A* **327**, p. 41.

Yoshida, Y., Ueda, T., Kobayashi T. and Tagawa, S. (1991). Radiation induced reactions of polyethylene model compounds, *J. Photopolym. Sci. Tech.* **4**, p. 171.

Index